IEE TELECOMMUNICATIONS SERIES 45

Series Editors: Professor C. J. Hughes
Professor J. D. Parsons
Professor G. White

RADIO SPECTRUM MANAGEMENT

2nd edition

Other volumes in this series:

Volume 1   **Telecommunications networks** J. E. Flood (Editor)
Volume 2   **Principles of telecommunication-traffic engineering** D. Bear
Volume 3   **Programming electronic switching systems** M. T. Hills and S. Kano
Volume 4   **Digital transmission systems** P. Bylanski and D. G. W. Ingram
Volume 5   **Angle modulation: the theory of system assessment** J. H. Roberts
Volume 6   **Signalling in telecommunications networks** S. Welch
Volume 7   **Elements of telecommunications economics** S. C. Littlechild
Volume 8   **Software design for electronic switching systems** S. Takamura,
             H. Kawashima, N. Nakajima
Volume 9   **Phase noise in signal sources** W. P. Robins
Volume 10   **Local telecommunications** J. M. Griffiths (Editor)
Volume 11   **Principles and practice of multi-frequency telegraphy** J. D. Ralphs
Volume 12   **Spread spectrum in communication** R. Skaug andJ. F. Hjelmstad
Volume 13   **Advanced signal processing** D. J. Creasey (Editor)
Volume 14   **Land mobile radio systems** R. J. Holbeche (Editor)
Volume 15   **Radio receivers** W. Gosling (Editor)
Volume 16   **Data communications and networks** R. L. Brewster (Editor)
Volume 17   **Local telecommunications 2: into the digital era** J. M. Griffiths (Editor)
Volume 18   **Satellite communication systems** B. G. Evans (Editor)
Volume 19   **Telecommunications traffic, tariffs and costs: an introduction for
             managers** R. E. Farr
Volume 20   **An introduction to satellite communications** D. I. Dalgleish
Volume 21   **SPC digital telephone exchanges** F. J. Redmill and A. R. Valdar
Volume 22   **Data communications and networks 2nd edition** R. L. Brewster (Editor)
Volume 23   **Radio spectrum management** D. J. Withers
Volume 24   **Satellite communication systems 2nd edition** B. G. Evans (Editor)
Volume 25   **Personal & mobile radio systems** R. C. V. Macario (Editor)
Volume 26   **Common-channel signalling** R. J. Manterfield
Volume 27   **Transmission systems** J. E. Flood and P. Cochrane (Editors)
Volume 28   **VSATs: very small aperture terminals** J. L. Everett (Editor)
Volume 29   **ATM: the broadband telecommunications solution** L. G. Cuthbert and J.-
             C. Sapanel
Volume 30   **Telecommunication network management into the 21st century**
             S. Aidarous and T. Plevyak (Editors)
Volume 31   **Data communications and networks 3rd edition** R. L. Brewster (Editor)
Volume 32   **Analogue optical fibre communications** B. Wilson, Z. Ghassemlooy and I.
             Darwazeh (Editors)
Volume 33   **Modern personal radio systems** R. C. V. Macario (Editor)
Volume 34   **Digital broadcasting** P. Dambacher
Volume 35   **Principles of performance engineering for telecommunication and
             information systems** M. Ghanbari, C. J. Hughes, M. C. Sinclair and
             J. P. Eade
Volume 36   **Telecommunication networks, 2nd edition** J. E. Flood (Editor)
Volume 37   **Optical communication receiver design** S. B. Alexander
Volume 38   **Satellite communication systems, 3rd edition** B. G. Evans (Editor)
Volume 39   **Quality of service in telecommunications** A. P. Oodan, K. E. Ward and
             T. W. Mullee
Volume 40   **Spread spectrum in mobile communication** O. Berg, T. Berg,
             S. Haavik, J. F. Hjelmstad and R. Skaug
Volume 41   **World telecommunications economics** J. J. Wheatley
Volume 42   **Video coding: an introduction to standard codes** M. Ghanbari
Volume 43   **Telecommunications signalling** R. Manterfield
Volume 44   **Digital signal filtering, analysis and restoration** J. Jan

RADIO SPECTRUM MANAGEMENT

2nd edition

Management of the spectrum and regulation of radio services

David Withers

The Institution of Electrical Engineers

Published by: The Institution of Electrical Engineers, London,
United Kingdom

© 1999: The Institution of Electrical Engineers

The Institution of Electrical Engineers,
Michael Faraday House,
Six Hills Way, Stevenage,
Herts. SG1 2AY, United Kingdom

British Library Cataloguing in Publication Data

A CIP catalogue record for this book
is available from the British Library

ISBN 0 85296 770 5

Printed in England by TJ International, Padstow, Cornwall

Contents

Principal abbreviations used in the text		**x**
Preface		**xv**
1	**Introduction**	**1**
	1.1 Reference	6
2	**The International Telecommunication Union**	**7**
	2.1 The evolution of the ITU	7
	2.2 The Union, post-1994	13
	2.3 The Radiocommunication Sector	17
	2.4 References	23
3	**Frequency allocation**	**25**
	3.1 The international frequency allocation table	25
	3.2 International frequency allocations and changing needs	33
	3.3 National frequency allocation tables	37
	3.4 References	38
4	**Frequency assignment**	**41**
	4.1. The frequency assignment process	41
	4.1.1 Licensing a private microwave fixed link	41
	4.1.2 Frequency assignment for other kinds of system	47
	4.1.3 Choice of frequencies for assignment	50
	4.1.4 Non-mandatory international frequency coordination	55
	4.2 Efficiency in spectrum utilisation	58
	4.2.1 The concept of efficient spectrum use	58
	4.2.2 The bandwidth of emissions	61
	4.2.3 A clean radio environment	64
	4.2.4 Enhanced spectrum utilisation efficiency by regulation	67

4.3	Licensing and licence fee policy		69
	4.3.1	The requirement for licensing	69
	4.3.2	Licensing non-commercial radio users	69
	4.3.3	Conventional procedures for licensing commercial radio users	70
	4.3.4	Spectrum pricing	71
	4.3.5	The application of spectrum pricing	72
4.4	Identification of stations and emissions		74
4.5	Interference		75
4.6	References		78

5   Frequency band planning and mandatory frequency coordination   **81**
5.1	Introduction		81
5.2	Frequency coordination for satellite and terrestrial services		82
	5.2.1	Coordination between satellite networks	82
	5.2.2	The criterion for coordination between GSO networks	92
	5.2.3	Band sharing by GSO satellite networks and terrestrial stations	93
	5.2.4	Band sharing by quasi-GSO satellite networks and terrestrial stations	102
	5.2.5	Band sharing for non-GSO satellite systems and terrestrial stations under RR Resolution 46	103
	5.2.6	Acceptable levels of interference	105
	5.2.7	Geostationary network coordination in practice	108
5.3	Miscellaneous allocations requiring coordination		110
5.4	International planning of frequency band utilisation		110
5.5	References		113

6   The fixed service   **119**
6.1	The FS operating below 30 MHz		119
	6.1.1	VLF and LF systems	119
	6.1.2	MF and HF systems	120
6.2	The FS operating above 30 MHz		124
	6.2.1	Frequency allocations and sharing	124
	6.2.2	Line-of-sight systems, 30–1000 MHz	128
	6.2.3	Line-of-sight systems above 1 GHz	129
	6.2.4	Trans-horizon systems	131
6.3	References		133

7   The broadcasting service   **137**
7.1	Introduction		137
	7.1.1	The international regulatory background	137
	7.1.2	National regulation of broadcasting	138
7.2	Sound broadcasting below 2 MHz		141
7.3	Tropical broadcasting, 2–5 MHz		147
7.4	HF sound broadcasting, 5.9–26.1 MHz		148
7.5	Sound and television broadcasting above 30 MHz		152
	7.5.1	The frequency allocations	152
	7.5.2	Television broadcasting	155
	7.5.3	Sound broadcasting above 30 MHz	160
7.6	References		162

8   The mobile services   **165**

   8.1   Introduction   165
   8.2   The maritime mobile service below 30 MHz   166
   8.3   The aeronautical mobile service below 30 MHz   172
   8.4   The land mobile service below 30 MHz   174
   8.5   Mobile services above 30 MHz   174
         8.5.1 Frequency allocations and sharing   174
         8.5.2 The MMS above 30 MHz   176
         8.5.3 The AMS above 30 MHz   177
         8.5.4 LMS radiotelephone networks   178
         8.5.5 Miscellaneous narrow-band mobile systems above 30 MHz   183
         8.5.6 Wide-band transportable systems   184
   8.6   Licensing mobile radio stations   185
   8.7   Distress and safety communication for mobile stations   187
   8.8   References   189

9   The fixed-satellite service   **193**

   9.1   Frequency allocations and sharing   193
         9.1.1 Introduction   193
         9.1.2 The FSS allocations, applications and sharing   194
   9.2   Geostationary FSS networks   202
         9.2.1 Technical factors affecting spectrum utilisation efficiency   202
         9.2.2 Very small aperture terminals   210
         9.2.3 International spectrum management for the FSS in the GSO   211
   9.3   The spectrum and orbit allotment agreement   212
   9.4   Non-GSO FSS systems   215
   9.5   National regulation and spectrum management in the FSS   217
   9.6   References   218

10 The broadcasting-satellite service   **221**

   10.1 Introduction   221
   10.2 Spectrum management and regulation of satellite broadcasting   226
   10.3 The 12 GHz frequency assignment agreement for TV   228
        10.3.1 The agreement and the plans   228
        10.3.2 The implementation and evolution of the agreement   232
        10.3.3 Frequency allocation sharing under the agreement   234
   10.4 Satellite sound broadcasting   236
   10.5 References   237

11 The mobile-satellite services   **239**

   11.1 The services and their frequency allocations   239
        11.1.1 The services   239
        11.1.2 Frequency allocations and sharing   239
   11.2 The commercial geostationary MSS systems   245
   11.3 Non-GSO MSS systems   247
        11.3.1 Little LEO systems   247
        11.3.2 GMPCS systems   248
        11.3.3 Spectrum management and regulation for GMPCS   249

11.4   Satellite news gathering   250
11.5   References   251

12 The amateur services   **255**
12.1   References   259

13 The science radio services   **261**
13.1   The services   261
13.2   The use of passive sensors   262
    13.2.1 Protection of passive sensors from interference   262
    13.2.2 Radio astronomy   264
    13.2.3 Passive sensors for the SR and the EES   265
13.3   Active sensors for the SR and the EES   267
13.4   Communication bands for the science services   268
    13.4.1 The EES, the MetS, the SR and the SO   268
    13.4.2 The meteorological aids service   271
    13.4.3 The standard frequency and time signal services   273
13.5   Managing spectrum for the science services   273
13.6   References   274

14 The inter-satellite service   **277**
14.1   References   279

15 Radionavigation and radiolocation   **281**
15.1   Introduction   281
15.2   International management of radiodetermination allocations   283
15.3   National spectrum management   287
15.4   References   288

Appendix: Radio propagation and radio noise   **289**
A1   Introduction   289
A2   Propagation by the tropospheric wave mode   290
    A2.1 Propagation in free space   290
    A2.2 Tropospheric phenomena   293
        A2.2.1 Refraction   293
        A2.2.2 The effects of solid obstructions   297
        A2.2.3 The effects of rain and other airborne particles   301
        A2.2.4 Absorption by atmospheric gases   304
        A2.2.5 Effects of the ionosphere on transmitted waves   305
A3   Ground wave propagation   307
A4   Ionospheric wave propagation   309
    A4.1 The ionosphere and its behaviour   309
    A4.2 Ionospheric wave propagation below about 2 MHz   314
    A4.3 Ionospheric wave propagation between 2 and 30 MHz   317
A5   Tropospheric scatter propagation   323
A6   Meteor-burst propagation   324
A7   The irregular propagation modes   324
    A7.1 Ionospheric cross-modulation   325
    A7.2 Irregular ionospheric propagation at HF   325

	A7.3	F region propagation at VHF	325
	A7.4	Sporadic E region ionisation	326
	A7.5	Ionospheric forward scatter at VHF	326
	A7.6	Tropospheric ducts	326
	A7.7	Reflection from aircraft	327
	A7.8	Precipitation scatter	327
A8	Radio noise		328
	A8.1	Introduction	328
	A8.2	Predicted external noise levels	330
		A8.2.1 Atmospheric noise	330
		A8.2.2 Man-made noise	331
		A8.2.3 Noise from the environment and beyond	331
A9	Prediction of radio link performance		335
A10	References		338

Index 343

Principal abbreviations
used in the text

AM	amplitude modulation.
AMS	aeronautical mobile service.
AMS(OR)	the AMS (off-route).
AMS(R)	the AMS (route).
AmS	the amateur service.
AMSS	the aeronautical mobile-satellite service.
AMS(OR)S	the AMSS (off-route).
AMS(R)S	the AMSS (route).
AmSS	the amateur-satellite service.
API	advance publication of information.
APP-92	The Additional Plenipotentiary Conference (of ITU), held in 1992.
ARNS	the aeronautical radionavigation service.
ARNSS	aeronautical radionavigation-satellite service.
ARQ	automatic error correction by repetition.
ATSC	the (United States) Advanced Television Systems Committee.
BDT	Telecommunications Development Bureau (of ITU).
BER	bit error ratio.
bit/s	bits per second.
BS	the broadcasting service.
BSS	the broadcasting-satellite service.
C	Celsius.
CB	Citizens' Band.
CCIF	International Telephone Consultative Committee.
CCIR	International Radio Consultative Committee.
CCIT	International Telegraph Consultative Committee.
CCITT	International Telegraph and Telephone Consultative Committee.
CDMA	code division multiple access.
CEPT	European Conference of Posts and Telecommunications Administrations.
C/I	carrier-to-interference ratio.
C/N	carrier-to-noise ratio.

CMTT	Joint Study Group for Television and Sound Transmission.
DAB	digital audio broadcasting.
dB	decibel.
dBi	antenna gain in decibels relative to an isotropic antenna.
dBW	decibels relative to 1 W.
$dB(W/m^2)$	dBW per square metre.
$dB(W/m^2/1\,MHz)$	dBW per square metre in 1 MHz sampling bandwidth.
$dB(W/m^2/4\,kHz)$	dBW per square metre in 4 kHz sampling bandwidth.
$dB(x)$	decibels relative to x.
DBS	direct broadcasting by satellite.
DSB	double sideband.
DSC	digital selective calling.
DTH	direct to home (satellite broadcasting).
EES	the Earth exploration-satellite service.
EHF	the frequency range 30–300 GHz; also called Band 11.
e.i.r.p.	equivalent isotropically radiated power.
EMC	electromagnetic compatibility.
ENG	electronic news gathering.
EPIRB	emergency position-indicating radio beacon.
ERC	European Radiocommunications Committee (of CEPT).
ERMES	European radio messaging system.
ERO	European Telecommunications Office (of CEPT).
ETSI	European Telecommunications Standards Institute.
FDM	frequency division multiplex.
FEC	forward error correction.
FM	frequency modulation.
FPLMTS	Future Public Land Mobile Telecommunication System (subsequently renamed IMT-2000).
FS	the fixed service.
FSK	frequency shift keying.
FSS	the fixed-satellite service.
GHz	gigahertz (1 GHz = 1000 MHz).
GMDSS	Global Maritime Distress and Safety System.
GMPCS	global mobile personal communications by satellite.
GPS	Global Positioning System.
GSM	Global System for Mobile Communication.
GSO	geostationary satellite orbit.
G/T	a symbol representing the figure of merit of a receiver/antenna combination.
HDTV	high definition television.
HF	the frequency range 3–30 MHz; also called Band 7.
HLC	The High Level Committee (set up by ITU in 1989).
HRC	Hypothetical Reference Circuit.
Hz	hertz (basic unit of frequency).
IARU	International Amateur Radio Union.
ICAO	International Civil Aviation Organization.
IEC	International Electrotechnical Commission.
IFL	International Frequency List.

IFRB	International Frequency Registration Board.
ILS	instrument landing system.
IMO	International Maritime Organization.
IMT-2000	International Mobile Telecommunications-2000 (formerly the FPLMTS).
ISB	independent sideband.
ISDN	integrated services digital network.
ISL	inter-satellite link.
ISM	industrial, scientific and medical.
ISO	International Standards Organization.
ISS	the inter-satellite service.
ITU	International Telecommunication Union.
ITU-D	ITU Telecommunication Development Sector.
ITU-R	ITU Radiocommunication Sector.
ITU-R x.y	ITU-R Recommendation or Report, series 'x', number 'y'.
ITU-T	ITU Telecommunication Standardization Sector.
J	joule.
K	degrees kelvin.
kbit/s	kilobits per second.
kHz	kilohertz (1 kHz = 1000 Hz).
km	kilometre.
kW	kilowatt.
LAN	local area network.
LEO	low Earth orbit.
LF	the frequency range 30–300 kHz; also called Band 5.
LMS	the land mobile service.
LMSS	the land mobile-satellite service.
LUF	lowest usable frequency.
m	metre.
MAC	multiplex analogue components (a TV standard).
Mbit/s	megabits per second.
MetA	the meteorological aids service.
MetS	the meteorological-satellite service.
MF	the frequency range 300–3000 kHz; also called Band 6.
MHz	megahertz (1 MHz = 1000 kHz).
MIFR	Master International Frequency Register.
MLS	microwave landing system.
MMS	the maritime mobile service.
MMSS	the maritime mobile-satellite service.
MPEG	Motion Pictures Expert Group (of ISO/IEC).
MRNS	the maritime radionavigation service.
MRNSS	the maritime radionavigation-satellite service.
MS	the mobile service.
ms	millisecond (1000 ms = 1 s).
MSS	the mobile-satellite service.
MUF	maximum usable frequency.
mV	millivolt (1000 mV = 1 V).
MVDS	multipoint video distribution system.
mW	milliwatt (1000 mW = 1 W).

NBDP	narrow-band direct printing (telegraphy).
NBPSK	narrow-band PSK.
non-GSO	non-geostationary satellite orbit.
ns	nanosecond (1000 ns = 1 μs).
NTSC	National Television Standards Committee (a TV chrominance encoding system standard).
OFDM	orthogonal FDM
OWF	optimum usable frequency.
PAL	phase alternate line (a TV chrominance encoding system standard).
PAMR	public access mobile radio.
PDA	predetermined arc.
PFD	power flux density.
PM	phase modulation.
PMR	private mobile radio.
PP	Plenipotentiary Conference (of ITU; subsequent digits show the year).
PSK	phase shift keying.
PSTN	public switched telecommunications network.
PTO	public telecommunications operator.
pW	picowatt.
pW0p	pW, psophometrically weighted, at a point of zero reference level.
QPSK	four-phase PSK.
quasi-GSO	an almost geostationary satellite orbit.
RA	Radiocommunication Assembly (of ITU; subsequent digits show the year).
RARC	Regional Administrative Radio Conference (of ITU; Subsequent digits show the year).
RAS	the radio astronomy service.
RDS	the radiodetermination service.
RDSS	the radiodetermination-satellite service.
RF	radio frequency.
RLS	the radiolocation service.
RLSS	the radiolocation-satellite service.
RNS	the radionavigation service.
RNSS	the radionavigation-satellite service.
RR	The (ITU) Radio Regulations.
RRB	Radio Regulations Board (of ITU).
RRx.y	Radio Regulations, edition of 1998, Article 'x', paragraph 'y'.
RTCE	real time channel evaluation.
RRC	Regional Radiocommunication Conference (of ITU; subsequent digits show date).
s	second.
SART	search and rescue radar transponder.
SCPC	single channel per carrier.
S-DAB	satellite DAB.
SECAM	séquential couleur avec mémoire (a TV chrominance encoding system standard).
SETI	Search for Extra-terrestrial Intelligence.
SF	Swiss Francs.
SG	study group.

SHF	the frequency range 3–30 GHz; also called Band 10.
SNG	satellite news gathering.
SO	the space operation service.
SOLAS	(The International Convention for the) Safety of Life at Sea.
SSB	single sideband.
SR	the space research service.
T-DAB	terrestrial DAB.
TDM	time division multiplex.
TDMA	time division multiple access.
TEC	total electron content.
TFS	the standard frequency and time signal service.
TFSS	the standard frequency and time signal-satellite service.
TFTS	terrestrial flight telecommunication system.
TV	television.
UHF	the frequency range 300–3000 MHz; also called Band 9.
UMTS	universal mobile telecommunication system.
V	volt.
VF	voice frequency.
VGE	Voluntary Group of Experts (set up by ITU in 1989).
VHF	the frequency range 30–300 MHz; also called Band 8.
VLF	the frequency range 3–30 kHz; also called Band 4.
VOR	VHF Omnidirectional Range system.
VSAT	very small aperture terminal (of the FSS).
VSB	vestigial sideband.
W	watt.
WARC	World Administrative Radio Conference (of ITU; subsequent digits show the year).
WARC HFBC-87	a WARC, held in 1987, to consider HF broadcasting.
WARC Orb-85	a WARC, held in 1985, to consider planning for the BSS and the FSS.
WARC Orb-88	a WARC, held in 1988, to consider planning for the BSS and the FSS.
WARC SAT-77	a WARC, held in 1977 to consider planning for the BSS.
WMO	World Meteorological Organization.
WRC	World Radiocommunication Conference (of ITU; subsequent digits show date).
WTPF	World Telecommunication Policy Forum.
λ	wavelength
μs	microsecond (1000 μs = 1 ms).

Preface

This book aims to provide for radio users a good understanding of the methods used by governments for managing the use of the radio spectrum. For clarity, and to keep the size of the book within bounds, the treatment concentrates on the principles of the methods used, references being provided for the technical and administrative details where appropriate. For the same reasons, minor details of the internationally agreed frequency allocation table and of national departures from the international table have also been omitted; reference should be made to the international frequency allocation table and national licensing authorities where precise information is necessary.

Radio spectrum management was first published in 1991 but much has changed since then. This has been a period of rapid growth in the use of radio, much innovation and spreading spectrum congestion. In particular, there has been a great increase in the use of mobile radiotelephones and this has necessitated new frequency allocations between 1 and 3 GHz. Growth in the use of radio for short terrestrial links is being made possible by extensive use of frequencies, not previously used, above 25 GHz. The introduction of low-orbit satellite systems serving mobile earth stations has made necessary many new frequency allocations between 100 and 2500 MHz and new techniques for coordinating the operation of these new systems with the systems already established in these bands.

In the past, the fees that radio users paid for their licences were usually calculated to cover the cost of licensing. Nowadays, fees are increasingly being determined by other criteria, for example to apply pressure on licensees to minimise their requirement for bandwidth, especially in locations and in frequency bands which are particularly congested, or to raise revenue. In some circumstances fees have been set by auction.

There has been a far-reaching change in the kinds of organisation that own and operate radio systems providing public telecommunications services and broadcasting. Up to the 1980s, with few exceptions, these systems were operated by publicly owned agencies and commercial monopolies, the latter being subject to strict government oversight of tariffs and the quality of service. Since about 1985 the trend has been towards liberalisation and commercialisation, including privatisation of the publicly owned operators and licensing many new commercial operators. One objective of this change has been the replacement of direct government action to restrain tariffs and optimise quality of service by market pressures brought about by competition. This has created a new function of government, since decisions as to which operators should be given access to spectrum, and to how much spectrum, must often be determined by the need to secure effective competition within the available spectrum. This new function is called regulation.

Finally, the International Telecommunication Union (ITU), which has a vital role in the management of the spectrum, had in 1991 the same basic organisational framework as it was given by the International Telecommunication Convention of 1947. The ITU Radio Regulations, which are the rule book for international spectrum management, had evolved from the version drawn up in 1947 to accommodate new kinds of radio system but with little retrospective revision; consequently they had become needlessly complex and awkward to use. Towards the end of the 1980s the view was emerging that the Union was not performing as well as modern conditions required; reform had become necessary. A new constitutional structure was adopted in 1992 and new, extensively revised, regulations were published in 1998. The basis of international spectrum management was not changed in principle by these developments but there were major innovations to deal with new situations and very many changes in detail. Some practices that had proved to be ineffective or inessential were abandoned.

As a result of all these changes, the 1991 edition of *Radio spectrum management* has become out of date. This new edition, a total revision, is up to date as of 1999.

Acknowledgements

To clarify some key issues in the international management of radio spectrum the author has selected several passages from publications of the International Telecommunication Union for inclusion in the text, with the prior authorisation of the Union, the copyright holder. The source of these texts is identified in each case. References to many other ITU publications are also provided. All ITU current publications are available from:

International Telecommunication Union, Sales and Marketing Service, Place des Nations, CH-1211 Geneva 20, Switzerland
Telephone: + 41 22 730 61 41 (English)/ + 41 22 730 61 42 (French)/ + 41 22 730 61 43 (Spanish)
Telex: 421 000 uit ch/Fax: + 41 22 730 51 94
X.400: S = Sales; A + 400 net; P = itu; C = ch
E-mail: Sales@itu.int
Website: http://www.itu.int/publications

Chapter 1

Introduction

Radio is the most versatile and flexible of the telecommunication and broadcasting transmission media; for satellite services and most applications involving mobile radio stations it is the only feasible medium and radio is often the cheapest medium for other communication applications. The operation of many navigation systems depends on the predictability of radio wave propagation. The use of radio for all of these purposes falls within the internationally agreed usage of the term 'radiocommunication'.

Because of these valuable properties, radio is already being used for a vast number of purposes. New uses and many new users are emerging every year. Fortunately the capacity of the radio spectrum is very great and, unlike most other natural resources, it is occupied but not consumed by use. Nevertheless, the various parts of the spectrum are not equally useful. Some parts are good for some purposes, other parts are technically preferable for other purposes, but a wide range of the highest radio frequencies has yet to be found economically useful for any application. As a result, the more readily usable parts of the spectrum tend to be heavily loaded with radio emissions, especially in extensive, densely populated areas. Heavy loading without efficient spectrum management leads to interference between systems and reduces the total realisable capacity of the medium. Thus where radio usage is heavy it is vitally important that spectrum management should be efficient.

National spectrum management

Radio spectrum management is a responsibility of governments but it has important international aspects; radio links, mobile radio stations and interference cross frontiers. If a country joins the International Telecommunication Union (ITU), and virtually all countries have done so, its government undertakes to ensure that radio stations within its jurisdiction do not cause harmful interference to radio stations in other countries that are operating in accordance with international agreements. Each country provides for the discharge of this responsibility and more generally for the management of the use that is made of the spectrum within its territory.

These duties do not necessarily become the function of a single administrative unit. For example it is not unusual for one unit to be responsible for the use of radio frequencies by the armed services and for another unit to manage radio use for non-military systems, while the management of some specialised application of radio may be devolved to an expert agency. Nevertheless, the work of these various units will be coordinated nationally, with an explicit chain of responsibility, to form what is termed an 'Administration'. An upper case initial is used for 'Administration' in this book.

In general principle, an Administration manages the spectrum as follows.

- The various kinds of radio station are classified into radio 'services'. The radio spectrum is divided into frequency bands and the Administration 'allocates' bands for each of the various services, producing a national table of frequency allocations. If the demand for a service is high, the Administration allocates additional bandwidth for the service or requires the use of more spectrum-efficient equipment requiring less bandwidth per user. Wherever possible, national frequency allocation tables are harmonised with an international table which has been agreed through the ITU.

- Transmission of radio signals is usually conditional on the issue of a licence by the Administration. Permission to use the radio spectrum for an authorised purpose takes various forms, depending on the intended use. For example, for a radio link between fixed points, the Administration would 'assign' a carrier frequency in an appropriately allocated frequency band and basic technical parameters would be determined that would ensure that the performance of the link would be adequate and that interference, to or from the new link, would be acceptably low. A ship's radio station might be licensed to use any of a large number of frequencies within frequency bands allocated for the maritime mobile service to communicate with the shore, the frequency to be used on any particular occasion being arranged with the coast station concerned. And so on.

- The Administration maintains surveillance of the spectrum to detect and stop unlicensed use and to ensure that licensed users abide by the terms of their licences. If it is found that a licensee is not making use of a frequency assignment, perhaps thereby denying the use of a scarce resource to another applicant, the licence may be withdrawn.

The effort required to carry out the functions of an Administration varies from country to country. One important factor is demographic; a sparse population is unlikely to use the spectrum intensively, reducing the need for highly efficient management. A second key factor is the regulatory regime governing the provision of public telecommunication services. In some countries the public telecommunications operator is a publicly owned agency or has a government-regulated monopoly; indeed until recently this was the usual arrangement. This kind of regime tends to minimise the use of radio and allows a substantial part of the spectrum management function to be delegated by the Administration to the service provider. However, the current trend is towards the provision of telecommunication services by competing, commercial, service providers. In addition, many new facilities which depend on radio have been introduced lately. These trends increase the pressure on the spectrum considerably, and spectrum management is made more complex by the need for the Administration to treat the competing service providers equitably.

Thus, some Administrations need to invest no great amount of effort or skill in managing the spectrum and have little need to ensure that radio users employ spectrum-efficient equipment. Other Administrations must manage the spectrum efficiently and place stringent technical constraints on radio use; both raise the costs falling on the Administration. However, where the need is acute, a well-managed, heavily used spectrum is of very great economic value. For example, recent studies [1] estimate that the use of radio added about 13 billion £UK to the economy of the United Kingdom in 1995/6, about 2% of the Gross Domestic Product, and the benefit is rising. Clearly, the cost of good management and efficient equipment can be amply rewarded.

Regulation of the provision of services by radio

In an economy where telecommunication and broadcasting services are supplied to the public by commercial undertakings, the government will ensure that the quality of the service provided is satisfactory. The prices charged for the facilities provided are also a matter of concern for governments, typically discharged by ensuring that there is effective competition between service providers. The achievement of effective competition is likely to be a responsibility of a regulating body, which licenses service providers, called the Regulator, with an upper case initial, in this book. Thus, in the United Kingdom the supply of telecommunication services to the public is regulated by the Office of Telecommunications (OFTEL), and commercial television and sound radio broadcasting are regulated by the Independent Television Commission and the Radio Authority respectively.

Where the provision of these services involves the use of radio, it is clear that there will be an overlap of the responsibilities of the Administration and the Regulator. It is pointless for the Administration to license the use of a frequency assignment to an organisation that will not be licensed to provide that service. Similarly, competition between licensed service providers may be made ineffective if the Administration refuses, or is unable, to license the necessary frequency assignments. Coordinated action by the Administration and the Regulator will often be necessary.

The basis in law for national spectrum management and regulation of services

Administrations and Regulators need legal powers for the exercise of their functions. The situation in the United Kingdom is summarised here as an indication of what may obtain in other countries.

The first United Kingdom law dealing with spectrum management was the Wireless Telegraphy Act of 1904. This was succeeded by the Wireless Telegraphy Acts of 1926 and 1949, the latter being the present basis, though much amended since, for spectrum management. It should be noted that the term 'wireless telegraphy', whilst appropriate in its literal sense in 1904, is applied in a much broader sense for the purposes of the Act of 1949, where it is defined to cover the use of radio for a comprehensive range of communication, broadcasting, remote control, radiolocation and research purposes. The same definition applies in the subsequent laws referred to below and corresponds with international practice.

With a few specified exceptions, a licence had to be obtained from the Postmaster General before any radio station was established or used within, or over, the United Kingdom, within its territorial waters or on any ship or aircraft of British registration.

This licence was subject to whatever terms the Postmaster General saw fit to apply and a fee was usually charged for it. The level of the licence fee for any specific use of radio was to be determined administratively but the total of fees received was to be related to the fully allocated cost of issuing licences. Most of the yield of broadcasting receiving licence fees was used to finance the production of public service broadcasting. A licence was operative for a specified period of time, normally one year, and it might be revoked before that time had expired. The Postmaster General also had a responsibility to locate electrical installations that cause interference to the operation of radio systems and to get the interference suppressed. Radio stations that operated without a licence or outside the terms of the licence were to be identified and the owners prosecuted. The Postmaster General might also regulate the use of radio equipment on board ships and aircraft of foreign registration within the United Kingdom or its territorial waters.

In the 1960s the unlicensed use of radio equipment, often operating in inappropriate frequency bands, was causing substantial interference to authorised radio systems. The Wireless Telegraphy Act of 1967 gave the government powers to make orders to restrict the manufacture, importation, sale or possession of radio equipment that was likely to cause such interference. Unlicensed 'pirate' broadcasting from ships was also causing interference to licensed spectrum users in the 1960s and it was found that the powers provided by the Act of 1949 were not sufficient to deal with the problem. The Marine Etc Broadcasting (Offences) Act of 1967 forbids broadcasting from ships and aircraft of British registration wherever they may be and provides comprehensive powers for the suppression of broadcasting within or over British territorial waters from off shore platforms and from ships and aircraft of foreign registration.

Until 1969 the Post Office, a government department, headed by its Minister, the Postmaster General, functioned as the Administration and also provided a virtual monopoly of postal and telecommunications services in the UK. By the Post Office Act of that year the office of Postmaster General was abolished and the function of the Administration was transferred to the newly formed Ministry of Posts and Telecommunications and subsequently to the Home Office, then to the Department of Trade and Industry. At the time of writing the responsibility rests with the Radiocommunications Agency, an executive agency of that Department. In that year the Post Office became a publicly owned corporation, still providing postal and telecommunication services as virtual monopolies. The two businesses were subsequently separated, a new publicly owned corporation called British Telecommunications being formed to provide telecommunication services.

The Telecommunications Act of 1984 initiated the sale of British Telecommunications to the public, created the Office of Telecommunications ('OFTEL') to act as the Regulator and began the licensing of other public telecommunications operators (PTOs) to compete with British Telecommunications PLC. The Wireless Telegraphy Act of 1949 was amended to allow licences to be issued to companies running telecommunication systems on terms which give better security of tenure to the licensee. A more liberal policy in the licensing of private telecommunications facilities began to be implemented by the Administration.

The Broadcasting Act of 1990 had little direct impact on the process of assigning frequencies by the Radiocommunications Agency for broadcasting. Its main purpose was to revise the processes of regulating commercial television and sound radio to ensure more effective competition between broadcasters. The Act established the Independent Television Commission and the Radio Authority to act as the Regulators for television and sound broadcasting, respectively. The Act also introduced the auctioning by the Regulators of licences for the right to use the frequencies that were assigned by the Administration, a practice that is closely parallel to the auctioning of frequency assignments but which the Act of 1949 does not allow. The Broadcasting Act of 1996 followed; the main purpose of that Act was to make regulatory provision for multiplex systems of digital terrestrial television and sound broadcasting. The Acts of 1990 and 1996 are reviewed in more detail in Sections 7.1.2 and 10.2 below.

Finally, the Wireless Telegraphy Act of 1998 makes major changes to the terms on which the Administration licenses frequency assignments. In particular:

- The fees charged for most licences are to be determined administratively in the future, as in the past, as provided for in the Act of 1949. However, there is specific provision for setting fees higher than would be required to recover the cost of issuing licences if, for example, the demand for spectrum for a specific purpose may exceed the bandwidth that is likely to be available or in order to achieve efficient management of the spectrum.
- In appropriate cases, licences may be auctioned.
- There is provision for licences to be issued offering better tenure to the licensee.

In addition, there is provision for making grants of money in order to promote the more efficient use of, or management of, the radio spectrum.

International collaboration in spectrum management

There must be a considerable degree of international uniformity in the allocation of spectrum to services. Ships, aircraft and road vehicles move from country to country and their radio equipment may need to be used wherever the mobile station may be. Radio links from one country to another should be assigned frequencies that are in frequency bands that are appropriately allocated in both countries. Furthermore, radio emissions used for systems that are wholly within one country may, nevertheless cross frontiers and cause interference. Spectrum can only be used efficiently if allocations are harmonised and, where necessary, if assignments are coordinated between neighbouring countries. Whilst spectrum management is a national responsibility, it cannot be carried out effectively without collaboration between Administrations.

Certain frequency bands are used exclusively, world-wide or regionally, by a single group of users operating under the oversight of an international body. In a few instances the choice of frequencies within those bands for specific stations has been devolved from the Administrations to that international body, although the function of formally assigning the frequencies remains with the Administrations. For example, frequency assignments in some bands that are used world-wide for communication with civil airliners, primarily for air traffic control, are managed by the International Civil Aviation Organization (ICAO).

More generally, Administrations collaborate with neighbouring Administrations, informally and usually bi-laterally, to resolve potential and actual interference problems. Multi-lateral collaboration between the countries in a region is found to be valuable in resolving broader problems. Within Europe the European Conference of Posts and Telecommunications Administrations (CEPT), in which almost all European countries participate, has this among its functions. Similar bodies serve other regions, such as the Inter-American Telecommunications Conference, the Asia-Pacific Telecommunity and the Arab Telecommunication Union.

The CEPT has a committee, the European Radiocommunications Committee (ERC), which has responsibilities in this area. ERC has set up a permanent body, the European Radiocommunications Office (ERO), located in Copenhagen and staffed by experts seconded from the Administrations of CEPT members, to deal with the long term spectrum management problems that arise in Europe. One important task of the ERO has been to prepare a draft European Common Allocation Table to form a basis for the harmonisation of the national allocation tables of the European countries. Within the European Union the Commission of the European Communities also has some involvement in the planning of spectrum use, in particular where issues of freedom of competition arise.

However, the ITU, which is involved with various aspects of telecommunication, is the forum for global collaboration on the use of the radio spectrum. Furthermore, through the ITU, the Administrations consult with other global organisations with a concern for radio, including ICAO, the International Maritime Organization (IMO), the World Meteorological Organization (WMO), the International Amateur Radio Union (IARU) and many others. The structure and functioning of the Union are reviewed in Chapter 2.

1.1 Reference

1 'Economic impact of the radio spectrum in the UK'. A report by National Economic Research Associates and Smith System Engineering Ltd for the Radiocommunications Agency, London, May 1997

The International Telecommunication Union

2.1 The evolution of the ITU

The predecessor international organisations

The first public telegraph services began operation in the 1840s. By 1860 extensive networks were in use, including international links and by 1870 a network of submarine cables was providing the service with global reach. A need emerged early for a forum for the resolution of the problems that were arising among international telegraph service providers, and the International Telegraph Union was set up for this purpose in 1865. The 'Berne Bureau', managed by the Swiss government, provided secretariat services for the Union. Public telephone services then emerged, the first in 1878, and international telephone links following soon afterwards. The same forum served for the resolution of international telephony problems.

Commercial radiotelegraph service between ships and the shore began in the first years of the twentieth century. Interference was common and competition between the operating companies was sharp, sometimes obstructing the flow of traffic. However, it was realised that radio could fill a vital role in reducing the risk to human life arising from disasters at sea if the service was operated in a disciplined manner. An inter-governmental conference in Berlin in 1903 prepared the way for the establishment of the International Radiotelegraph Union in 1906, partly to provide that discipline. Regulations were agreed and implemented. A register of radio stations was published and kept up-to-date, showing the carrier frequencies assigned to each station and the date on which each assignment was first taken into use. In general, the right to use a radio frequency without unacceptable interference was conceded to the station with prior registered use. The Berne Bureau provided secretariat functions for radio also.

Early in the 1920s electronic devices began to be used to enhance telecommunication systems, extending their effective range, making higher quality of performance feasible and increasing system complexity. Governments, as Administrations,

Regulators and/or service providers, needed to collaborate in developing the still very immature technology of telecommunications and in standardising systems to the extent necessary to enable national networks to be effectively interconnected with one another. International technical forums were found to be necessary for these purposes. The International Telephone Consultative Committee (CCIF) was set up in 1923, the International Telegraph Consultative Committee (CCIT) followed in 1925, and the International Radio Consultative Committee (CCIR) in 1927. (The abbreviations for these bodies were formed from their names in the French language.) The work of these consultative committees was complementary to that of the two Unions and there was collaboration between them.

The emergence of the International Telecommunication Union

Thus by around 1930, five international bodies were meeting separately from time to time to deal with international telecommunications problems, technical, operational, commercial and administrative. Rationalisation was seen to be desirable and in 1932 a rather loose federation of the five organisations was created. The new body was called the International Telecommunication Union (ITU). Codding [1] provides a more detailed account of the ITU and its predecessors.

Substantial changes to the organisational structure of the ITU were agreed at the International Telecommunication Conference at Atlantic City in 1947. For the new structure, see the revised International Telecommunication Convention which the conference produced [2]. A permanent headquarters in Geneva with a general secretariat and a Secretary General came into being. The three consultative committees were more closely integrated with the administrative core. A semi-autonomous group of elected experts, called the International Frequency Registration Board (IFRB) took charge of a specialised secretariat that dealt with the international registration of radio frequency assignments. The Board was also to provide technical and regulatory guidance to Administrations on request. The International Radio Conference, which followed on from the International Telecommunication Conference at Atlantic City in 1947, prepared and agreed a new edition of the Radio Regulations [2]. Finally, the Union became a specialised agency of the United Nations.

In the 45 years that followed 1947 there were organisational changes in the ITU. The absorption into the public telephone network of the transmission facilities used for the public telegraph and related services led in 1956 to the merging of the CCIF and CCIT to form the International Telegraph and Telephone Consultative Committee (CCITT). By 1965 the flood of frequency assignment problems arising from the reconstruction activities that followed World War II had receded and the number of members forming the IFRB, initially eleven, was reduced to five. New functions were added to the Board's responsibilities from time to time. Codding [3] provides a history of the Board. Developments in technology and changing service requirements made it necessary for the CCIR and CCITT to set up new study groups from time to time and other study groups, no longer essential, were discontinued. But basically the organisation of the Union remained until 1993 much as the 1947 Convention had left it.

Criticisms of the ITU in the 1980s

However, shortcomings were perceived in the performance of the Union in the 1980s. The main criticisms were as follows:

1   The regulations and procedures involved in the international management of the spectrum had become very complex and it was thought that their sheer complexity might be impeding optimum spectrum use.

2   Too little was being done by the Union to assist developing countries to modernise and develop their telecommunications facilities.

3   Too much of the time at the periodical meetings of the Plenipotentiary Conference (PP), the supreme organ of the ITU, was being spent on consideration of minor amendments proposed for the International Telecommunication Convention, the 'basic instrument' of the Union. This, it was argued, diminished the time available for more important business and tended to deny the Union the constancy of objectives that was desirable.

4   The drafting and agreeing of system standards for global application was slow at a time when the development of new technology and new systems was fast and accelerating.

5   The running costs of the Union were rising and they were thought to be too high.

These criticisms led to several parallel streams of activity between 1983 and 1995, seeking improvement.

Technical assistance to developing countries

Until recently most of the assistance given by the ITU to developing countries for the development of their telecommunications networks was funded through the United Nations Development Programme. In the late 1980s the level of this funding was about four million Swiss Francs (SF) per annum. There was sustained pressure from developing countries for substantial additional funding to be provided for this purpose from the ITU budget. The Plenipotentiary Conference that met in 1982 (PP-82) had set up an 'Independent International Commission for World-wide Telecommunications Development', its members to be drawn from the highest decision-makers of administrations, operating agencies and industry and from both developed and developing countries; see Resolution 20 attached to the International Telecommunication Convention as revised by PP-82 [4]. The mandate of the Commission was, in brief, to consider all of the ways in which countries might assist one another in telecommunications development, to suggest a range of effective options for achieving progress and to consider how best the ITU could participate in that assistance.

The report of the Commission, which was chaired by Sir Donald Maitland, a senior British diplomatist, was published in 1984 [5]. It made many recommendations, addressed to the developing countries themselves, developed countries, international agencies and operating organisations, financial institutions and manufacturing industry. Of particular relevance in the present context was a proposal that the ITU take urgent action to set up a centre for telecommunications development, comprising a Development Policy Unit, a Telecommunications Development Service and an Operations Support Group. The new centre was indeed set up and PP-89 gave additional emphasis to this aspect of the Union's work by establishing a Telecommunications Development Bureau (BDT) as part of the ITU secretariat. Provision was made for BDT expenditure from ITU funds of up to SF 15 million in 1990, rising to 22.4 million SF in 1994; see Decision 1 in the Final Acts of PP-89 [6]. Management by the ITU of United Nations Development Fund resources for telecommunications development continues.

The basic instrument of the ITU; the Convention

Until 1994 the International Telecommunication Convention was the basic instrument of the Union, setting out its purposes, its organisational structure, its membership, its methods of working and much else. It could be amended by a simple majority of the votes of the Member States at a meeting of the PP. Other specialised agencies of the United Nations have basic instruments that require a larger majority of votes, typically two thirds, to effect an amendment. PP-82 came to the view that the contents of the Convention should be divided between two documents, a constitution containing provisions which are of a fundamental character and a convention containing the other provisions. It was foreseen that a majority greater than 50% should be required for amending the constitution. PP-82 asked for drafts for a two-part text to be prepared for consideration at the next meeting of the PP; see Resolution 62 [4]. However, other major issues had emerged by 1989, as follows.

The preparation of technical standards

The preparation of standards for systems and equipment, to enable equipment from any manufacturer to interface with the public network and corresponding equipment from other manufacturers, was one of the tasks of the CCITT and, to a lesser degree, the CCIR. These consultative committees and their study groups always had a wide range of studies in progress, and they normally worked in a four-year cycle, with completed recommendations being published after approval at one of the quadrennial plenary assemblies of the committees. This procedure worked well enough for the technical studies required to guide the drafting of new regulations, but it was too slow to serve as a basis for system design when the pace of technical innovation accelerated in the mid-1980s. In default of timely global standards from the ITU, manufacturers often designed systems to standards drawn up by national or regional standardising organisations, leading to a lack of global uniformity.

It is arguable that the inadequacy of the Union's performance in this respect was inevitable in a body constituted like the ITU; see, for example, Besen and Farrell [7]. However, PP-89, recognising the importance of global uniformity in telecommunication standards and conscious that the ITU was the only relevant institution with global representation and responsibilities, was determined to reorganise ITU procedures to enable the Union to carry out this vital function effectively.

The High Level Committee and reorganisation of the ITU

PP-89 agreed with PP-82's view that the contents of the Convention of the Union should be divided between two texts, a constitution and a convention. It concluded that the inadequacies in the preparation of technical standards were caused, in large part, by weaknesses in the working methods of the consultative committees and by insufficient participation in the work by manufacturing industry. It also considered that running the Union could be made more cost-effective by changes of organisation, in particular changes that would reduce the duration of major conferences, which are a heavy burden on Union funds.

Accordingly, PP-89 initialled a constitution and a convention, based on the drafts which PP-82 had called for; the texts are to be found in the Final Acts of PP-89 [6]. These new texts changed little of substance of the Convention agreed at PP-82 and subsequently ratified by governments, other than requiring a two-thirds majority for amendment of the constitution. However, the new texts were never ratified by a quorum of Member States and so never entered into force because PP-89 also decided to establish a High Level Committee (HLC) to make an in-depth review of the structure and functioning of the Union and to recommend changes to make the Union more effective and cost-efficient; see Resolution 55 of PP-89 [6].

The main business of the Additional Plenipotentiary Conference in 1992 (APP-92) was to consider the report of the HLC. The Committee had recommended substantial changes to the structure of the Union and these recommendations were, in general, adopted by APP-92. A Constitution and a Convention were agreed, incorporating the substance of the HLC recommendations; see the Final Acts of APP-92 [8]. These new texts entered into force on 1 July 1994 between countries that had ratified them. Various amendments were agreed at PP-94 and PP-98; see the Final Acts of PP-94 [9] and PP-98 [10]. PP-98 decided to publish the Constitution and the Convention, as amended, as a unified text.

The Radio Regulations and the Voluntary Group of Experts

The principal contents of the Radio Regulations (RR) are:

- a table of frequency allocations covering the whole of the radio spectrum that is in use at present;
- a set of procedures intended to enable the spectrum to be used without unacceptable interference from radio stations of one country to those of another;
- various measures to increase the efficiency with which the spectrum is used;
- rules for the day-to-day operation of the maritime and aeronautical mobile services; and
- resolutions and recommendations bearing on these matters.

Since 1947 the RR had grown extensively by the accretion of provisions made by successive administrative radio conferences. Substantial growth had, of course, been made inevitable by the great increase in the complexity of radio interference problems that had occurred during this period, but too little had been done retrospectively to rationalise the procedures and the text which sets them out. This made the RR awkward to use, needlessly so in some cases. Increasingly, their use demanded expertise. Furthermore, at a time when the technologies of telecommunication transmission were converging and many new systems were multi-purpose, the appropriateness of the established frequency allocation system was in question. By its Resolution 8 [6], PP-89 set up a Voluntary Group of Experts (VGE) to review the RR and make proposals for changes.

The VGE proposed extensive amendments to the text of the RR, in particular to simplify the application of the articles concerned with the examination and registration of frequency assignments. It suggested that there was no need to retain within the RR some detailed technical and operational material, since it could be accommodated elsewhere, being incorporated by reference into the RR if necessary. Discontinuation was proposed for some procedures which the IFRB had found ineffectual. A few minor changes were proposed for the international table of frequency allocations and a number of ways of further simplifying the table were suggested for subsequent consideration.

The report of the VGE was considered at the World Radiocommunication Conference in 1995 (WRC-95); see the Final Acts of that conference [11]. The conference adopted most of the Group's proposals for amending the version of the RR then current; namely the edition of 1990, revised in 1994 [12] to incorporate amendments agreed at the World Administrative Radio Conference in 1992. A few points of substance was left for WRC-97 to determine. WRC-95 itself also made substantial amendments to the RR, the outcome of its own agenda. Some of the revised articles of the RR came into force on 1 January 1997 and these were published, together with the resolutions and recommendations agreed at WRC-95, in 1996 [13]. A complete revised edition of the RR, published in 1998 [14], the so-called simplified RR, includes the rest of the WRC-95 amendments plus others agreed by WRC-97. All further reference to the RR in this book is to the edition of 1998 if no earlier edition is identified.

Volume 1 of the 1998 edition of the RR contains the 59 Articles of the Regulations, grouped into six Chapters and subdivided into Sections as required. Every paragraph can be identified by an alpha-numeric group in the form S9.52, where 52 is the number of the paragraph within Article 9, the prefix 'S' (for 'Simplified') being used to distinguish numbering provided by the VGE from previous numbering schemes, now obsolete. Where it has been necessary to insert a new paragraph into the numbering scheme, it has been given the same number as was given to the previous paragraph numbered by the VGE, with the addition of a distinguishing suffix in the form of one or more upper case letters. Thus the numbers of four new paragraphs inserted immediately after paragraph S9.52 are S9.52A to S9.52D.

Volume 2 of the RR contains the 25 Appendices, their numbers also prefixed with 'S'. Volume 3 contains all the Resolutions and Recommendations of past WARCs and WRCs that are still in force; there has been no change in the numbering scheme for these texts and so no prefix is required.

Among the proposals of the VGE that were adopted by WRC-95 was the principle of incorporating texts into the RR by reference. This device enabled the mandatory authority of the RR to be conferred upon a text, typically an operational or technical standard, without adding to the bulk of the regulations. The principle is set out in RR Resolution 27, which also lists the texts, all at present recommendations of the ITU-R, which have in effect been incorporated into the RR in this way. Volume 4 of the 1998 edition of the RR contains the current texts of these recommendations.

The ITU in 1998

Thus the ITU has been changed substantially in organisation and in the form, if not the substance, of its basic instrument and the RR. Much has been done to meet the criticisms of the former organisational structure. The new organisation appears to be functioning well, although it is recognised that further refinement of working methods will probably be necessary. The main features of the new organisation are outlined in Section 2.2 below and the parts of the organisation which are principally involved in the management of spectrum are described in more detail in Section 2.3.

2.2 The Union, post-1994

The purposes of the Union

Article 1 of the ITU Constitution defines the purposes and functions of the Union. The items that relate in particular to the management of the radio spectrum are as follows:

The purposes of the Union are

3   to maintain and extend international cooperation among all its Member States for the improvement and rational use of telecommunications of all kinds;

3A   to promote and enhance participation of entities and organizations in the activities of the Union and foster fruitful cooperation and partnership between them and Member States for the fulfilment of the overall objectives as embodied in the purposes of the Union;

5   to promote the development of technical facilities and their most efficient operation with a view to improving the efficiency of telecommunication services, increasing their usefulness and making them, as far as possible, generally available to the public;

8   to harmonize the actions of Member States and promote fruitful and constructive cooperation and partnership between Member States and Sector Members in the attainment of these ends.

To this end, the Union shall in particular:

11   effect allocation of bands of the radio-frequency spectrum, the allotment of radio frequencies and registration of radio-frequency assignments and, for space services, of any associated orbital position in the geostationary-satellite orbit or of any associated characteristics of satellites in other orbits in order to avoid harmful interference between radio stations of different countries;

12   coordinate efforts to eliminate harmful interference between radio stations of different countries and to improve the use made of the radio-frequency spectrum for radiocommunication services and of the geostationary-satellite and other satellite orbits;

13   facilitate the world-wide standardization of telecommunications, with a satisfactory quality of service;

15   coordinate efforts to harmonize the development of telecommunication facilities, notably those using space techniques, with a view to full advantage being taken of their possibilities;

17   promote the adoption of measures for ensuring the safety of life through the cooperation of telecommunication services;

It should be noted that paragraph 3A is new, having been added to the Constitution at the ITU Plenipotentiary Conference in 1998, and paragraphs 3, 8, 11 and 12 were amended at that conference. These paragraphs enter into force in this form on 1 January 2000 between Member States, at that time Parties to the Constitution and Convention of the International Telecommunication Union (Geneva, 1992), that have deposited before that date their instrument of ratification, acceptance or approval of, or accession to, the amending instrument adopted by the 1998 Plenipotentiary Conference.

Paragraph 1012 of the ITU Constitution defines Telecommunication as

- any transmission, emission or reception of signs, signals, writing, images and sounds or intelligence of any nature by wire, radio, optical or other electromagnetic systems.

and paragraph 1009 of the Constitution defines Radiocommunication as

- telecommunication by means of radio waves.

Paragraph 1005 of the ITU Convention sets an arbitrary upper limit of 3000 GHz on the frequency range covered by the Union's radio regulatory activities, although paragraph 1006 withholds any frequency limit from the scope of the work of the radiocommunication study groups.

Membership, participation and voting

Any state that had been a member of the Union before the entry into force of the new Constitution and any other member of the United Nations that accedes to the new Constitution is automatically a Member State of the Union. Any other state, even though not a member of the United Nations, may become a member of the Union if it accedes to the new Constitution provided that two-thirds of the members approve. In fact, virtually all independent states are Member States of the ITU.

Various other kinds of organisations are encouraged to participate in many of the activities of the Union; see Article 19 of the Convention. These organisations include telecommunications operating agencies that are recognised by the responsible government, scientific and industrial organisations, financial and development institutions dealing with telecommunications, and the United Nations and its specialised agencies. In some cases participation is conditional on the approval of the relevant Member State. Participants, other than Member States, are termed 'Sector Members'.

If it is necessary to resolve an issue by voting at a conference or by correspondence, each Member State has one vote. Delegates of recognised operating agencies may be authorised by their Member State to cast one vote at certain conferences if the state is not represented directly. In no other circumstances may a Sector Member vote.

The 'instruments' of the Union

The instruments of the Union are the Constitution (the basic instrument), the Convention and the Administrative Regulations, the latter consisting of the International Telecommunication Regulations and the Radio Regulations. These instruments have the status of international treaties. By acceding to the Constitution and the Convention, a state binds itself to observe the provisions of the Administrative Regulations also. Thus, for example, a state undertakes to observe the terms of all of the instruments at any radio stations which it operates, and to require the same of any radio stations that it permits others to operate within its jurisdiction if these stations are involved with international links or could cause harmful interference to radio stations outside the state's jurisdiction. These constraints do not apply to national defence services.

The structure of the Union

The main elements of the structure of the Union, illustrated in Figure 2.1, are defined, with their functions, in the Constitution and the Convention.

The *Plenipotentiary Conference* met for the first time under the new instruments of the Union in 1994 and then again in 1998, and its future meetings will normally be at intervals of four years. The principle functions of the PP are:

- to consider proposals for amending the Constitution and the Convention;
- to review recent activities of the Union;

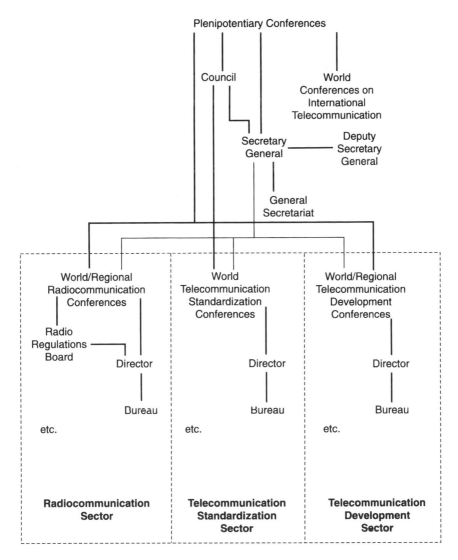

Figure 2.1   The main elements of the structure of the ITU, post-1994

- to determine the main elements of Union policy for the future, including the preparation of a five-year strategic plan;
- to consider relationships between the Union and other international organisations;
- to examine accounts and authorise budgets;
- to elect the states that are to appoint a person to serve on the ITU Council;
- and to appoint the persons who are to serve as the principal officers of the Union (that is, the Secretary General, the Deputy Secretary General and the Directors of the three Bureaux) and as the members of the Radio Regulations Board.

The decisions of the PP are published as the Final Acts of the conference. Final Acts consist mainly of formal resolutions which define action required in any area of the Union's activities, together with changes that have been agreed to the Constitution or the Convention and a Final Protocol recording qualifications that participating Member States have applied to their acceptance of the outcome of the meeting. The minutes of the plenary sessions are also published.

The *ITU Council* meets, usually once per year, to act as the governing body of the Union in the four-year intervals between meetings of the PP. PP-98 chose 46 Member States and invited each to provide a delegate to participate in meetings of the Council, up to the time when a subsequent PP should make another selection. The procedure of the election is designed to ensure that there is an equitable distribution of seats on the Council among all regions of the World.

The *Secretary General*, assisted by the Deputy Secretary General, is responsible for the management of the headquarters organisation, under the direction of the PP and the Council.

The *Sectors*: the Union's primary activities are carried out in three 'sectors', namely the Radiocommunication Sector (ITU-R), the Telecommunication Standardization Sector (ITU-T) and the Telecommunication Development Sector (ITU-D). Subject to the ultimate authority of the PP, the business of the sectors is directed by periodical World Radiocommunication Conferences (WRCs), World Telecommunication Standardization Conferences and World Telecommunication Development Conferences, respectively. Each sector has study groups, the successors of the study groups of the CCIR, the CCITT and the BDT. And each sector has a secretariat, called a Bureau, in the charge of a Director. There are other organs within the sectors, according to their particular requirements; for example in the ITU-R there is the Radio Regulations Board (RRB) that meets up to four times a year; in some respects the RRB is the successor to the IFRB. The structure and functions of the ITU-R are considered in more detail in Section 2.3.

When necessary, the PP convenes a World Conference on International Telecommunications to review and amend the International Telecommunication Regulations. In recent times the intervals between these conferences have been long.

Finance

The ITU budget is expected to be of the order of SF 333 million per annum at 1998 prices between 1999 and 2003. This cost is met by the Member States and, to a limited extent, by the Sector Members.

In most of the specialised agencies of the United Nations, each Member State contributes to the cost of running the agency in proportion to the Gross Domestic Product of the state. However, the ITU still uses the method that it used before the United Nations was formed. Each state chooses freely the number of contributory units that it is willing to pay on a scale ranging from 40 units to one-quarter of a unit. For the least developed countries there is the option of contributing an eighth or a sixteenth of a contributory unit. The total funding requirement is divided by the aggregate number of contributory units that have been offered, and the states are billed accordingly. At present a typical major developed state contributes between SF 10 and 30 million per annum and one of the least developed states contributes about SF 60 000 per annum. A similar scheme applies to Sector Members; for a major Sector Member, a contribution of several hundred thousand Swiss Francs per annum would be typical.

PP-98 decided that the ITU would, in future, recover from Administrations requiring them the cost of some of its products and services. It was foreseen that this policy would yield several advantages. For example, it would ensure that the cost to the ITU of some of its services, such as the frequency coordination of proposed new satellite networks, would fall on Administrations in proportion to the use that they made of the service. Frivolous use of ITU services might also be discouraged. Cost recovery will, no doubt, reduce in time the sums that must be recouped by the voluntary contributions from Members.

If the contributions of a Member State are seriously in arrears, the right to vote may be lost, but this does not deny the Member the right to participate in Union business.

Other ITU activities

The ITU supports a number of activities, meeting temporary, occasional or on-going requirements, in addition to those that are specified in the Constitution or the Convention. These include several series of global forums for the examination of long term issues of importance to global telecommunication for which a formal occasion would be unduly constraining. One of these is the World Telecommunication Policy Forum (WTPF), set up under Resolution 2 of PP-94 [9] and maintained by PP-98. In addition, the quadrennial Telecom events combine a trade fair of global importance with forums and seminars for the discussion of major current issues and the presentation of views on the prospects for the future.

2.3 The Radiocommunication Sector

The ITU-R has four main elements:

1   *World Radiocommunication Conferences* (WRCs) have the primary function of maintaining the effectiveness, in a rapidly changing world, of the RR as an instrument for ensuring the rational, equitable, efficient and economic use of the radio frequency spectrum by all kinds of radio services. More generally, WRCs have ultimate responsibility for all activities in the Radiocommunication Sector, including the Radiocommunication Bureau. Regional Radiocommunication Conferences (RRCs) may be convened to consider matters of concern to only one of the ITU's three geographical Regions.

2   *The Radiocommunication Assembly* (RA). The RA provides briefing, mainly technical, for WRCs. See below.
3   *The Radio Regulations Board.* The twelve members of the RRB are elected by the PP from among nominees of Member States but they are required to be very knowledgible in radiocommunications and spectrum management and to serve on the Board as independent experts. The Board meets up to four times a year to consider matters referred to it by the Bureau that cannot be resolved through the application of the Radio Regulations. Other problems or duties may also be referred to the Board by a competent conference or the Council.
4   *The Radiocommunication Bureau,* under its Director, provides secretariat services for all activities within the ITU-R. However, it has a particular responsibility, under the oversight of the RRB:

- to process frequency assignments notified by Administrations,
- to provide assistance on request to Administrations when the Bureau's processing of notifications of frequency assignments, or operational experience, reveal interference problems, and
- to publish the outcome of this work, in a variety of forms and media, for the information of all Administrations and radio operating organisations.

The basic structure of the Sector is illustrated in Figure 2.2.

World Radiocommunication Conferences

A WRC is normally convened every two to three years. The agenda is normally based on proposals outlined four years in advance, developed in more detail at the next WRC and finalised by the Council in consultation with the Member States. It usually covers a broad range of proposals, mostly having the direct effect or the ultimate objective of amending the RR. On technical, operational, regulatory and procedural matters the WRC has the guidance of the Radiocommunication Assembly (RA). The first WRC was in 1993. RRCs are arranged when the need arises.

Before 1993, the functions that WRCs and RRCs perform were carried out by World Administrative Radio Conferences (WARCs) and Regional Administrative Radio Conferences (RARCs). The agendas of these conferences were usually restricted to one, or in some cases two, major topics, although other matters were also included if they were urgent and limited in scope. In the 1980s there was a substantial number of WARCs, sometimes two in the same year, and the change to regular, multi-purpose conferences that was introduced by APP-92 was intended, in part, to reduce costs.

Before PP-98 the interval between meetings of WRCs was set by the Constitution as two years and the next conference was planned for 1999. Accordingly, many of the texts produced at WRC-97 refer to the subsequent meeting as WRC-99. However, WRC-97 found that its workload was very high, with much preparation to be done before the meeting and much that was new to be assimilated into practice after the meeting. RR Resolution 50, agreed at WRC-97, foresaw that similar problems would occur again and invited PP-98 to reconsider the rigid requirement for biennial WRCs. PP-98 amended the Constitution, which now requires a WRC meeting every two or three years. The next meeting has been fixed for May/June 2000. In this book all references to that meeting take the form WRC-00.

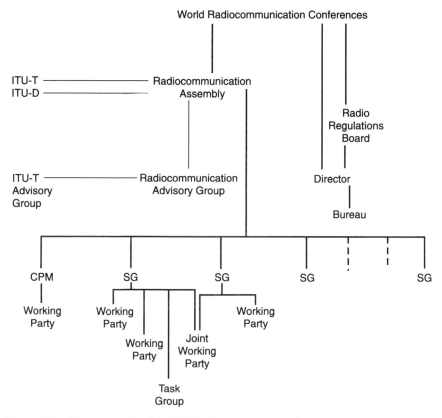

Figure 2.2   The organisation of the ITU Radiocommunication Sector

The outcome of a WRC is published as its Final Acts, consisting mainly of:

- A schedule of additions and amendments to the RR and its appendices.
- New and amended resolutions, typically urging Member States, inviting the PP or the Council, or instructing the RRB, the RA or the Secretary General to implement a course of action. Some resolutions set out a new regulatory procedure which is to be taken into use on a trial basis, pending incorporation into the RR by a subsequent WRC after any amendments found to be desirable have been made.
- New and amended recommendations, typically inviting action or studies, usually in preparation for decisions by future WRCs.
- A Final Protocol recording qualifications that participating Member States have applied to their acceptance of the text.

Every few years, subject to need, the RR are revised in accordance with the amendments agreed by recent WRCs, new Resolutions and Recommendations are annexed to the RR, old Resolutions and Recommendations that a WRC has decided are no longer useful are omitted and the revised text is republished.

Radiocommunication Assemblies

The RA meets immediately before a meeting of the WRC and at the same place. The primary functions of the RA are to provide a basis, mainly technical, for the decisions of WRCs and RRCs and to publish recommendations on the use of the radio spectrum that will promote the purposes defined in the Constitution. Mention should be made here of the recommendations which take the form of standard system specifications for radio equipment, often prepared jointly with ITU-T. The working methods of the RA are set out in Resolution ITU-R 1-1 [15].

A number of Radiocommunication Study Groups (SGs) work under the direction of the RA, studying problems and preparing texts in response to questions defined by the RA. See Resolution ITU-R 5 [15] for the current list of questions. A recommendation prepared by an SG gives specifications, data or guidance on the preferred way of undertaking a specified task or a preferred procedure for a specified application which is considered to be sufficient to serve as a basis for international cooperation. An ITU-R recommendation normally requires the approval of a meeting of the RA before it is published, although there is a procedure for getting speedy approval for urgent recommendations; see below. These recommendations are the successors to the corresponding texts of the CCIR.

The study groups are concerned mainly with technical and, to a lesser degree, operational problems but the RA and WRC also need advice on regulatory and procedural issues. Under Resolution ITU-R 38 [15] RA-95 set up the Special Committee on Regulatory/Procedural Matters to study these matters.

The RA maintains liaison with ITU-D; see Resolution ITU-R 7 [15]. One of the main objectives of this liaison is to ensure that the work programme set out in the questions which the RA puts to its study groups takes fully into account the requirements perceived by developing countries. The RA also seeks to facilitate the involvement of developing countries in the work of ITU-R. One task of ITU-D is the preparation and up-dating of handbooks on telecommunication systems, in particular to assist developing countries in the development and maintenance of their networks; the ITU-R study groups assist in this work.

Close collaboration between ITU-R and ITU-T is essential, especially in the development of standards for systems of concern to both sectors. Resolution ITU-R 6 [15] sets out the basis for this collaboration.

Thus, the output of a meeting of the RA consists mainly of:

- resolutions and opinions, the former directing the work of the RA and its dependent groups and the latter containing suggestions or requests addressed to bodies outside the RA's authority; and
- recommendations prepared by the study groups and approved by the RA for publication.

The Radiocommunication Advisory Group

With the authority of Resolution 3 of APP-92 [8], RA-93 set up the Radiocommunication Advisory Group; see also Resolutions 16 and 17 of PP-94 [9] and Resolution ITU-R 3-1 [14] of RA-95. The function of this Group is to provide broadly based advice on an informal basis to the Director of the Radiocommunication Bureau on the business of the Radiocommunication Sector. As an aid to the discharge of this function, the Group maintains direct liaison with the Telecommunication Standardization Advisory Group, the corresponding body in the ITU-T.

The Radiocommunication Study Groups

The SGs working within ITU-R are semi-permanent, new groups being set up and existing groups being discontinued as need and economy dictate. At present there are eight SGs, their fields of activity being as follows:

SG 1   Spectrum management.
SG 3   Radiowave propagation.
SG 4   The fixed-satellite service.
SG 7   The science services.
SG 8   The mobile, radiodetermination and amateur services, including the associated satellite services.
SG 9   The fixed service.
SG 10   Sound broadcasting, terrestrial and by satellite.
SG 11   Television broadcasting, terrestrial and by satellite.

In addition, the ITU-R organises the Coordination Committee for Vocabulary, which liaises with the study groups of all three sectors and with interested bodies outside the ITU, with the objective of standardising telecommunications terminology.

Study Groups set up working parties which meet as required, each being assigned one or more of the questions put by the RA to the SG, to study and draft a response, usually in the form of a draft recommendation or amendments to an existing recommendation. Task groups are also set up to work on a single urgent question against a stringent timetable. Where a question falls within the responsibilities of more than one SG, a joint working party or a joint task group may be set up to deal with it. Each SG meets once or twice in the two-year period between meetings of the RA to deal with questions and other business which have not been devolved to dependent groups and to consider the draft texts produced by the working parties and task groups.

When a formal response to a question is necessary but the state of work is not sufficiently advanced for a recommendation to be drafted, a 'report' may be prepared. In general, ITU-R reports are not published, and in this respect the practice of the ITU-R differs from that of the CCIR, which published all its reports. However, a report may be published if there is a need for information on the state of the work to be made generally available.

Reports and handbooks may be approved for publication by the study group that produces them. Most recommendations are referred to the next meeting of the RA for approval and publication.

However, special procedures arise in two cases:

- If publication of a recommendation is urgent, there are procedures whereby, without waiting for a meeting of the RA, a study group meeting may initiate through the Director of the Bureau a process of direct consultation with the Member States, approval of the text being dependent on 70% of replies being favourable; see Resolution ITU-R 1 [15].
- If an existing recommendation has been incorporated into the RR by reference, and has thereby been given mandatory status, a revision of that recommendation

may be approved by the RA and published in the usual way, but the previous version retains its mandatory status until the revised version has been approved by a WRC; see RR Resolution 28 [14].

Conference Preparatory Meetings

A consolidated report on regulatory, technical, operational and procedural aspects of the agenda of each WRC is drawn up under the authority of the RA and distributed to all Member States before the meeting. This report is prepared by the Conference Preparatory Meeting, drawing on the expertise of the ITU-R study groups, the Special Committee on Regulatory/Procedural Matters, any other ITU groups engaged in relevant work, and on documents submitted to it directly by Member States; see Resolution ITU-R 2-1 [15]. In order that this time-critical activity should not be constrained by the timing of the RA meetings, responsibility for approving the report is devolved by the RA to the Conference Preparatory Meeting itself.

Numbering and publication of ITU-R recommendations and reports

Every fourth year the ITU publishes all ITU-R recommendations that are in force at that time, about 750 in 1997, including CCIR recommendations that are still in force. The texts are bound in a number of 'series', some comprising several volumes or 'parts', each containing texts covering a broad topic area. In 1997 for example there were 17 series totalling 25 volumes. Volumes are identified by the year of issue, one or more letters identifying the series and a part number if necessary. New or amended recommendations approved in the years between the quadrennial issue are published in interim volumes with similar identifying codes. For a few recommendations of the CCIR, still in force but not yet integrated into the ITU-R documentation, it is necessary to refer to the last series of volumes of the CCIR recommendations, published in 1990 [16].

A new recommendation is numbered, using an alpha-numerical group in the form ITU-R xxx.yyyy, where xxx is the identifying code of the series in which the recommendation is to be found and yyyy is the number assigned to the recommendation. The first version of a recommendation approved in 1993 or since is numbered in a single series starting from 1001. Recommendations approved by the CCIR have numbers below 1000, also in a single series. When a recommendation is revised, it retains its former number, with the addition of a hyphen followed by a number to identify this, the latest of perhaps a series of revisions. Thus, let a current text, Recommendation ITU-R P.453-5, be taken as an example; the original version of this text would have been approved by the CCIR as Recommendation 453 but the current version is the fifth revision of it; it is to be found in Series P of the current (1997) quadrennial set of ITU-R recommendations. It so happens that Series P, in the 1997 set, is published in two volumes and a complete reference to that recommendation would include the fact that it is in Part 1 of Series P.

Reports are numbered and some are published in a similar way, although without the automatic quadrennial reprinting which applies to recommendations. It may be noted that useful information is still to be found in some of the more recent CCIR reports, which are contained in the annexes to the volumes of CCIR recommendations [16].

2.4 References

1 CODDING, G.A.: 'Evolution of the ITU', *Telecommunications Policy*, 1991, **15**, (4), pp. 271–285
2 'The International Telecommunication Convention and the Radio Regulations' included in the Final Acts of the International Telecommunication and Radio Conferences, Atlantic City, 1947 (The ITU, 1947)
3 CODDING, G.A.: 'The 1989 ITU Plenipotentiary Conference and the IFRB', *Telecommunications Policy*, 1988, **12**, (3), pp. 234–242
4 International Telecommunication Convention; Nairobi 1982 (ITU, Geneva, 1982)
5 'The Missing Link' (the report of the International Commission for World-wide Telecommunications Development) (ITU, Geneva, 1984)
6 'Final Acts of the Plenipotentiary Conference', Nice, 1989 (ITU, Geneva, 1989)
7 BESEN, S.M., and FARRELL, J.: 'The role of the ITU in standardisation', *Telecommunications Policy*, 1991, **15**, (4), pp. 311–321
8 'Final Acts of the Additional Plenipotentiary Conference', Geneva, 1992 (ITU, Geneva, 1992)
9 'Final Acts of the Plenipotentiary Conference', Kyoto, 1994 (ITU Geneva, 1994)
10 'Final Acts of the Plenipotentiary Conference', Minneapolis, 1998 (ITU, Geneva, 1998)
11 'Final Acts of the World Radiocommunication Conference' (WRC-95) (ITU, Geneva, 1996)
12 'Radio Regulations', Volumes 1, 2 and 3. 1990 edition, revised 1994 (ITU, Geneva, 1994)
13 'Radio Regulations', Volume 4 (ITU, Geneva, 1996)
14 'Radio Regulations', Volumes 1, 2, 3 and 4. 1998 edition (ITU, Geneva, 1998)
15 'Radiocommunication Assembly, Book of Resolutions and Opinions' (ITU, Geneva, 1998)
16 'XVIIth CCIR Plenary Assembly, Düsseldorf 1990'. 15 Volumes plus Annexes (ITU, Geneva, 1990)

Chapter 3

Frequency allocation

3.1 The international frequency allocation table

The designation of frequency ranges

The radio spectrum that has been taken into use for radio systems extends from about 10 kHz to about 80 GHz, and every year sees an extension into higher frequencies. The international table of frequency allocations in RR Article S5 divides this vast range, together with an extension upwards to 275 GHz that is not yet in substantial use, into many hundreds of precisely defined frequency bands for the purpose of allocation. However, there is also a need for means for designating broad frequency ranges within the radio spectrum and the standard ways of doing so are set out in RR Article S2 and, in more detail, in Recommendation ITU-R V.431-6 [1]. The most commonly used terms are shown in Table 3.1. Other sets of designations are sometimes used, in particular in broadcasting and satellite communication; see Sections 7.5 and 9.1 below, respectively.

The terms 'microwave' and 'millimetre wave' are often used in the literature although they have no generally accepted definition. Where used in this book, microwave implies the range 1–20 GHz and millimetre wave implies frequencies above 10 GHz.

Table 3.1   Designations of frequency ranges

Frequency band	Band number	Symbol
3–30 kHz	4	VLF
30–300 kHz	5	LF
300–3000 kHz	6	MF
3–30 MHz	7	HF
30–300 MHz	8	VHF
300–3000 MHz	9	UHF
3–30 GHz	10	SHF
30–300 GHz	11	EHF

The international table of frequency allocations

Frequency allocation involves the division of the radio spectrum into frequency bands, wide or narrow according to need, each band to be used, ideally, for one specified category of radio system, called a radio service. The international table fills many pages but Figure 3.1 shows a sample page. The principal features to note are:

- the radio services for which the bands are allocated;
- some allocations are world-wide but for some frequency bands there are regional differences in the allocations;
- most frequency bands are in fact allocated, not for one service but for several; these bands are said to be 'shared';
- many allocations are qualified by footnotes.

The radio services

There are about 35 radio services in all, the exact number depending on the criteria used in counting them. Formal definitions of the services are to be found in RR S1.19 to S1.58. However, they fall conveniently into a few groups, as follows.

The fixed service (FS) consists of radio stations in specified terrestrial locations and the radio links between them. Stations at fixed, specified locations on Earth and links between them routed via one or more satellites form the fixed-satellite service (FSS). Feeder links, which connect satellites of certain other services, such as the mobile-satellite and broadcasting-satellite services, to earth stations at specified, fixed locations may also be included in the FSS.

The broadcasting service (BS) comprises terrestrial transmitters and their emissions for direct reception by the general public. Corresponding transmissions from satellites form the broadcasting satellite service (BSS). As mentioned above, the feeder links between broadcasting satellites and the earth stations that supply the programme signals are part of the FSS.

The mobile service (MS) consists of mobile radio stations, stations at fixed locations on land that communicate directly with them, and the radio links that they use for that purpose. These mobile stations may be on ships, aircraft, land vehicles or hand-portable. Direct communication between one mobile station and another is also part of the MS. Similarly, links between a mobile station and a satellite form the mobile-satellite service (MSS), although the feeder links between these satellites and stations at fixed locations on land through which the mobile station gets access, for example, to the public switched telephone network (PSTN) are usually treated as part of the FSS. In addition to these two general services, the MS and the MSS, which can cover any kind of mobile station, six more services have been defined to enable frequency bands to be allocated more specifically, namely the maritime mobile service (MMS), the aeronautical mobile service (AMS), the land mobile service (LMS), the maritime mobile-satellite service (MMSS), the aeronautical mobile-satellite service (AMSS) and the land mobile-satellite service (LMSS).

Among the frequency bands allocated for the AMS and the AMSS, some are reserved for air traffic control and safety messages for aircraft flying on civil air routes. These allocations are designated AMS(R) or AMS(R)S, the letter R signifying 'Route'. Other AMS or AMSS allocations may be qualified by the indicator (OR), signifying 'Off-route'; these bands are used for aircraft on other missions.

2170–2520 MHz

Allocation to services		
Region I	Region 2	Region 3
2170–2200	FIXED MOBILE MOBILE-SATELLITE (space-to-Earth) S5.388 S5.389A S5.389F S5.392A	
2200–2290	SPACE OPERATION (space-to-Earth) (space-to-space) EARTH EXPLORATION-SATELLITE (space-to-Earth) (space-to-space) FIXED MOBILE S5.391 SPACE RESEARCH (space-to-Earth) (space-to-space) S5.392	
2290–2300	FIXED MOBILE except aeronautical mobile SPACE RESEARCH (deep space) (space-to-Earth)	
2300–2450 FIXED MOBILE Amateur Radiolocation S5.150 S5.282 S5.395	**2300–2450** FIXED MOBILE RADIOLOCATION Amateur S5.150 S5.282 S5.393 S5.394 S5.396	
2450–2483.5 FIXED MOBILE Radiolocation S5.150 S5.397	**2450–2483.5** FIXED MOBILE RADIOLOCATION S5.150 S5.394	

2483.5–2500
FIXED
MOBILE
MOBILE-SATELLITE
  (space-to-earth)
Radiolocation
 	
S5.150 S5.371 S5.397 S5.398
S5.399 S5.400 S5.402
	2483.5–2500
	FIXED
	MOBILE
	MOBILE-SATELLITE
	  (space-to-earth)
	RADIOLOCATION
	RADIODETERMINATION-SATELLITE
	  (space-to-Earth) S5.398
	
	S5.150 S5.402
		2483.5–2500
		FIXED
		MOBILE
		MOBILE-SATELLITE
		  (space-to-earth)
		RADIOLOCATION
		Radiodetermination-satellite
		  (space-to-Earth) S5.398
		
		S5.150 S5.400 S5.402

2500 2520
FIXED S5.409 S5.410 S5.411
MOBILE except aeronautical
  mobile
MOBILE-SATELLITE
  (space-to-Earth) S5.403
S5.405 S5.407 S5.408 S5.412
S5.414
	2500 2520
	FIXED S5.409 S5.4.11
	FIXED-SATELLITE (space-to-Earth) S5.415
	MOBILE except aeronautical mobile
	MOBILE-SATELLITE (space-to-Earth) S5.403
	
	S5.404 S5.407 S5.414 S5.415A	

Figure 3.1 A sample page from the international table of frequency allocations (RR, Article S5) showing the framed part of the table covering the frequency range 2170–2520 MHz

Two special categories of MMS traffic are known as the port operations service and the ship movement service. Channels are allotted for these purposes in the plan for the MMS band around 156 MHz (see RR Appendix S18, note k) but there are no separate allocations for them and they are treated as part of the MMS in this book.

Allocations for the amateur service (AmS) and the amateur-satellite service (AmSS) are for the use of radio amateurs.

The technical and scientific services use radio in the course of space exploration, surveying the Earth and for similar activities. This group comprises the space research service (SR), the Earth exploration-satellite service (EES), the meteorological-satellite service (MetS), the meteorological aids service (MetA), the radio astronomy service (RAS), the standard frequency and time signal service (TFS) and the standard frequency and time signal-satellite service (TFSS). It is convenient to include the space operation service (SO) in this group, although it is used for links that are needed, not directly for a scientific purpose but for establishing and maintaining spacecraft and satellites in their intended trajectories and orbits.

There are also the radiodetermination service (RDS) and the radiodetermination-satellite service (RDSS), in which radio waves are used for measuring the location, distance, relative motions, or other physical characteristics of objects. Within the RDS, systems are categorised into the radionavigation service (RNS) or the radiolocation service (RLS), the former covering systems used in the navigation of ships or aircraft and the latter covering other RDS systems. More selective allocation within the RNS has been provided for by the definition of the maritime radionavigation service (MRNS) and the aeronautical radionavigation service (ARNS). Differentiation within the RDSS is provided for by the radionavigation-satellite service (RNSS), the maritime radionavigation-satellite service (MRNSS), the aeronautical radionavigation-satellite service (ARNSS) and the radiolocation-satellite service (RLSS).

In almost every case where a frequency band is allocated for a satellite service, the permitted direction of transmission, space-to-Earth or Earth-to-space (downlinks and uplinks, respectively), is specified in the table. A few bands allocated, for example, for the SR may be used for direct links between spacecraft of that service. However, bands allocated for the inter-satellite service (ISS) may be used for direct links between satellites, in most cases regardless of their service.

There is no standard code of abbreviations for the names of radio services. Those used above and throughout this book are widely used, but other abbreviations may also be encountered in the literature.

The ITU Regions

Uniform frequency allocations world-wide are essential for some purposes and would be desirable for all. However, agreement has not always been reached to allocate a frequency band everywhere for the same services. When it is important for one country or a few countries to have allocations in a band that differ from those of the rest of the world, a departure from the world-wide standard has sometimes been agreed in that limited area, the terms of the agreement being recorded in a footnote to the table. In other circumstances a more suitable solution has been to divide the world into regions which have different allocations in that band. In such a case, the ITU divides the world into three Regions which are defined in RR S5.3 to S5.9; see Figure

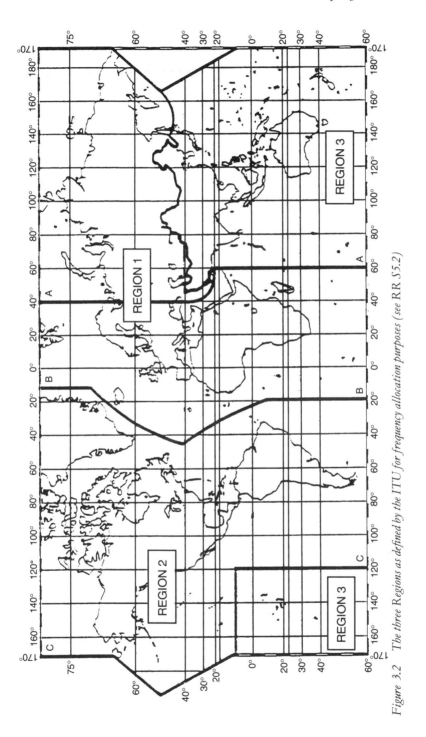

Figure 3.2 The three Regions as defined by the ITU for frequency allocation purposes (see RR S5.2)

3.2. Where reference is made in this book to these Regions, the word is printed with an upper case initial.

Frequency band sharing

It is particularly desirable for some services to be allocated the exclusive use of a frequency band. This is true, for example, of the principal allocations for the MMS, the AMS and the BS in the parts of the spectrum where interference may occur at great distances and there are a number of exclusive allocations in, for example, the HF band. Elsewhere in the spectrum exclusive allocations are relatively rare; thus there is none in the frequency range shown in Figure 3.1, where all the frequency bands are allocated for at least three services. This joint use of bands is called 'sharing'.

Often the allocation of a band for two or more relatively compatible services raises no major international spectrum management problems, and may give valuable flexibility to spectrum use at a national level. In other cases, particularly where stations of the services involved may cause or suffer interference over great distances, sharing has been made feasible by the application of sharing constraints or mandatory frequency coordination procedures, or both. Sharing constraints are limitations placed on the parameters of systems in order to reduce the likelihood of significant interference. Mandatory frequency coordination is a procedure, involving a joint computation of prospective interference levels by the Administrations concerned before a radio station starts to use a new frequency, to ensure that interference will not be unacceptably high. These methods are considered in some detail later in this book; see in particular Sections 5.2 and 5.3.

It has not always been possible to secure international agreement for bands to be shared by different services on equal terms. Accordingly, there are two categories of allocation, primary and secondary. An emission from a transmitting station of a service with a secondary allocation may not cause harmful interference at a receiving station in another country of a service which has a primary allocation. Also, the RR offer no remedy if a receiving station of a secondary service suffers harmful interference from a transmitting station belonging to a primary service. However, a number of other factors also affect the relative status of frequency assignments in the event of interference, and these factors are brought together in Section 4.5 below.

Primary service allocations are printed in the table in upper case letters, whereas only the initial letter is in the upper case for a secondary service; see, for example, the allocation to the AmS, a secondary allocation, in the band 2300–2450 MHz in Figure 3.1.

The footnotes to the table

The international table of frequency allocations takes two forms, one consisting of entries 'within the framed part of the table' (as shown in Figure 3.1) and the other in the form of footnotes, each treated as a paragraph of the RR. These footnotes have been agreed by a WRC or a WARC, no less than the entries within the frame and the two parts have equal status. No fewer than 33 footnotes apply to the frequency range covered by Figure 3.1 and Figure 3.3 shows a selection of them. Footnotes serve various purposes, most commonly the following.

1   *Additional allocations.* A footnote may make an allocation, primary or secondary, for a service in the band, or a stated part of the band, that is additional to the services

S5.388   The bands 1885–2025 MHz and 2110–2200 MHz are intended for use, on a worldwide basis, by administrations wishing to implement International Mobile Telecommunications-2000 (IMT-2000). Such use does not preclude the use of these bands by other services to which they are allocated. The bands should be made available for IMT-2000 in accordance with Resolution **212 (Rev.WRC-97)**. (WRC-97)

S5.391   In making assignments to the mobile service in the bands 2025–2110 MHz and 2200–2290 MHz, administrations shall not introduce high-density mobile systems, as described in Recommendation ITU-R SA.1154, and shall take that Recommendation into account for the introduction of any other type of mobile system. (WRC-97)

S5.392A   *Additional allocation:* in Russian Federation, the band 2160–2200 MHz is also allocated to the space research service (space-to-Earth) on a primary basis until 1 January 2005. Stations in the space research service shall not cause harmful interference to, or claim protection from, stations in the fixed and mobile services operating in this frequency band.

S5.393   *Additional allocation:* in the United States; India and Mexico, the band 2310–2360 MHz is also allocated to the broadcasting-satellite service (sound) and complementary terrestrial sound broadcasting service on a primary basis. Such use is limited to digital audio broadcasting and is subject to the provisions of Resolution **528 (WARC-92)**. (WRC-97)

S5.395   In France, the use of the band 2310–2360 MHz by the aeronautical mobile service for telemetry has priority over other uses by the mobile service

S5.396   Space stations of the broadcasting-satellite service in the band 2310–2360 MHz operating in accordance with No. **S5.393** that may affect the services to which this band is allocated in other countries shall be coordinated and notified in accordance with Resolution **33 (Rev.WRC-97)**. Complementary terresrial broadcasting stations shall be subject to bilateral coordination with neighbouring countries prior to their bringing into use.

S5.397   *Different category of service:* in France, the band 2450–2500 MHz is allocated on a primary basis to the radiolocation service (see No. **S5.33**). Such use is subject to agreement with administrations having services operating or planned to operate in accordance with the Table of Frequency Allocations which may be affected.

S5.399   In Region 1, in countries other than those listed in No. **S5.400**, harmful interference shall not be caused to, or protection shall not be claimed from, stations of the radiolocation service by stations of the radiodetermination satellite service.

S5.400   *Different category of service:* in Angola, Australia, Bangladesh, Burundi, China, Eritrea, Ethiopia, India, the Islamic Republic of Iran, Jordan, Lebanon, Liberia, Libya, Madagascar, Mali, Pakistan, Papua New Guinea, Dem. Rep. of the Congo, Syria, Sudan, Swaziland, Togo and Zambia, the allocation of the band 2483.5–2500 MHz to the readiodetermination-satellite service (space-to-Earth) is on a primary basis (see No. **S5.33**), subject to agreement obtained under No. **S9.21** from countries not listed in this provision. (WRC-97)

S5.407   In the band 2500–2520 MHz, the power flux-density at the surface of the Earth from space stations operating in the mobile-satellite (space-to-Earth) service shall not exceed $-152\,\mathrm{dB}$ $(\mathrm{W/m^2/4\,kHz})$ in Argentina, unless otherwise agreed by the administrations concerned.

S5.409   Administrations shall make all practicable efforts to avoid developing new tropospheric scatter systems in the band 2500–2690 MHz.

S5.412   *Alternative allocation:* in Azerbaijan, Bulgaria, Kyrgzstan, Turkmenistan and Ukraine, the band 2500–2690 MHz is allocated to the fixed and mobile, except aeronautical mobile, services on a primary basis. (WRC-97)

Figure 3.3   A selection of the footnotes to the international table of frequency allocations, applicable to the frequency range 2170–2520 MHz (See Figure 3.1)

listed in the framed part of the table. There are footnotes which add an allocation to a band world-wide (see, for example, RR S5.479) but they are few. Most footnote allocations are limited to named countries; see, for example, RR S5.392A and RR S5.393 in Figure 3.3. Sometimes agreement to the addition of an allocation in this way has been made conditional upon a limitation of the rights of stations of foot-note service with regard to interference to or from stations of services with alloca-tions in the framed part of the table, as in RR S5.392A. Sometimes sharing constraints are imposed on the stations of the footnote allocation. In other cases, a mandatory coordination procedure is imposed, as in RR S5.396. For the rights obtained by means of a footnote of this kind, see RR S5.35 to S5.37.

2   *Alternative allocations.* Instead of the allocations in the framed part of the table, a foot-note may state that there are different allocations in named countries in the band, or a stated part of the band. For example, RR S5.412 (see Figure 3.3) has the effect of omitting the MSS allocation in the band 2500–2520 MHz in the five countries named. For the rights obtained by means of a footnote of this kind, see RR S5.39 to S5.41.

3   *Different category of service.* These footnotes have the effect of changing, for the coun-tries named, the category of allocations in the table from primary to secondary or vice versa. See, for example, RR S5.397 and RR S5.400 (see Figure 3.3). It may be noted that these footnotes, together with RR S5.399, include protection of the status of established stations in other countries that might be affected by this change of category.

4   *General footnotes.* These may be technical, operational or administrative but the three general footnotes shown in Figure 3.3 all impose, or seek to impose, restrictions on the use of a frequency band for one allocated service in order to facilitate sharing with another allocated service. Thus, RR S5.391 requires that the allocation for the MS in the band 2200–2290 MHz should not be used for high density land mobile systems, such as major cellular radiotelephone systems, in order to protect inter-satellite links and downlinks of the SO, the EES and the SR, which share the band. RR S5.407 applies a sharing constraint on the power of MSS satellite emissions in Argentina in the band 2500-2520 MHz to protect other services with allocations in that country. RR S5.409 calls for a world-wide halt to the growth of tropospheric scatter systems in the same band to protect line-of-sight systems of the allocated services.

5   *Advisory footnotes.* These footnotes have no regulatory force. Thus RR S5.395 provides information on the use made of the band 2310–2360 MHz in France. RR paragraph S5.388 recalls that the bands 1885–2025 MHz and 2110–2200 MHz are intended for the implementation of International Mobile Telecommunications-2000 (IMT-2000) systems (formerly called the future public land mobile telecommunication systems (FPLMTS)), on a world-wide basis.

There is a tendency for the footnotes to become more numerous and more complex. RR Resolution 26 calls for standardisation and restraint in their use.

Industrial, scientific and medical use of radio frequency energy

In addition to radiocommunication services which depend for their functioning on the radiation of radio frequency (RF) energy and its reception at a distance, there are other applications for RF energy that do not involve deliberate radiation outside the limits of the apparatus in which it is used. The most important of these involve industrial, scientific or medical (ISM) apparatus. A considerable amount of RF energy may be radiated unintentionally from such installations, capable of causing severe interference at radio stations nearby. Some of this unwanted radiation is wide-band and noise-like but some is narrow-band in nature, the designer of the apparatus having close control over the frequency band within which the energy is generated. A number of frequency bands have been designated for use by ISM apparatus, details being given in 'general' footnotes to the table; see also Section 4.2.3 below. These bands are also allocated for radio services, but radio stations assigned a frequency within one of these designated bands must tolerate interference from ISM apparatus.

3.2 International frequency allocations and changing needs

The case for international frequency allocation

The mixture of radio services for which the Administration must provide assignments at any point in time is not the same for all countries. It depends on such factors as:

- the stage of development that the country has reached,
- the density of the population,
- the impact of the climate on radio propagation,
- the history of the frequency assignment and telecommunications liberalisation policies of that country,
- the mixture of telecommunications transmission facilities, cabled, by terrestrial radio or by satellite radio, that is currently available and required.

Furthermore, the mixture changes, at a different rate for different countries, as users' needs develop and technical innovation provides new options.

Frequency band sharing in the framed part of the international table provides flexibility which helps Administrations to meet their different requirements. Additional flexibility is provided by differences in the allocations for the three Regions and by footnotes to the table which vary the allocations within more limited geographical areas. Furthermore, as discussed in Section 3.3 below, Administrations have reserved to themselves freedom to use the radio spectrum within their jurisdiction in any way they wish, provided that international interference problems do not arise.

Nevertheless there must be occasions when the adoption of a frequency allocation impedes the making of new frequency assignments. Also, current developments in the technology of telecommunications, in particular the trend towards digital transmission, are creating anomalies in the application of the principle of frequency allocation. The technologies, for example, of broadcasting and mobile communication, of broadcasting and data communication, and of mobile communication and radiolocation are tending to converge and the same emissions are sometimes used to provide facilities that belong to more than one service. Whether a comprehensive, internationally agreed, scheme of frequency allocations is, on balance, beneficial to the users of the spectrum nowadays is a question for consideration.

Experience has shown that international frequency allocation is virtually essential for some services and beneficial for others. The most compelling reasons for taking this view are as follows:

1   Ships and aircraft need to operate their communication and radionavigation equipment wherever they may be, and this would not be practicable unless the frequencies that they may use are virtually uniform world-wide. A need for world-wide harmonisation of some land mobile allocations is emerging.

2   Much of the HF frequency range is used for fixed links and broadcasting, much of it being international. A large fraction of the SHF frequency range and a growing fraction of the UHF range are used for satellite systems, most of which are multinational in use. The choice of frequency bands to be used for such purposes must be acceptable to all the Administrations involved if these systems are to operate effectively.

3   The total number of radio systems that can operate satisfactorily within given geographical limits and in a given bandwidth is increased if the range of the parameters of the various systems is minimised. Grossly inefficient use of spectrum may arise, for example, if mobile stations share a band with fixed or broadcasting stations, if systems which use very sensitive receivers share with systems with very insensitive receivers, or if systems which are continuously operating share with systems that are used intermittently, in the same broad geographical area.

4   The bandwidth required in the UHF and SHF ranges for both terrestrial and satellite services is very great. Some terrestrial systems can share frequency bands with some satellite systems without a significant loss of usable spectrum from either service, subject to the application of constraints on the parameters of emissions and the implementation of mandatory frequency coordination procedures. However, these constraints are disadvantageous for other systems and the coordination procedures are labour-intensive; without uniformity of frequency allocation the disadvantages would be spread unnecessarily.

5   By allocating a frequency band for a narrowly defined purpose, it becomes feasible, and may be found expedient, to devolve spectrum management functions for international services to a recognised, expert international agency. Thus, for example, a number of bands allocated for AMS(R) are managed world-wide by ICAO.

6   Countries are not all alike in the priority they give to the various new applications that can be foreseen for radio, and development and investment for some important applications has lagged behind perceived needs. International agreement on a table of frequency allocations provides a means for establishing a negotiated division of the available spectrum between services, enabling bandwidth to be reserved for such needs.

There seems to be little doubt that the principle of frequency allocation is useful, indeed necessary, although there might be a need for change in the ways in which it is implemented. The VGE was asked by PP-89 to consider the need for changes to the service definitions and more general changes to the way in which the spectrum is allocated, in order to maximise the efficiency of spectrum use and to provide more satisfactorily for multifunctional radio systems. However, the VGE recommended no significant changes in this respect in the short term, a view that was accepted by WRC-95. It may be supposed that this is a matter that will continue to receive attention. Perhaps, for example, as the trend towards the use of digital transmission for combinations of convergent applications progresses further, some frequency bands will be allocated, not for the BS, the MS, the FS or the RDS, but for wide-band multiplexed emissions with area coverage transmitting a range of facilities which would at present be classified with all of these services.

Accommodating change of international requirements

New services need allocations, sometimes of very large bandwidth. For example, by stages since 1963 about 55% of the spectrum between 1 and 20 GHz has been allocated for satellite services. In the 1980s, a substantial number of new LMS allocations were made between 800 and 2000 MHz to accommodate the explosive growth in radiotelephone systems. In the 1990s it is the turn of LMSS systems using satellites in non-geostationary orbits (non-GSO). Almost all of these new allocations share with the previous allocations of the bands. In addition to virtually new requirements like these, the requirements for some established services have grown considerably, above all wide-band FS systems, although the demand for other services has shrunk. Despite the various forms of flexibility referred to above, major national spectrum availability problems often arise from developments like these, especially in situations where interference can be propagated over great distances.

An Administration may find that the demand for a radio service within its jurisdiction is growing, so that the bandwidth that is allocated for it in the international table within the frequency range that is technically suitable is becoming exhausted. Solutions to the immediate problem may be found.

- The efficiency of spectrum management that the Administration provides may be improvable.
- The bandwidth per emission required for new systems may be reduced by technical means.
- If the allocations used for the growing service are shared locally, it may be possible to make room for the growth by transferring elsewhere the assignments to stations of the sharing service.
- Frequencies to meet the growing demand may be assigned in a frequency band in which there is no international allocation for the service, provided that harmful interference to and from the stations of other countries does not arise; see Section 3.3.
- The Administration may continue for a while to assign frequencies to meet new demands in frequency bands that are already fully loaded, so increasing interference levels. Alternatively, it may raise licence fees or refuse to license new stations, suppressing growth of traffic or diverting it to other media.

But if none of these options enables a long term solution to be reached, the final option is to secure the agreement of a WRC to an amendment of the international table which provides more bandwidth for the growing service.

It may be desirable for the new allocation to be exclusive; this is seldom achievable but success is not unknown. Thus between 1947 and 1979, most of the HF range, which is uniquely useful for long distance ionospherically propagated links, was allocated in narrow exclusive bands for the FS, the BS and the MMS. These bands, more specifically those between 4 and 23 MHz, were very heavily loaded with assignments and interference was rife. There were repeated proposals for bandwidth to be transferred from one of these services to another to relieve the pressure on the latter but no transfers were agreed. However, the use of repeatered submarine cables or satellites for long fixed links eventually reduced the pressure on the HF FS allocations and it was agreed at WARC-79 that some of the FS spectrum could be reallocated for other services, given a period of time for FS systems still operating in this medium to be transferred to other HF bands. After 10 years for some parts of the spectrum, 15 years for others, a number of bands from which the FS had withdrawn became new exclusive allocations for the BS or the MMS.

There is one other set of circumstances which sometimes allows new exclusive allocations to be made. The international table has been extended upwards in frequency from time to time, providing allocations to be taken into use when the technical means for operating at higher frequencies became available. Thus in 1947 the upper limit of the allocated spectrum was raised from 200 MHz to 10.5 GHz. In 1959 the upper limit was raised to 40 GHz, and it was further extended to 275 GHz in 1971 for satellite services and in 1979 for terrestrial services. By 1998 much of the spectrum below 25 GHz had been taken into use in many countries and substantial use is now being made of bands between 25 and 50 GHz and even higher. At times when allocations are being made in advance of the availability of the technical means for using them, an opportunity arises to make exclusive allocations, and such opportunities have sometimes been seized, although often only to be lost again to new sharing arrangements when new pressures develop.

A need perceived for a new allocation for a new service or an established, growing service is not likely to be unique to one country. If care is taken to develop with other Administrations in advance of a WRC a proposal for amending the international table that meets the needs of the countries that are also interested in the new allocation and causes the fewest problems to countries that are not interested, the new allocation may well be agreed at the conference. It is, however, very difficult to get agreement to the cancellation of an allocation that is already in substantial use. Large investments and important facilities in many countries may depend on the maintenance of existing frequency allocations. Consequently, most new frequency allocations take the form of shared access to bands that are already allocated for other services. Agreement may be given willingly, often in the form of a sharing primary allocation in the framed part of the table, world-wide or Regional. In some cases, provision is made for the orderly introduction of a new allocation by means of a resolution inserted into the RR.

If, on the other hand, other Administrations foresee the possibility that planned growth of the use already made of the band in their own country may be jeopardised by a new allocation, they may insist on measures to protect these established uses. The measures might consist of limiting the new allocation to the secondary category, or limiting geographically its applicability by means of a footnote to the table. In addition, a sharing constraint may be imposed, for example one limiting the power radiated by transmitting stations of the new service. Alternatively or additionally, a requirement may be imposed that every new frequency assignment, before international registration, should be coordinated with other Administrations with frequency assignments that are already registered; see Section 5.3 below.

Amending the international table in these ways has enabled many new needs to be met and most of the benefits claimed above for the principle of frequency allocation remain intact. However, the table and the spectrum management function as a whole have become more complex. This complexity was the subject of some of the strongest criticisms of the ITU raised in the 1980s. The revisions of the RR agreed at WRC-95 clarified the text but did little to reduce the complexity, and new complexities are already being introduced to meet more new requirements. No doubt there will be further efforts to simplify spectrum management but simple procedures which meet major new requirements without disturbing established ones may be hard to find.

3.3 National frequency allocation tables

National obligations

When a government ratifies the Instruments of the ITU, it does not undertake to implement the internationally agreed frequency allocations in their entirety. Its basic obligation is to conform with the allocation table in RR Article S5 and the other provisions of the RR in assigning frequencies to stations which are capable of causing harmful interference to stations of another country; see RR S4.2.

It may be argued that an Administration might reasonably make an 'unconforming assignment' (that is, an assignment in a frequency band that is not allocated in the international table for the service to which the station belongs) if it knew of no foreign station with an assignment to which its own new assignment could cause harmful interference. RR S4.4 accepts that such an assignment may be made, but only on condition that an express undertaking is given that the Administration will not allow the unconforming emission to cause harmful interference to a foreign, conforming assignment. Military radio systems are excepted from these regulations by Article 48 of the ITU Constitution [2].

National tables of frequency allocations

It might be said that a Member State's obligation is to respect the right of other Member States to implement the international table, albeit without commitment with regard to military systems. Each Administration draws up a national table of frequency allocations to meet as closely as possible its own national needs, current and foreseen. However, those national needs will include the adoption of many elements from the international table, for example for international mobile services, for bands used for international fixed and fixed-satellite systems and for services, like broadcasting, for which bands have been planned internationally. More generally, it would be perverse to depart from the ITU table without good reason.

To a large extent the national table will in fact be based on the international table, but developed in two ways. On the one hand, the Administration will consider the many options that sharing has provided in the international table. Some options will be adopted nationally, others will be discarded, to produce a table that meets well the national requirements. On the other hand, the national table will be elaborated to show in some detail how the allocations will be used nationally, and how it is foreseen that this use will evolve in the course of the near and medium term future.

Taking the LMS as an example of this detailed planning of the allocations for a service, some bands will be reserved nationally for military systems and others for civil use. Some of the latter will be used for vehicle-borne or hand-portable radiotelephones and other systems that may be operated whilst in motion, whereas other bands will be used for transportable equipment that is operated while at rest. Some of the road vehicle system bands will be designated for cellular systems while others will be for private mobile radio or paging systems. Bands for these various separate LMS purposes, and others, will need to be identified in the national table. Furthermore, in each case some of the bands will be used for base station transmitters only and the others will be for mobile transmitters only.

This consideration of the national planning process can be taken a stage further, still using the various land mobile applications as the example. The demand for land mobile services, currently growing vigorously, can be expected to continue to grow. Indeed some commentators, for example Sung [3] and Forge [4], foresee massive growth of spectrum requirements in the medium term for mobile radio aids to business activities. An Administration would seek to make a realistic estimate of the bandwidth that will be needed say 5, 10 and 15 years ahead for existing systems and foreseen new systems. In making these estimates, expert opinion would be sought from service providers, equipment manufactures, users' representatives and the Regulator. The experience and intentions of the Administrations of other countries would also be studied.

The feasibility of accommodating the forecast levels of traffic 5, 10 and 15 years into the future in the bands already in use for the same purposes, assuming no changes of spectrum management practice and no major changes of equipment technique other than those already firmly planned, would be considered. Even if no probable need for more bandwidth can be foreseen, it might nevertheless be desirable to consider whether some rearrangement of band usage would be desirable; for example, if none of the bands planned for the IMT-2000 system deployment (see RR Resolution 212, and RR S5.388, quoted in Figure 3.3) are already in use for LMS, it might be thought desirable to prepare for such use. If, however, the bandwidth already available for these facilities will become exhausted soon if technology is not improved, it will be necessary to consider how best to provide spectrum for further growth. It might be feasible to obtain the necessary scope for growth by ensuring that more efficient systems are taken into use; the means of achieving this improvement would then be planned. Otherwise, more radical action would have to be considered.

This more radical action may take several forms. New frequency bands, suitable for the purpose and allocated internationally for LMS but remote from the bands already used nationally for LMS, may be set aside for the purpose. It may, of course, be necessary to make other provision for the services for which the band is already allocated. Any such decision should be harmonised if possible with the usage and plans of the Administrations of neighbouring countries. Alternatively it may be possible to increase significantly the amount of traffic that a given bandwidth can carry, without adding unacceptably to interference, by improving the effectiveness of an Administration's own frequency assignment procedures. Ways may be found to encourage users of the service to transfer their traffic to other systems that use spectrum more efficiently. If none of these options can meet the forecast needs, it may be necessary to set in motion the process, discussed in Section 3.2, of seeking changes in the international table to entrench a new national allocation. Whatever way ahead is determined, it should be recorded in the national table.

The LMS has been used as the example of national spectrum planning because it is a complex service and currently in a state of vigorous growth, but it is also necessary to consider nationally the future of some other services in the same way. With this scope, broad in range of applications and long in its reach of time, the national frequency allocation table is an important document of national economic development.

3.4 References

1 'Nomenclature of the frequency and wavelength bands used in telecommunications'. Recommendation ITU-R V.431-6. ITU-R Recommendations Series V (ITU, Geneva, 1997)

2 'Final Acts of the Additional Plenipotentiary Conference', Geneva, 1992 (ITU, Geneva, 1992)
3 SUNG, LICHING: 'WARC-92: setting the agenda for the future', *Telecommunications Policy*, 1992, **16** (8), p. 624–634
4 FORGE, S.: 'The radio spectrum and the organization of the future', *Telecommunications Policy*, 1996, **20**, (1), pp. 53–75

Chapter 4

Frequency assignment

4.1 The frequency assignment process

4.1.1 Licensing a private microwave fixed link

The assignment of a frequency to a radio station and the issue of the associated licence to the owner of the station gives authority for transmitting or receiving at that frequency for stated purposes using stated emission parameters. The process followed in making a frequency assignment and issuing a licence depends on the service to which the station belongs. For some services the process is quite simple, but for others it is complex and particularly so if circumstances require that spectrum should be used very efficiently, so as to accommodate a large number of assignments in a limited frequency band. The licensing of a private two-way microwave fixed link in a country with a liberal regulatory regime may be taken as an example of the latter and an outline of the important elements of the process follows. Some of these elements are considered in more detail elsewhere in the book.

Stage 1. Dialogue between applicant and Administration

Assume that a company has premises at two locations, 10 km apart, with substantial staff at both. The internal telephone exchanges at the two sites are linked by private channels leased from a public telecommunications operator (PTO) but a radio equipment supplier has estimated that a private two-way multichannel radio link would cost less and could also provide desirable new facilities economically, such as a conference video link and a high speed digital link for computer-to-computer file transfer.

The propagation path between the two premises, at A and B, has been surveyed and is expected to provide a clear line of sight under all normal conditions of atmospheric refraction, with no likelihood of troublesome multipath propagation. The climate is temperate. The equipment supplier recommends digital equipment operating at 11 GHz, using antennas 1 m in diameter, 4-condition phase shift keyed (QPSK) carriers with a transmitter power of 1 W, baseband signals of nominally 4 Mbit/s and a bit error ratio (BER) objective not exceeding 1 in 10^9 for 99.9% of the time.

The company applies to the Administration for licences and frequency assignments for this link. The Administration might ask for evidence to support the request for a link of such considerable capacity but, if satisfied on this point, it might agree that a link would be licensed, provided that the equipment to be used meets the appropriate type-approval specifications. However, the Administration might refuse to assign frequencies in the 11 GHz band for the link; the transmission loss due to rainfall would be relatively low at 11 GHz, and assignments in this band might be reserved for systems with longer hops. Furthermore, the Administration would probably not accept the proposed BER objective; whilst taking the view that performance of this order would probably be obtained for much of the time, it might consider that such a stringent objective was not essential for a link that was primarily for telephony and its adoption would increase the difficulty of maintaining a sufficiently low interference level for the new assignments.

The Administration might offer to assign frequencies in the 18.1–18.4 GHz band and a BER objective of 1 in 10^7 for 99.0% of the time. This performance would be ample for a private telephone network; if a better BER was essential, for example for a bit-reduced video system or for computer file transfer, it could be obtained by automatic error correction. To further reduce the interference potential, and after having made calculations of the transmission loss and received power requirement, the Administration might ask for the 1 m antennas to be retained, despite the increase in the frequency, with the carrier power limited to a nominal 0.1 W.

The applicant would discuss the Administration's response with its technical advisors and the equipment supplier. If it was concluded that the company's objectives could be met in the way proposed by the Administration, the applicant would inform the Administration accordingly and would obtain any necessary town planning consent for the radio installation; this might include demonstrating to the planning authority that the antennas would not be aesthetically objectionable and would pose no radiation hazard to staff or the public.

Documents confirming that the equipment would be type-approved would then be supplied to the Administration. Specifications might be required covering such basic matters as carrier frequency stability and spurious emission levels. Information might also be required on the response of the receivers to interference at frequencies just outside the necessary bandwidth, the performance of the demodulators in the presence of low level noise or interference, the control of transmitter output power level, and the sidelobe gain and cross-polarisation discrimination of the antennas. Protection of staff from electric shock or radiation hazard would be shown to be adequate.

Stage 2. Search for a short list of potential assignments

Meanwhile, the Administration would be searching for suitable frequencies, frequency F_{AB} for assignment for transmission from station A to station B and frequency F_{BA} for transmission from station B to station A. The band 18.1–18.4 GHz is allocated world-wide for the FS, the MS, the FSS and (between 18.1 and 18.3 GHz) for the MetS (downlinks). The FSS allocation is exclusively for feeder links for the BSS in the Earth-to-space direction and for both GSO and non-GSO satellites of the FSS in the downlink direction. All of these allocations have primary status. For the present purpose it is assumed that all of these allocations are included with primary status in the national frequency allocation table.

For FS assignments in the band 18.1–18.4 GHz the Administration would use a radio channel plan (see Section 4.1.3 below, and Figure 4.1) for assignments for emissions having the bandwidth (about 2.5 MHz) that these new links would require. There might be provision in this channel plan for the sharing services; thus two blocks of four adjacent radio channels (2 × 10 MHz) might be reserved nation-wide for temporary assignments to wide-band single-hop MS links. Two radio channels might be reserved nation-wide for short-range mobile links used for hand-portable wireless data terminals. The remainder of the radio channels would be available for assignment to links of the FS, subject to verification that unacceptable interference would not arise. These channels would probably be divided into two groups of equal size; when assignments are made for the 'GO' and 'RETURN' legs of a two-way link, one channel would be taken from each of these groups.

Probably many, perhaps all, of these plan channels would have been assigned already for other systems within the jurisdiction of the Administration. The first task would be to draw up a list, possibly a long list, of channels which might be assigned for the new links, neither causing unacceptable interference to established stations operating in the same radio channel or the adjacent channels, nor suffering unacceptable interference from them. The probability of interference to or from domestic earth stations of the satellite services with allocations in the band and foreign earth stations that had been coordinated previously would also be investigated.

Potential interference levels will have to be derived in almost all cases using calculations of the transmission loss over the propagation path. It will be seen from the Appendix that the dominant propagation mode for links operating in the 18.1–18.4 GHz band is the tropospheric wave, although long distance interference by tropospheric scatter may also occur. The Appendix reviews data and methods that are available for calculating transmission loss. The Appendix also discusses the prediction of noise levels due to natural causes and the impact of noise and interference on link performance.

As a minimum, the process of calculating the impact of interference between stations A and B and other stations licensed to operate in these various channels would take into account station location, transmitter power and antenna characteristics, including gain in the relevant direction, antenna height above ground and polarisation, assuming transmission over smooth spherical terrain. Limited in this way, the calculations could be done manually. If large numbers of assignments have to be taken into account, the task could be made easier by eliminating from it those stations

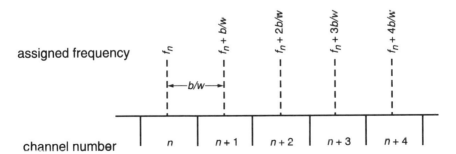

Figure 4.1   A group of adjacent radio channels of uniform bandwidth (b/w)

that are so remote from A and B that the risk of interference can be ignored, using methods such as those discussed in the last paragraph of Section 4.1.4.

However, the use of these minimalist methods leads to imprecise, usually pessimistic, estimates of interference levels and so to inefficient spectrum utilisation. Computer software is available for carrying out this task much more precisely. Given the necessary databases, such programs can take into account the irregularities of the surface of the Earth over which the transmission path passes, known meteorological characteristics and significant indirect propagation modes for interference. This provides a basis for considerably more efficient spectrum utilisation. The databases would desirably include:

- details for the relevant frequency band of all entries in the national register of frequency assignments and of the areas covered by successful coordination negotiations for foreign earth stations,
- programs for calculating wanted-carrier-to-interfering-carrier ratios and signal-to-interference ratios,
- local climatic data, and
- a large scale digital relief map of the terrain.

The list that this process would provide of channels that could be assigned to stations A and B without unacceptable interference arising would then be scrutinised. A short list would be abstracted from it of the channels, the use of which for the new system, considered in conjunction with the assignments already made, could be expected to lead towards efficient utilisation of the frequency band as a whole.

The RR limit the power of FS transmitters and the equivalent isotropically radiated power (e.i.r.p.) of FS stations in the band 18.1–18.4 GHz in order to keep acceptably low the interference to FSS uplinks (feeder links) at the receivers of broadcasting satellites. The low power FS assignments under consideration, however, would be well below these constraints. Also the power of the emissions from transmitters on satellites of the FSS and the MetS in this band will be kept acceptably small by a limit on their power flux density (PFD) at the Earth's surface.

Stage 3. *Coordination with foreign frequency assignments*

In making assignments, Administrations are guided by the rules set out in RR Article S4. See in particular RR S4.3, which requires that a new assignment is not made if it would cause harmful interference to stations receiving frequency assignments that are in accordance with the regulations and registered in the Master International Frequency Register (MIFR). The MIFR is a database, maintained by the Radiocommunication Bureau of the ITU, the essential details from which are published as the 'International Frequency List' (IFL) [1]. The IFL, up-dated by the Radiocommunication Bureau's 'Weekly Circular' [2], gives warning to Administrations of stations in neighbouring countries that have registered assignments that may cause or suffer interference as a result of new assignments to stations A and B.

If station A or station B lies within the coordination area, and uses a frequency band, already coordinated for reception at a foreign earth station of the FSS or the MetS, formal frequency coordination must be carried out to ensure that whatever frequencies may be assigned to stations A or B for transmission will not cause unacceptable interference at the earth station. Similarly, formal coordination of receiving frequencies is necessary if station A or B lies within the area and the frequency band previously coordinated for the transmissions of a feeder link earth station operating in this band.

However, the Administration should consider whether its short list of possible assignments for stations A and B should be discussed with the Administrations of neighbouring countries to establish whether the assignment of any of them is likely to raise interference problems with terrestrial stations abroad, for example of the FS, using frequency assignments that have not yet been notified to the ITU. There is no obligation upon an Administration to do this, but time will be wasted if interference arises after a new assignment has been taken into use. Informal frequency coordination methods have been evolved to ensure that such discussions are not initiated unless there is some significant risk that interference will arise; see Section 4.1.4.

Stage 4. *Assignment of frequencies and issue of licence*

Frequencies F_{AB} and F_{BA} having been selected in this way and shown to be unlikely to cause or suffer interference under current conditions, they are assigned by the Administration to stations A and B. Licences for transmitting and receiving these frequencies are issued to the owner of the stations, subject to various conditions and usually subject to the payment of a fee; see Section 4.3. The details of the assignments are entered in the national register of frequency assignments.

Stage 5. *International registration of frequency assignments*

Administrations are required to notify their assignments to the Radiocommunication Bureau of the ITU, for inclusion in the MIFR, if:

- their use is capable of causing harmful interference to systems under the jurisdiction of another Administration, or
- they are to be used for international communication, or
- they are subject to mandatory coordination procedures (see Sections 5.2 and 5.3 below),

and in certain other special circumstances; see RR S11.2 to S11.14. The meaning attached to the term 'harmful interference' is set out in RR S1.169, quoted in full in Section 4.5 below.

The information to be notified is listed in RR Appendix S4. Among other items to be notified for FS assignments are:

- the assigned frequency, which is at the centre of the assigned frequency band,
- the assigned frequency band, which is the necessary bandwidth plus twice the frequency tolerance (see Section 4.2.2),
- the locations of the transmitting and receiving stations,
- a description of the emission, in coded form, which shows the necessary bandwidth (see Section 4.2.2), the type of modulation employed and in some cases other information bearing on the spectrum of the emission; the code that is used is given in RR Appendix S1,
- the power transmitted and the chief characteristics of the antennas, and
- the date of bringing into use.

For FS assignments, notifications should be sent to the Bureau not more than three months before, nor one month after, the assignment has been brought into use.

The Bureau publishes the notification in its 'Weekly Circular' [2]. Then the Bureau scrutinises the notification to verify that all the relevant provisions of the Radio Regulations have been observed. Thus, for F_{AB} and F_{BA}, the Bureau would check

- that there is an allocation for the FS in the international table of frequency allocations at the frequencies that have been notified,
- that the parameters of the emissions are in accordance with any constraints that apply to the FS at these frequencies, and
- that any mandatory international frequency coordination processes have been completed satisfactorily.

If all is well, the assignments would be entered in the MIFR, together with their dates of bringing into use; if not, the Bureau would return the notification to the Administration for further action.

Stage 6. Action after the assignments have been taken into use, and monitoring

Interference may arise in use, at stations A or B or at other stations using the same frequencies. Low level interference that causes no degradation of service must be tolerated; such interference is characteristic of the heavily used radio medium. If harmful interference arises to or from other stations within the jurisdiction of the assigning Administration, this is a problem that the Administration must solve. If foreign stations cause or suffer harmful interference, see Section 4.5.

A licensee's need for the use of frequency assignments may come to an end. If so, it is important that the licences and the assignments should be promptly withdrawn and the entries deleted from the national and international registers.

However, it is not unknown for a licensee to fail to advise the Administration when the use of an assignment has ended. Also applicants for licences to operate a radio system tend to exaggerate the scale of the requirement and therefore the radio bandwidth needed. Both of these practices may needlessly deny access to the spectrum to other applicants with real present needs. Moreover, radio systems are sometimes operated contrary to the terms of the licence, with the carrier frequency outside tolerance, excessive carrier power and so on, possibly causing interference to other radio systems. In response to RR S3.14 and to fulfil their own functions, Administrations monitor the use that licensees make of their transmitters and their frequency assignments in order to get errors corrected and abuses curbed, licences being withdrawn and licensees being prosecuted for the more serious violations of the terms of the licence. There is a good general introduction to monitoring in an ITU handbook [3]; see also Recommendation ITU-R SM.1050 [4].

There is an international dimension to radio signal monitoring. RR Article S16 and RR Recommendation 32 encourage Administrations, individually or in cooperation, to establish high performance monitoring stations. The services of these stations may be offered to the ITU, creating a network of international stations to assist in the resolution of difficult interference problems; see also RR S15.44 and Recommendation ITU-R SM.1139 [5]. See also RR Recommendation 36, which foresees particular benefit from reports from monitoring stations that will enable the GSO to be used more effectively. The ITU publishes a list [6] of the stations which form the network; see RR S20.12.

4.1.2 *Frequency assignment for other kinds of system*

Frequency assignment to stations of the FS in the band 18.1–18.4 GHz, described in Section 4.1.1, involves most of the processes that also arise elsewhere in the radio spectrum and for assigning frequencies to stations of other services. However, there are important differences in other bands and for some other services. The main points of difference between the various services with regard to the assignment of frequencies are outlined below and in Chapter 5. For some applications of radio in a country with a liberal regulatory regime, the Regulator will play a major role, not in how spectrum is managed, but as to which of competing applicants will be licensed; see Section 4.3. However, the ways in which systems are operated in the various ranges of the spectrum affect many details of spectrum management and the intricate pattern of allocation sharing leads to differences in practice from band to band; these differences are examined in Chapters 6–15 below.

A general point to note is that the various procedures set out in the RR and the good offices of the ITU are used to solve the trans-border interference problems that may arise. An Administration must find its own solutions to problems arising wholly within its own jurisdiction, although the methods used for solving international problems will often be found helpful in a domestic context also.

1 Satellite services

Special conditions apply to the assignment of frequencies for satellite broadcasting and for assignments to stations of other services that share these frequency bands; see item 4 below.

For the frequencies assigned to the satellites of new networks of other services, there is a mandatory coordination procedure to limit interference to and from existing foreign satellite networks; see Section 5.2.1 below. This procedure, which is intended to identify and resolve interference problems before the satellite is launched, must be satisfactorily completed before frequency assignments for reception or transmission at the satellite are notified to the Radiocommunication Bureau for registration in the MIFR.

The frequencies that are assigned to a new satellite are also assigned, as appropriate, to the earth stations that will use them for communicating via that satellite. Virtually all frequency bands allocated for satellite services are shared with terrestrial services, so a different set of potential interference situations arises in the vicinity of a new earth station, some of which may involve foreign terrestrial stations. In relatively rare situations, foreign earth stations may also have to be coordinated. There is, therefore, another mandatory coordination procedure to be followed to identify and resolve potential interference problems at earth stations at fixed locations before the satellite starts service. A third coordination procedure is applied where the earth stations are mobile. These procedures are also reviewed in Section 5.2 below.

2 Terrestrial services sharing bands with satellite services

Item 1 above and item 4 below refer to the coordination procedures that are used when a new earth station is planned, to ensure that there will not be interference between frequencies to be assigned to it and foreign terrestrial stations that may be operating in the same frequency band. After an earth station has been coordinated in this way, a new frequency subsequently assigned to a foreign terrestrial station within interference range of the earth station has to be coordinated with the earth station before it is registered.

3 *Mandatory frequency coordination in specified shared bands*

It has sometimes happened that a proposal to add a new sharing allocation to a frequency band has been agreed on condition that every assignment for the new service, before being notified for registration in the MIFR, is coordinated with any foreign assignment of the other services sharing the band within possible interference range. See Section 5.3.

4 *International planning of frequency band use.*

For the allocations of several services, many of the processes described in Section 4.1.1 apply, but they are carried out not by Administrations acting in isolation but by all the concerned Administrations meeting together in advance at a planning conference. This planning process is outlined in Section 5.4. It is applied most commonly in the mobile services, broadcasting and some radionavigation services. Perhaps the most complex of these planning exercises were those for the principal allocations for direct broadcasting by satellite (DBS), which are in the vicinity of 12 GHz, and the associated uplinks, called feeder links. A special procedure is prescribed for taking these broadcasting assignments into use and there are special coordination procedures for assignments to stations of the other services which share these frequency bands; see Section 10.3. There is another set of procedures for satellite broadcasting assignments in unplanned BSS allocations; see Section 10.2.

5 *Delegated management of blocks of spectrum*

As indicated in Section 3.2, it has sometimes been agreed, globally or regionally, that the use of a frequency band should not be managed by the Administrations acting independently. Instead, management has been devolved to an expert international body. The devolvement to the ICAO of the management of HF and VHF bands allocated exclusively for the AMS(R) is one such case.

An analogous arrangement may sometimes be employed, although with the Administration retaining close surveillance of spectrum use, on a national basis. For example, a major PTO may need the use of a substantial number of new microwave FS frequency assignments every year in developing its trunk network, or for linking isolated private subscribers or business subscribers with large bandwidth requirements into the public network. A similar arrangement might be applied in the LMS, for example for a major cellular radiotelephone system. The Administration may decide to allot the sole use of a block of spectrum to a service provider, enough to meet the foreseen demand for one or two years given careful management. Under strict guidelines the service provider would then carry out many of the functions that would otherwise fall to the Administration, the latter remaining, if only nominally, the frequency assigning authority and the channel for the notification of assignments to the ITU.

6 *HF assignments for FS stations*

For an FS assignment in the HF frequency range, the procedure will be much as is described for a microwave assignment in Section 4.1.1, but there is one important difference. There is no systematic way for a radio user or an Administration to select a frequency for assignment to an HF link with a high probability of prolonged satisfactory operation.

It might be thought that reference to the IFL, to confirm that a proposed new assignment would be compatible with any assignment that had already been registered, would provide satisfactory guidance on this point. Alternatively, an Administration might notify a tentative assignment to the ITU Radiocommunication Bureau, in the expectation that the publication of the notice in the 'Weekly Circular' would lead to an objection by an Administration with an earlier registration. However, this frequency range is congested, the IFL contains many out-of-date entries, and the vagaries of ionospheric propagation ensure that the frequencies that systems use change erratically and from hour to hour, from season to season and from year to year. It may be virtually impossible to forecast whether the same frequency will be used simultaneously by two stations on any but rare occasions, and if so, whether harmful interference will occur.

Thus, when predictions have been made of the optimum frequency ranges for a new link, it is usually necessary to monitor spectrum usage in a frequency band with the appropriate predicted propagation characteristics, find an unoccupied frequency band wide enough for the proposed emission, make an assignment, notify the ITU and begin test transmissions. Complaints of interference may well follow quickly, or after the lapse of months or years, in which case the process has to be repeated in another apparently unoccupied frequency band. Recently the introduction of frequency adaptive systems (see Section 6.1.2) promises to provide a more satisfactory solution to this difficult problem.

7 The terrestrial mobile services

One rule that is common to all the terrestrial mobile services is that the frequency assignments, transmitting and receiving, made to land stations, which are at fixed locations (see RR S11.2 and S11.9 to S11.11) may be registered in the MIFR but those made to mobile stations are neither notified to the ITU nor registered. Also it is characteristic of most mobile systems that several or many frequencies are assigned to each station, at a fixed location or mobile, in accordance with some kind of plan, the frequency to be used on any particular occasion being selected for operational reasons.

8 Broadcasting at HF

A number of frequency bands between 5950 and 25 670 kHz are allocated exclusively for terrestrial broadcasting. They are used for sound broadcasting, mainly international. The frequency used for each broadcast must be selected having regard to predictions of ionospheric propagation conditions, but the total bandwidth available is not sufficient to provide for interference-free operation for all broadcasters wishing to operate at HF. A procedure has been developed in which Administrations, or broadcasters acting for them, collaborate twice per year to plan how to use the channels which are available to best advantage with least interference. The Radiocommunication Bureau is informed of the plans and publishes them, but this use of frequencies is not recorded in the MIFR. See RR Article S12 and Section 7.4 below.

9 Radio amateurs

The terrestrial stations of radio amateurs are not licensed, nor are frequencies assigned to them or registered in the MIFR. A much simplified version of the usual procedure for licensing and assigning frequencies for earth stations and satellites is used for the AmSS. Amateur licences are issued to persons who have shown that they are competent to operate a station, and who are thereby licensed to operate any amateur station.

Amateurs are given access to the frequency bands allocated for AmS and AmSS in the national frequency allocation table, subject to certain constraints. Discipline in the use of spectrum within the bands is exercised primarily by the amateur operators themselves and their national and international societies.

10 Citizens' Band
A frequency band allocated for the LMS may be allotted for 'Citizens' Band' radio and divided in accordance with a radio channel plan. Users are licensed and their equipment is required to meet technical standards, but frequency assignments are not made, choice of channel within the band being left to the user.

11 Broadcasting receiving licences
Licences are sometimes required for domestic broadcasting receiving stations. The fee that is charged contributes to the cost of eliminating interference where it arises. In addition, in countries with public service broadcasting, there may be a component in the fee to cover the cost of programme-making and transmission.

12 Low power devices
Various narrow sub-bands, typically in frequency bands allocated for the MS or LMS, are set aside for low power radio devices having very short range and low sensitivity to interference. A wide variety of devices fall into this category, such as cordless telephones, burglar alarms and remote control systems for toys. These bands are used without formality for devices which conform to appropriate guidelines issued to manufacturers by Administrations.

13 Non-conforming assignments
Frequency assignments made to radio stations in bands in which there is not an appropriate allocation in the international table are said to be non-conforming. If such an assignment is notified to the ITU for registration it will be registered in the MIFR, but only for information purposes, and only if the notifying Administration gives specific undertakings on two points. The first is that protection will not be claimed for the station from interference from any station operating in conformity with the Radio Regulations. Secondly, operation of the non-conforming assignment will cease immediately if it causes harmful interference to a conforming station; see RR S4.4.

4.1.3 Choice of frequencies for assignment

Radio channel plans

A frequency assignment to a radio station states the centre frequency of the licensed emission and the width of the band symmetrically around that frequency within which the sidebands of the emission are to be contained. These figures are termed the 'assigned frequency' and the 'assigned frequency band'. The relationship between the assigned frequency band, the nature of the signal to be transmitted and the type of modulation used is considered further in Section 4.2.2 below.

In the HF allocations of the FS, there are various kinds of emission with bandwidths that range from less than 100 Hz to about 12 kHz. The arrangement of frequency assignments for these emissions makes no regular pattern in the spectrum, narrow-band emissions being next to wide-band emissions in any order, wherever a band of sufficient width, sufficiently free from interference, can be found. In transponders of satellites of the FSS which are operated in the frequency division multiple access mode a similar non-uniformity of bandwidth also arises.

These irregular situations of the FSS and the FS at HF are, however, exceptional. In most frequency bands the emissions tend to be uniform in bandwidth and it is convenient for assignments to be uniformly spaced in frequency at intervals approximately equal to the bandwidth that each single emission requires. This practice leads to the concept that each emission occupies an imaginary channel through space, called a radio channel, with the assigned frequency at the centre and the sidebands of the emission filling the channel; see Figure 4.1. Managed with care, the use of radio channels in frequency assignment can lead to convenient and efficient spectrum use. For some frequency bands a common radio channel plan has been agreed internationally.

The assignment of the same frequency to two or more links for simultaneous use for different purposes is called frequency re-use. This is feasible if some form of isolation, provided by geographical separation, physical obstruction of interference propagation paths, antenna directivity, polarisation discrimination or some combination of those factors ensures that interference between the links sharing the same channel will be sufficiently low. Frequency re-use is valuable wherever demand for access to the spectrum is substantial.

A variant of the basic channel arrangement shown in Figure 4.2 may be used to further enhance the scope for frequency re-use. Thus, if the method of modulation used produces a concentration of spectral power density near the centre of the emission spectrum (see Figure 4.9), as happens with frequency modulation (FM) and some forms of amplitude modulation (AM), an interleaved channel plan as shown in Figure 4.2(a) provides perhaps 60% more radio channels than the basic plan without a substantial increase in the requirement for isolation by geographical separation etc. If the gaps between channels are eliminated altogether as in Figure 4.2(b), the need for isolation will probably be increased significantly but with a net increase of the frequency re-use factor.

Cellular coverage plans

When choosing frequencies for assignment to radio stations distributed over an extensive region, the radio channel planning concept can often be usefully extended into the geographical domain. The method of linear frequency planning has been used for terrestrial broadcasting since 1961; see the ITU 'Handbook of national spectrum management' [7] and Recommendation ITU-R BT.1123 [8]. Applied to an extensive and uniform region, this method generates a pattern of hexagonal coverage cells. The concept of dividing a region into a lattice of hexagonal cells, each small enough to be served by a single station, is familiar in the context of land mobile systems but it can usefully be applied more generally, especially when the radio emissions involved have uniform parameters.

In the simplest concept a region, assumed to have smooth and level terrain, is divided into a lattice of uniform hexagonal cells, each of a size that can be covered from a centrally located base station; see Figure 4.3. Three channels, A, B and C are allotted to adjacent cells, as shown in the figure. The same channels are allotted to all the rest of the cells in the system, as shown by lower case letters in the figure, without the same frequency being made available in adjacent cells.

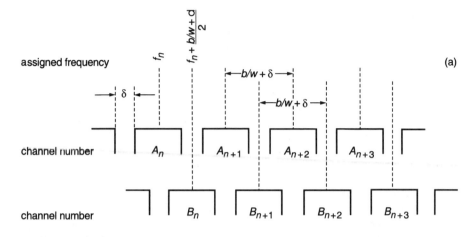

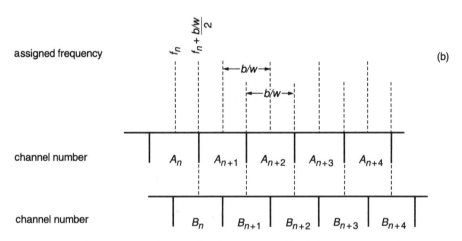

Figure 4.2 *Two sets of interleaved radio channels, A and B. Plan (a) shows an arrangement with a frequency gap δ Hz wide between adjacent channels, where the interference power would otherwise be greatest. Plan (b) maximises the number of channels by omitting the gaps*

Figure 4.4 sketches the variation of the field strength with distance from a base station transmitter. A uniform cellular system would function without unacceptable interference in the base-station-to-out-station direction between users of the same frequency assignments in different cells provided that both of the following conditions apply:

- The field strength at distance d_1, one of the most distant points within coverage area A shown in Figure 4.3, exceeds F_1, the minimum field strength for satisfactory operation.

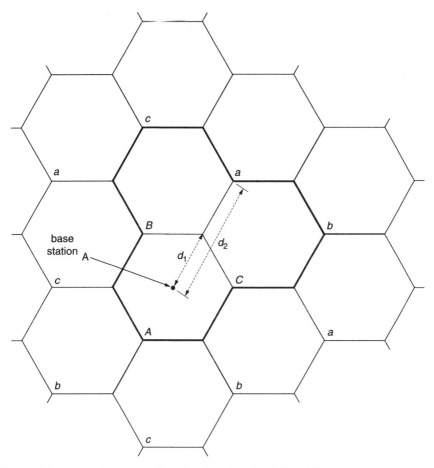

Figure 4.3   A cellular coverage plan using three channels, A, B and C

- The field strength at distance d_2, one of the nearest points in the coverage area of another cell with the same allotment of channels, does not exceed F_2, where the worst acceptable signal-to-interference ratio is F_1/F_2.

In a two-way system, similar conditions must also be achieved in the out-station-to-base-station direction.

Using only three channels and with the assumption that each base station is at the centre of its cell, d_2 equals $2d_1$. This ratio is likely to be too small for practical use at UHF and VHF, where LMS cellular systems and terrestrial broadcasting currently operate. It must also be expected that interference would be particularly high in some locations owing to obstructions to propagation of the wanted signal within a cell, or particularly favourable propagation of interfering signals from an unwanted base station using the same channel. A non-central base station location is another potential source of interference. Perhaps the only point in the spectrum where it might be expected that three channels could be used to form a practicable cellular

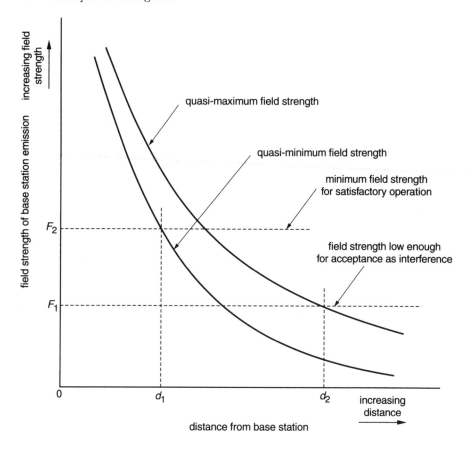

Figure 4.4   The field strength distribution determining the maximum size of cells and the minimum co-channel repetition distance in a cellular system

system is much higher in frequency, in one of the atmospheric absorption bands, and in particular around 60 GHz.

However, a better ratio of d_2 to d_1 can be obtained with a larger number of channels. Thus seven channels, as shown in Figure 4.5, provide a worst-case d_2/d_1 ratio for central base stations of 4.8 to 1. A system using 12 channels, as shown in Figure 4.6, provides a ratio of 6.0 to 1. With 16 and 19 channels the ratios obtained are 7.2 to 1 and 8.0 to 1.

Thus as the number of channels used rises, the geographical isolation of channels from interference from other cells using the same frequencies improves, although the available degree of frequency re-use declines. When the number of channels is high enough, there will be a margin of protection from unacceptably high interference levels due to unfavourable circumstances. This margin may enable the same scheme of frequency utilisation to be applied to regions where the terrain is less regular, the

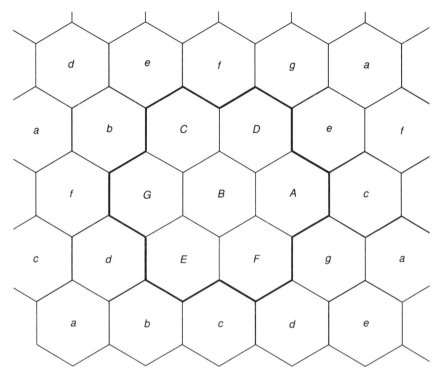

Figure 4.5 A cellular plan using seven channels, A to G

base stations are not necessarily in the centre of the cell or the cells are not all of the same size.

4.1.4 Non-mandatory international frequency coordination

Even where frequency coordination of new assignments with other Administrations is not mandatory, it may nevertheless be desirable to coordinate informally. RR S4.3 requires Administrations to assign frequencies with due care, to avoid causing harmful interference to conforming emissions that are recorded in the MIFR. Furthermore, time is lost in establishing new systems if an assignment, having been made, has to be withdrawn because of the interference it causes or suffers. But coordination is time-consuming and costly, and Administrations make many assignments; it is important to be able to decide which assignments need to be coordinated with which other Administrations.

   The main features of one systematic approach to minimising the need for coordination are as follows (see Recommendation ITU-R SM.1049-1 [9]). Two neighbouring Administrations, X and Y, decide that they make enough assignments each year in a frequency band to justify a special procedure to minimise the need for coordination.

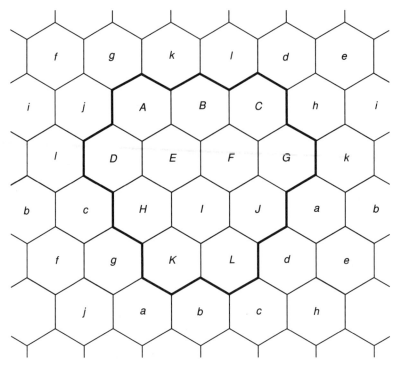

Figure 4.6   A cellular plan using 12 channels, A to L

So:

- An agreement is drawn up harmonising principal system parameters for radio systems used in the band for application in both countries.
- A coordination perimeter is defined at a suitable distance, perhaps 100 km, to either side of the common border, the distance being calculated to ensure that interference from a greater distance in one country is very unlikely to be harmful to stations anywhere in the other country. The territory between the coordination perimeters and the border becomes a coordination zone in each country; see Figure 4.7.
- It is agreed that country X shall have unrestricted use within its coordination zone of one half of the frequency band without coordination, and country Y shall have equivalent rights within the other half of the band.
- Each country may make use, within its coordination zone, of the half-band to which the other country has unrestricted access, subject to constraints on emission parameters that ensure that cross-border interference is unlikely to occur, it being agreed that such use will be stopped if interference does occur.
- Each country uses without prior coordination the whole band outside its coordination zone except for systems which are much more likely to cause or suffer interference than systems of the harmonised characteristics.

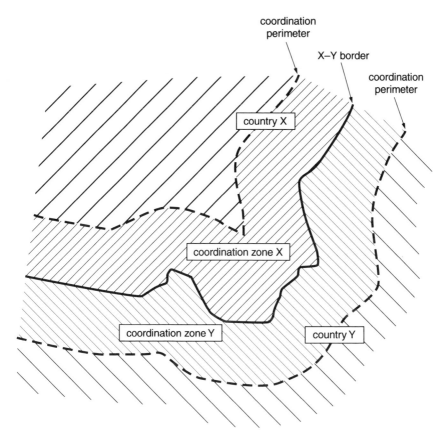

Figure 4.7   *The minimisation of the need for non-mandatory frequency coordination by spectrum division in coordination zones*

More complex arrangements are necessary where more than two countries are involved. Schemes of this kind are widely used in Europe, where they are known as 'Vienna agreements'.

Vienna agreements are helpful for frequency bands used for systems with relatively low transmitter power and antennas with relatively low gain, typical of the LMS. For systems with antennas having high gain, likely to cause and suffer interference at long distances if antenna beams are unfavourably oriented, such agreements would be undesirably restrictive. Thus, for example, for frequency bands used for radio relay systems, it is more satisfactory to determine the area within which each specific station with a proposed new assignment might cause or suffer interference with a foreign station if the orientation of the antenna of the latter were unfavourable. The level of interference can then be assessed for each station within this coordination area in consultation between the responsible Administrations; see Recommendation ITU-R F.1095 [10]. This is a simplified version of the method used in the mandatory frequency coordination between earth stations of a GSO satellite network and terrestrial stations; see Section 5.2.3 below.

4.2 Efficiency in spectrum utilisation

4.2.1 The concept of efficient spectrum use

The frequency ranges that are good for important services are becoming congested in most countries. If the full economic value of radio is to be realised, it is important that these congested frequency ranges should be used efficiently. Indeed, since parts of the spectrum which are lightly loaded today may well be in great demand in a few years time for requirements unforeseen at present, there is long term benefit in giving attention to the basic principles of efficient spectrum use in all parts of the spectrum. But can this efficiency be quantified and what means for achieving efficiency are available to Administrations?

Defining efficient spectrum use

There is an examination of concepts of spectrum utilisation and efficient spectrum use in Recommendation ITU-R SM.1046-1 [11] and in the ITU 'Handbook of national spectrum management' [7]. In brief, the spectrum utilisation (U) of a one-way radio link can be expressed by the measure

$$U = B \times S \times T \qquad (4.1)$$

where B is the occupied bandwidth,
  S is the volume within which the use of that bandwidth is denied to other would-be users, and
  T is the fraction of the time for which that bandwidth is denied to those would-be users.

The space factor, (S) can usually be regarded not as a volume, but as a geographical area. However, two areas may need to be considered for each emission, the area which the emission denies to receivers of other links (the 'transmitter space') and the area which the receiver denies to transmitters of other links (the 'receiver space').

A measure of spectrum utilisation efficiency (η) is given by the ratio of the quantity of information (M) transferred over the link to the amount of spectrum utilisation, thus

$$\eta = M/U = M/(B \times S \times T) \qquad (4.2)$$

Also a measure may be obtained of the relative spectrum utilisation efficiency of an actual link by dividing its spectrum utilisation efficiency (η_a) by that of a reference link (η_r), perhaps a standard system or an ideal system.

These basic concepts may be illustrated by a simple example. Thus, transmission from the base station of a VHF land mobile system A is illustrated in Figure 4.8. The base station antenna is omnidirectional in the horizontal plane, as are the antennas of the mobile stations it serves. There are similar systems, B, C and D, operating on the same frequency assignment in the same area and the antennas of these systems are also omnidirectional. The service area for the base-to-mobile leg of system A is circular, with radius (R_{sA}) centred on the base station. The area which system A base station emissions would deny to receivers of other systems is also circular, centred on the transmitter and with radius (R_{iA}). The areas which mobile stations X and Y of system A deny to base station transmitters of systems C and D when the system A base station transmitter is transmitting are sketched on the figure. Let it be supposed

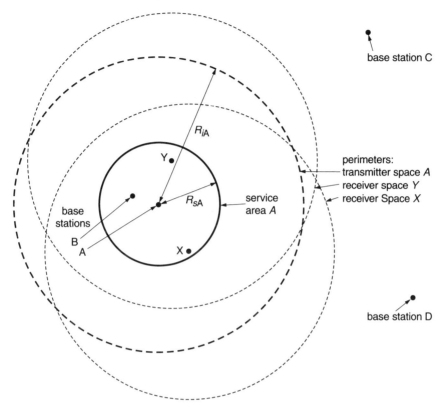

Figure 4.8   *The spectrum utilisation of LMS base station transmitter* A, *and two associated mobile receivers* X *and* Y. *Base station* B *time-shares the same channel. Base stations* C *and* D *use the same frequency and are potential sources of interference*

that it has been agreed that systems A and B will not operate simultaneously and accordingly, the mobile stations of system A do not deny the neighbourhood of system A to base station B. Then equation (4.1) can be rewritten

$$U_A = B \times \pi(R_{iA})^2 \times T_A \tag{4.3}$$

where (T_A), the fraction of the time that system A denies access to the channel to system B, is less than unity. Similarly (T_B), the fraction of the time for which system B denies access to the channel to other systems, is also less than unity. If systems A and B are the only systems sharing this frequency assignment with overlapping service areas, $(T_A + T_B)$ does not exceed unity.

The quantity of information (M) that is transmitted through these base stations can be expressed simplistically in terms of the fraction of the time that communication (a telephone conversation, a unidirectional data message or whatever other form the system's business may take) is in progress. This fraction will obviously depend on the pressure of demand and the nature of the traffic. If demand is high and traffic is between operators, perhaps the dispatchers and drivers of a taxi service, $(M_A + M_B)$

may approach unity. If the traffic is serving the users of automatic networks requiring a high grade of service, much will depend, despite high pressure of demand, on the number of channels covering the same service area to which the users of the network have access; if the number of channels is one or few, $(M_A + M_B)$ will be much less than unity, as Erlang has shown.

Changes to the signal characteristics and the modulation parameters of the LMS systems involved are easy to incorporate into this model. If a change were made, for example, to the modulation index of FM analogue systems or the number of significant states of PSK emissions and the transmitter power is adjusted to ensure that the size of the service area $(\pi(R_s)^2)$ remains unchanged, then the bandwidth (B) will change and the level of interference (R_i) will probably change also. If single channel LMS systems are replaced by multi-channel FS systems, (T) usually becomes effectively equal to unity, (B) usually increases, perhaps greatly, and the relationship between the amount of information that could be carried and is carried becomes more complex.

However, the inclusion of other factors, common to many systems in practice and very important if spectrum utilisation is to be economic, makes evaluation of the impact of the changes on spectrum use efficiency much more difficult to determine. The use of directional antennas is one such complicating factor. Should the length of the link be taken into account in quantifying the effectiveness of spectrum utilisation? How can allowance be made for the impact of the environment of a radio station on the optimisation of its characteristics? And what value can be given to the efficiency of spectrum utilisation of a station such as a radio beacon, which may provide navigation services of great value without conveying any information, as defined for telecommunications purposes?

The attainment of efficient spectrum use

Methods such as these can be used to derive expressions for the relative spectrum utilisation efficiency of single radio systems in isolation and thereby they help to maximise the capacity of limited frequency bands used for specific purposes. However, it is difficult to extend the analysis meaningfully to include a number of unrelated systems operating within a limited area and a limited frequency band.

Furthermore, there are likely to be limits to the power that Administrations have to control the use made of the spectrum. They may use regulatory powers to forbid the use of equipment with characteristics that are obviously seriously wasteful. On the other hand, the most efficient available equipment and system configurations are likely to involve initial cost penalties and perhaps subsequent revenue penalties for a service provider. There may also be a conflict between the objective of efficient spectrum use and the achievement of fair competition between service providers. Thus, the Administration may need to encourage, rather than compel, a change from a tolerable practice to good practice, delaying evolution to the best practice until a suitable opportunity arises.

How, then, can an Administration seek to ensure that efficient spectrum use is achieved? The main measures available fall into four categories.

- Obviously the procedures that are carried out by the Administration's own staff should be done efficiently; the functions outlined in Sections 4.1.1 and 4.1.2 above are indicative of what is necessary if there is a need for intensive spectrum use. There is further guidance in the ITU 'Handbook of national spectrum management' [7] and Recommendation ITU-R SM.1047 [12]. Computer methods for carrying out the necessarily complicated processes of spectrum management with reliability and economy are available; see Recommendations ITU-R SM.1048 [13] and SM.1370 [14]. There is also a particular need for the Administration to cultivate close liaison with radio users, service providers and equipment suppliers and with the Administrations in neighbouring countries.

- Secondly, while there may be little that can be done to limit the impact of nature on radio propagation, apart from using knowledge of the phenomena to ameliorate their effects, the same is not true of several kinds of man-made noise. Directly or indirectly the Administration can do much to ensure that the environment in which its radio services operate is as free from unwanted radio frequency noise as possible; see Section 4.2.3 below.

- Thirdly, Administrations should ensure that the systems that they license achieve certain technical standards and occupy no more spectrum than is necessary, having regard to operational requirements. Many of these standards are set out in the RR and the recommendations of the ITU-R.

- Fourthly, however, it is arguable that some of these objectives can be achieved, perhaps more easily, through the non-technical terms of licences, and in particular by the licence fees that are charged.

These factors are considered further in the sections that follow. The optimum balance between the use of technical regulations and licence terms to optimise spectrum use is debatable, and varies from application to application and perhaps from country to country.

4.2.2 *The bandwidth of emissions*

The necessary bandwidth

Figure 4.9 is a sketch of the envelope of the spectrum of a typical AM emission with an unsuppressed carrier. The spectrum of an FM emission is not greatly different. It will be seen that the power density, whilst relatively high at the centre of the band, tapers off towards its outer fringes, making the overall bandwidth rather indeterminate. If the bandwidth of the emission which reaches the demodulator is significantly less than the bandwidth at the output of the modulator, the received signal will be degraded. The least bandwidth that, given equipment of good design, permits satisfactory transmission of a signal is called the 'necessary bandwidth'; see RR S1.152.

RR S3.9 requires the bandwidth occupied by emissions to be kept to the lowest feasible value, to be taken in general as the necessary bandwidth. The necessary bandwidth of typical emissions can be calculated with sufficient accuracy; see Recommendation ITU-R SM.1138 [15], which is incorporated into the RR by reference in RR Appendix S1, paragraph 1(2). This recommendation does not cover multichannel digital radio relay emissions, but Recommendation ITU-R F.1191-1 [16] provides some guidance on this matter. For systems using spread spectrum modulation there is no simple relationship between the bandwidth occupied and the rate at which information is being transmitted.

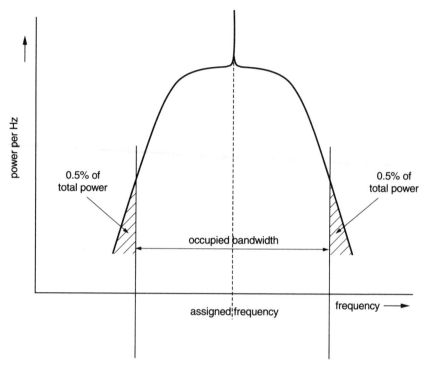

Figure 4.9   A sketch of the distribution of power in the spectrum of a typical emission, showing how the occupied bandwidth is typically defined

Frequency tolerance

The centre frequency of an emission, typically that of the carrier wave, may not be precisely equal to the assigned frequency. A margin for error, called the 'frequency tolerance', is allowed in recognition of this; see RR S1.151. The bandwidth that an Administration assigns to a station (the assigned frequency band) is equal to the necessary bandwidth plus twice the frequency tolerance.

A stringent frequency tolerance brings two important benefits. The bandwidth that an emission denies to other radio users is reduced. Secondly, given receivers tunable with comparable precision, it becomes feasible to use narrow-band emissions without the presence of a skilled operator at the receiver. The use that is now being made of spectrum-efficient radio channel plans for radiotelegraphy in the MMS at HF is a graphic example of these two benefits. The frequency tolerances set out in RR Appendix S2 (see also RR S3.5) are designed to secure one or both of these benefits, depending on the frequency range and the use made of the frequency allocations affected. Recommendation ITU-R SM.1045-1 [17] recommends that some of the tolerances given in RR Appendix S2 should be made more stringent, and recommends other, still more stringent, tolerances as long term design objectives.

The tolerances given in RR Appendix S2 are sufficiently stringent for the good operation of most radio systems but tighter tolerances are essential for some special applications. For example, narrow-band LMS systems in the UHF range need considerably better tolerance than the 20 parts in 10^6 which Appendix S2 requires. Thus it may sometimes be necessary for an Administration or a service provider to specify a tolerance which is more stringent than that required by the RR.

The occupied bandwidth

Necessary bandwidth is a regulatory term, a parameter that can and should be attained but which may, in particular cases, be no more than approached, typically because the parameters of the modulator, the post-modulator filters or power amplifiers of the transmitter are sub-optimal. There is sometimes a need for a measure of the actual bandwidth of an emission, for use for example in estimating prospective interference levels. The term 'occupied bandwidth' is defined for this purpose; see RR S1.153. The occupied bandwidth is the width of the band between upper and lower frequency limits, such that the mean power radiated above the upper limit and below the lower limit is a small percentage (normally 0.5%) of the total mean power of the emission; see Figure 4.9.

Out-of-band emission

'Unwanted emission' is the energy of an emission which is radiated outside the necessary bandwidth. It is divided into the 'out-of-band emission' (which is outside but in the immediate vicinity of the necessary bandwidth) and 'spurious emissions' which are more remote; see RR S1.144 to S1.146. No specific boundary has been defined to separate out-of-band and spurious emissions but RR Appendix S3 paragraph 11 indicates, as a broad guideline, that spurious emissions may be considered to begin at ± 2.5 times the necessary bandwidth relative to the centre frequency of an emission. RR Recommendation 66 calls for studies to be made that would lead, for example, to the adoption of a firm demarcation between out-of-band and spurious emissions, to recommended limits for out-of-band emissions and to a decision as to whether these limits should be given mandatory force, in general or where the need is greatest. There is further discussion of spurious emissions in Section 4.2.3 below.

The RR do not at present specify general limits for out-of-band emission levels, although limits are specified for some particular situations, especially where the ITU has accepted a measure of responsibility for an operational scheme. See for example Recommendation ITU-R M.1173 [18], which defines the out-of-band characteristics required of transmitters for operation in ITU radio channel plans in certain maritime HF allocations. This recommendation is incorporated in the RR by reference, for example, in RR S52.181. More generally, RR S15.10 rules that transmitters operating close to the edge of an allocated band shall not cause harmful interference to a conforming system with receivers of good design operating in the adjacent allocated band. An Administration may find it necessary to impose limits on the level of out-of-band emissions of specific kinds of transmitters operating within its jurisdiction to meet, for example, the needs of a national radio channel plan.

Receiver characteristics

The complement of narrow-band transmission is matching receiver selectivity, without which the full benefits of the former are not realised. RR S3.12 reminds Administrations of this need.

4.2.3   A clean radio environment

Spurious emissions

The carrier generation and the amplification processes in a transmitter produce, and may fail to suppress, spectral components more remote from the necessary bandwidth than the out-of-band emissions. These components are called spurious emissions; see RR S1.145. Spurious emissions typically include harmonics of the wanted emission, parasitic oscillations, intermodulation products and various unwanted components that are generated in frequency conversion processes in the transmitter.

RR Appendix S3 gives limits for spurious emissions from transmitters; see also RR S3.6 and Recommendation ITU-R SM.329-7 [19]. In addition, RR S15.11 requires special measures to be taken to eliminate harmful interference caused by spurious emissions even if their level is within the constraints of Appendix S3. The appendix was most recently revised by WRC-97 and it is in two sections. The provisions of Section I, applicable to transmitters installed in or before the year 2002, are intended to remain in force for those transmitters until 2012. Section II is applicable to transmitters installed in 2003 or later and for all transmitters after 2012.

The main Section I limits are summarised in Figure 4.10. They differ only in detail from those that applied previously to transmitters installed since 1985. However, they are not comprehensive. It is accepted that some system configurations, such as wideband HF transmitters and groups of transmitters sharing a common antenna may fall short of these limits. Particular problems arise in defining limits for radar transmitters and transmitters used with digital modulation. A number of types of transmitter have been exempted from these limits altogether, including those used on survival craft at sea but also all transmitters operating above 17.7 GHz, all microwave transmitters carrying digital signals and all satellite and earth station transmitters. These exempted transmitters are to have their spurious emission levels reduced as much as possible.

There are, however, a number of other studies on this subject that have not yet reached a sufficiently advanced state for their results to be incorporated into RR Appendix S3. Thus, Recommendation ITU-R M. 1177-1 [20] provides guidance on the measurement of spurious emissions from maritime radar transmitters; see also Section 15.1 below. Recommendation ITU-R F.1191-1 [16] recommends design objectives for limiting spurious emissions from digital radio relay systems. Recommendations ITU-R S.726-1 [21] and M.1343 [22] propose limits, not yet finalised, for very small aperture fixed earth stations (VSATs) and mobile earth stations used in non-GSO global networks operating between 1 and 3 GHz, respectively.

Section II of RR Appendix S3 puts more stringent constraints on spurious emissions for the future. They are applicable to all transmitters but some of the limits, in particular those for earth station and satellite transmitters, are shown at present as design objectives, and the constraint appropriate for digital modulation close to the

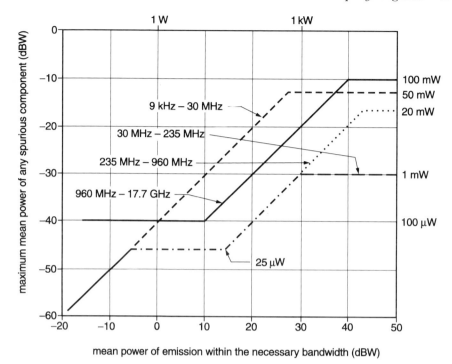

Figure 4.10   *Maximum permitted mean power of any component of spurious emission at the output of a transmitter as a function of the power within the necessary bandwidth (RR Appendix S3). These limits apply until 2012 to any transmitter installed before 2003, subject to various exclusions*

limit of out-of-band emission is not finalised. Firm limits are to be determined at WRC-00. See also RR Recommendation 66. RR Recommendation 506 urges that special measures be taken to limit second harmonic radiation of emissions from broadcasting satellites operating in the band 11.8–12.0 GHz, to protect radio astronomy in the band 23.6–24.0 GHz.

In addition to the harmonic and intermodulation products that are generated within transmitters, spurious products may also arise from the flow of radio frequency currents through non-ohmic contacts between the metallic structural components of transmitter buildings and cabins, especially those housing more than one transmitter. These spurious products may not cause interference to distant radio users but they may interfere with receivers operating nearby at many frequencies; this is particularly troublesome on ships. Similar problems may arise from intermodulation in wide-band antenna pre-amplifiers at receiving stations, especially those which are close to transmitters.

Radiation from the local oscillators of receivers may cause interference to other receivers; this may be particularly troublesome in the broadcasting and land mobile services, where receivers under independent control may be physically close together. Interference may also enter receivers at image frequencies. Very powerful interference may degrade reception of signals over a wide frequency band because of non-linear amplification in early stages of receivers. All of these phenomena may circumscribe the freedom of an Administration to assign frequencies for reception. However, they can to a large degree be eliminated by good equipment design, and weight should be given to this in the Administration's oversight of the equipping of stations.

Interference from industrial, scientific and medical sources

In addition to man-made noise-like interference (see below) there are many industrial, scientific and medical devices which make use of large amounts of radio frequency power in a form that is not noise-like but narrow-band, sometimes coherent in waveform, at frequencies that are determined by the design of the device. A considerable amount of RF energy may be radiated unintentionally from such installations, capable of causing severe interference at radio receivers nearby. Recommendation ITU-R SM.1056 [23] briefly reviews the principal industrial, scientific and medical (ISM) applications and lists some typical levels of leaked radiation.

These ISM applications are required to operate only in specific frequency bands in various parts of the radio spectrum. These frequency bands have been designated for use in such apparatus, details being given in footnotes to the international table of frequency allocations. The principal bands (see RR S5.150) are as follows:

13.553–13.567 MHz	2400–2500 MHz
26.957–27.283 MHz	5725–5875 MHz
40.66–40.70 MHz	24–24.25 GHz
902–928 MHz (Region 2 only)	

In addition, the following bands are available for ISM use subject to the special authorisation of the Administration concerned and with the agreement of other Administrations whose services might be affected (see RR S5.138):

6765–6795 kHz	122–123 GHz
61.0–61.5 GHz	244–246 GHz

In addition, the band 433.05–434.79 MHz is available for ISM use in ten European countries listed in RR S5.280 and, subject to the terms of RR S5.138, in other countries in Region 1.

These ISM bands are also allocated for radio services, and a radio station assigned a frequency within a designated band must tolerate the ISM interference. RR S15.13 calls on Administrations to seek to ensure that the radiation from ISM installations is minimal. There are, at present no agreed constraints on radiation from ISM installations. However, RR Resolution 63 initiated discussions between the ITU and the International Special Committee on Radio Interference of the International Electrotechnical Commission (IEC), having the objective of limiting interference to radio services.

Man-made noise

In addition to the relatively coherent ISM radiation, a large amount of wide-band noise-like interference at radio frequency is radiated from many kinds of electrical equipment. In many localities this 'man-made noise' is the dominant form of noise, external to the radio receiver, over much of the VHF and UHF frequency ranges; see Section A8.2.2 below. The main sources of man-made noise are electric traction systems, certain industrial processes and the ignition systems of petrol-driven vehicles. Much can be done to ensure that this noise is controlled at the source by the application of laws and regulations on electromagnetic compatibility (EMC). RR S15.12 directs Administrations to get involved in the implementation of such constraints, in the telecommunications service and generally.

4.2.4 Enhanced spectrum utilisation efficiency by regulation

Various measures have been adopted internationally, by means of the RR or through recommendations of the ITU-R, that constrain systems in minor ways in order that the radio spectrum shall yield a greater aggregate communications capacity or to facilitate efficient sharing of frequency bands between different services. The limitation of the bandwidth of emissions (see Section 4.2.2) is one such measure, but there are various others and this section outlines the main forms that these other measures take.

Man-made hazards in the space environment

There is a risk of serious long-term interference from satellite transmitters, for example when the orbit of the satellite can no longer by controlled, if the transmitters cannot be switched off. RR S22.1 requires all satellites to be equipped with devices to ensure that their transmitters can be switched off immediately by telecommand if the need should arise.

There is a growing risk that discarded body-sections of launchers, debris from the launch process and satellites that have passed the end of their operational life, remaining in orbit indefinitely, will become a significant physical hazard to operational satellites. Collisions have already occurred and will become more frequent, and the additional debris from collisions will aggravate the problem. The danger is seen to be greatest for geostationary satellites but it is likely that the hazard will be intensified in lower orbits soon. Recommendation ITU-R S.1003 [24] provides a preliminary analysis of the problem in the geostationary context and recommends measures to check its growth.

Constraints on geostationary satellites: (a) Station-keeping

Ideal geostationary satellites revolve in a circular orbit with a radius of 42 164 km in the Earth's equatorial plane in a west-to-east direction. Their orbital period is one sidereal day and they remain stationary relative to a point on the surface of the rotating Earth. However, a satellite in the GSO experiences perturbing forces which tend to change the period and the shape of the orbit and the angle of inclination of the orbital plane relative to the equatorial plane, causing the satellite to cease to be perfectly geostationary. This leads to various disadvantages, including a reduction in the efficiency with which the GSO can be used. Consequently, limits are placed on the departure of a nominally geostationary operational satellite from an ideal geostationary orbit, especially in the east–west direction.

Satellites of the BSS operating between 11.7 and 12.7 GHz must be maintained within $\pm 0.1°$ of their nominal longitude; see RR Appendix S30, Annex 5, Section 3.11. This standard of station-keeping is recommended in the north–south plane also.

With specified exceptions, other geostationary satellites operating in other frequency bands allocated for the BSS or for the FSS must be capable of maintaining station within ±0.1° of their nominal longitude. It is not required that this precision be maintained in normal operation unless this is necessary to prevent unacceptable interference to another satellite network. However, note that the term 'unacceptable interference' may denote a relatively small amount of interference, much less that the 'harmful interference' defined in RR S1.169 and quoted in full in Section 4.5 below; see also Section 5.2.6.

Exceptionally, experimental geostationary satellites and satellites which were put into service before 1987, the 'advance publication of information' having been made before 1982, are not required to achieve these standards. For these satellites the east–west station-keeping capability required is ±0.5° and ±1.0°, respectively. See RR S22.6–S22.18. See also Recommendation ITU-R S.484-3 [25].

There are no mandatory requirements for north–south station-keeping for satellites. However, the status of a quasi-GSO satellite with a daily north–south excursion exceeding ±5° is unclear. It may be termed an inclined geosynchronous satellite, not geostationary, and the longitudes at which it crosses the equatorial plane going north and going south (that is, at the ascending node and descending node, respectively of its orbit) must both occur within the range required for the station-keeping of a geostationary satellite. See the footnote to the title of Section III of RR Article S22.

Constraints on geostationary satellites: (b) Precision of antenna beam pointing

The earthward beams of satellites of the BSS operating between 11.7 and 12.7 GHz must be maintained within ±0.1° of their nominal direction. For the corresponding feeder link beams, the required pointing accuracy is ±0.2°. Limits are also placed on the angular rotation of beams. See RR Appendix S30, Annex 5, Section 3.14 and RR Appendix S30A, Annex 3, Section 3.7.4.

For other geostationary satellites it must be feasible to maintain the pointing direction of the maximum radiation of an earthward beam within ±10% of the half-power beamwidth of the nominal direction, or within ±0.3° if that is greater. As with station-keeping, it is not required that this precision be implemented except to prevent unacceptable interference to another system. See RR S22.19–S22.21. As before, 'unacceptable interference' may denote a relatively small amount of interference. Recommendation ITU-R S.1064-1 [26] recommends that a more stringent beam-pointing tolerance, namely ±5% or ±0.2° of the half-power beamwidth, should be adopted as a design objective for satellites of the FSS.

Other technical standards and constraints which contribute to efficiency of spectrum use

The frequency tolerances set by the RR (see Section 4.2.2 above) apply to all radio services. The station-keeping and beam-pointing standards (see above) apply to geostationary satellites of several services. Various other measures are applied by international agreement to promote efficient use of spectrum. Some of these measures take the form of the rules that are used in drawing up frequency assignment and frequency allotment plans. Sharing constraints may be used where two services have been allocated the same frequency band if it is found that systems of one service can accept a limited curtailment of their freedom to use the band and thereby facilitate the use of the band by systems of the other service. Examples of sharing constraints are to be found in Section 5.2.3; these enable frequency bands to be shared efficiently by terrestrial and satellite services. There are also many relevant recommendations of ITU-R and ITU-T; some of these are made mandatory by reference in the RR and others, for the present, are only advisory but widely implemented. Many of these measures are reviewed in later chapters of this book.

The basic objective of these various measures is to ensure that radio systems can achieve adequate standards of performance whilst tolerating a significant but controlled level of interference from other systems, so enabling the spectrum to be used by a very large total number of systems. However, the application of these internationally approved principles is not in itself sufficient to ensure that efficient spectrum utilisation is achieved in an economical way. This is in part because many of the problems which Administrations must overcome if spectrum is to be used efficiently arise in the licensing of domestic systems, and these are of deliberately limited concern to the ITU. The licensing process also has an important role in the achievement of efficient spectrum use; see Section 4.3 below.

4.3 Licensing and licence fee policy

4.3.1 The requirement for licensing

RR S18.1 requires all transmitting stations established or operated by private persons or enterprises to be licensed by the Administration. Other paragraphs of RR Article S18 deal with basic responsibilities of the licensee and provide guidance for Administrations on various problems that may arise in licensing mobile stations which may enter the jurisdiction of other Administrations.

All licences will, no doubt, identify the licensee, the location of the transmitting station or stations or the area within which they move, the purpose for which the frequency assignment(s) may be used and the period of validity of the licence. In most cases the assigned frequency, the parameters of the emission and essential characteristics of the transmitting antenna will be stated, although the authority to transmit may have to be defined in another form where a single document authorises the use of more than one transmitter or more than one assignment. A licence may also be issued for a receiving station, recording the assigned frequency etc. of emissions which have been registered for reception at the station. A fee will usually be charged for a licence. Key issues that arise in licensing are who should be licensed to transmit and how is efficient spectrum use to be achieved.

4.3.2 Licensing non-commercial radio users

Non-commercial users are one of the main categories of radio user. These organisations use radio because it is a resource that is essential for the pursuit of a primary activity which is of public concern. The communications and navigation requirements of ships and aircraft, of the police, the emergency services and certain public utilities come into this category. It is also convenient to include here the users of the science radio services reviewed in Chapter 13 below. Almost all of the radio services are used by non-commercial users to some extent and for some services they are the sole users.

While the Administration may insist on the collaboration of non-commercial users in ensuring that spectrum assigned to them is used efficiently, the case for providing for the essential radio needs of these organisations, to the extent that it is feasible, is not usually at issue. A licence fee or some equivalent transfer of funds will usually be charged, sufficient to meet some appropriate fraction of the Administration's costs.

4.3.3 Conventional procedures for licensing commercial radio users

Commercial users of transmitters fall into two basic categories: service providers and private radio users. Service providers set up systems for other parties to use; thus PTOs provide telecommunications facilities to users and broadcasters deliver entertainment and information to the public. Private radio users typically use their own radio systems in the course of their own business. Disregarding non-commercial use, service providers are virtually the sole users of the BS and the BSS and they are the dominant users of the FSS and the MSS. Private users are the sole users of the AmS and the AmSS. Both categories of user compete for commercial use of the FS and the LMS, although the share taken by service providers can be expected to be dominant in each case. Commercial use of the other radio services is small.

The products that service providers supply and private radio users obtain from the FS and the LMS are similar. However, in some circumstances private systems can be less costly and more flexible than facilities leased from a service provider, but the latter can be expected to make more economical use of spectrum. Thus, where spectrum is not abundant, it may be supposed that the Administration will be more willing to license service providers to use it than private radio users but the Regulator is likely to prefer the maintenance of competition between them.

In the conventional licensing procedure for commercial radio users in a liberal regime, the Administration follows the principles outlined in Sections 4.1 and 4.2 to ensure that an applicant has needs or opportunities that justify the grant of a licence and has equipment that would use spectrum efficiently. If satisfactory frequencies can be found for all applicants, a licence is issued and a licence fee is charged. The fee is typically set at a level that will cover an appropriate share of the Administration's costs. If satisfactory frequencies cannot be found for all applicants of both categories of radio user, the Administration and the Regulator must find some basis for deciding which applications are to be rejected, pending the allocation of more spectrum. This decision will involve a choice between service providers and private users and between one service provider and another.

This procedure is a laborious one, applied fairly effectively in the past but increasingly difficult to implement as the pressure of demand increases, in particular for systems of the BS, FS and LMS. Particularly difficult and embarrassing problems arise when it is necessary to choose which of several service providers should be given access to a limited amount of spectrum. Other ways of optimising spectrum use and of choosing between applicants for licences have been considered. The application of economic solutions to both problems, known as spectrum pricing, has been favoured in some countries.

4.3.4 Spectrum pricing

Economists have considered the application of market forces to the allocation and assignment of spectrum. Levin [27] for example, points out that the radio spectrum is exceptional among factors of substantial economic importance in that it is made available virtually without charge. In consequence, he argues, there is no satisfactory mechanism for ensuring that spectrum is distributed between users in an economically optimum way. He recognises that the establishment of a market in spectrum would present special problems, probably making a completely free market impracticable, but he concludes that some substantial application of market forces in frequency management would be beneficial. It could be expected to ensure that spectrum would always be available to organisations that would extract the greatest economic value from it. Existing licensees would minimise their holding of spectrum, relinquishing what they no longer needed, if spectrum cost its full market value. Furthermore, he concludes, the cost of managing the spectrum might be reduced by such a change. See also References [28–33].

The implementation of a free market in spectrum, in which a party could buy a frequency band to use for any purpose or to lease or sell to another party, whole or sub-divided by frequency or geographically, with the buyer having the same freedom of use or disposal, seems likely to present difficult domestic problems and formidable international problems. It may be supposed that the availability of spectrum for the most profitable applications would be increased, but there seems to be little doubt that the efficiency of use of the medium would be reduced, and that many of the less profitable applications would be squeezed out altogether.

Nevertheless, there are various options between a spectrum management regime based on the executive application of technical principles and a totally free market ruled by price, and many radio users would see advantages in some of those options. For a private mobile telephone user, a one-year licence, renewable at the Administration's pleasure, is too insecure; many licensees would pay a higher fee for better tenure. Some governments find inappropriate and objectionable their commitment to manage a complex administrative and technical institution serving vast numbers of citizens and companies, especially where telecommunications and broadcasting services have been privatised and liberalised. Governments are embarrassed by the need to choose between applicants for, for example, a limited number of assignments for broadcasting stations and would be glad to replace choice by auctions of assignments. Higher licence fees might, for example, encourage private radio users to transfer to leased channels in a service provider's system which uses spectrum more efficiently. And it may be supposed that governments would welcome a new source of revenue that would be seen, not as a tax, but as a contribution to the public purse for leasing the use of spectrum, a public asset.

It seems likely that spectrum pricing in some form could help to resolve some important non-technical problems that confront Administrations and Regulators. This might include eliminating trivial applications for assignments, discouraging inflated applications, encouraging the use of narrow-bandwidth systems and resolving choice between service provider applicants. Furthermore, these benefits might be obtained selectively, wherever geographically and where in the spectrum they were most needed. Spectrum pricing is not, however, an alternative to technical requirements designed to increase the efficiency with which spectrum is used, such as those reviewed in Section 4.2 above.

4.3.5   The application of spectrum pricing

A scheme of spectrum management that draws heavily on free market principles has been established in New Zealand under the Radiocommunications Act of 1989. Nationwide 'management rights' for selected frequency bands, valid for 20 years, have been leased by the Ministry of Commerce by auction. These management rights can be traded, sub-divided or aggregated. A holder of management rights leases 'licence rights' of stated duration, to itself or to other would-be users, the licence rights holder being entitled to set up radio transmitters of specified carrier frequency, power and type of emission at specified locations, to be used for whatever purpose the licence rights holder chooses. Licence rights holders pay an annual fee to cover the costs of the Ministry and are responsible for ensuring that radiation beyond the frequency limits of the management rights holder's lease does not exceed fixed levels. These licence rights are also tradable. Frequency bands which remain outside this scheme, including bands in which international interference problems are thought likely to be troublesome, continue to be managed by the Ministry of Commerce.

The radical course taken by New Zealand has not been followed elsewhere so far. Indeed the geographical isolation of New Zealand probably enables practices to be implemented there that would not be feasible elsewhere. However, a variety of much more limited spectrum pricing initiatives, linking economic pressures with more conventional methods of spectrum management, have been introduced in several countries, some of which are outlined below.

Spectrum pricing for private radio systems

Fees for private mobile radio licences are being increased in some metropolitan areas to discourage inessential use or to divert traffic to cellular systems, which use spectrum more efficiently. Recognising that considerable progress has been made in reducing the bandwidth per channel of LMS systems in recent years, this practice might be extended by relating licence fees to the bandwidth occupied, so encouraging the replacement of out-of-date equipment, See further in Section 8.6.

Licence fees for multichannel fixed links might be made proportionate to the necessary bandwidth or increased in frequency bands which have particularly favourable radio propagation characteristics. See References [34] and [35]. Here again, where there is a shortage of spectrum, higher licence fees might give relief by encouraging the use of broad-band facilities leased from service providers, whose radio systems may use spectrum more efficiently and who have more ready access to the alternative of cabled connections.

Licensing policy for service providers

An Administration is responsible for managing assignments for many radio services but by far the most extensive are the FS, LMS and BS. Where telecommunications and broadcasting are monopolies, many of the problems of managing the frequency bands allocated for these services can be delegated from the Administration to the service providers. However, in a country where competition between the suppliers of these services is a significant element in the regulation of quality and prices, the Administration must retain a strong grip on the use of the spectrum involved, in collaboration with its Regulator colleague, who will have different objectives in controlling access to spectrum for these three services. Licensing for competition in these services is considered below, and policy is likely to differ from service to service.

Licensing fixed links for telecommunications service operators

PTOs use the FS for three main purposes. Line-of-sight inter-city links, typically spanning distances of 10–50 km, often formed into long radio relay chains, occupy a large part of the SHF spectrum. Short SHF and EHF links are used within cities for the prompt provision of wide-band connections to subscribers, avoiding the delays that installing new underground cables might involve. Radio links such as these are also being used extensively to connect LMS base stations together and into the PSTN. Thirdly, short distance multiple access systems, typically at UHF, are being brought into use to connect subscribers into the telephone network in rural areas.

In each of these applications the objectives of an Administration serving a liberal regime are likely to be:

- to maintain conditions of fair competition between operators,
- to ensure that an efficient pattern of assignments is employed but without needless duplication or delay of the planning done by the operators' own planning staff and
- to ensure that the industry as a whole makes constructive use of both cable and radio if the usable range of the radio medium should approach exhaustion.

Above about 20 GHz the FS allocations may not be in heavy use. These bands, up to 55 GHz at least, are very suitable for short distance wide-band connections between the PSTN and the premises of both major subscribers and the base stations of land mobile networks. It may be found practicable to delegate the detailed management of an allotment of spectrum intended for these purposes, under terms that ensure efficient use. However, the frequency bands allocated for the FS below about 10 GHz which are necessary for long links without intermediate relays and long radio relay chains, are likely to be crowded already. The Administration will probably assign frequencies in these bands to operators that apply for them in the manner described in Section 4.1.1 above for private fixed links. In the frequency bands between 10 and 20 GHz that are usable for the shorter inter-city links, one of these solutions or the other will be appropriate, depending on local circumstances.

An allotment made to a PTO for delegated management would be the subject of a formal agreement. The allotment would consist of a defined band of frequencies for use within a defined area; it would desirably be free from assignments to other radio users and capable, given good management, of providing enough assignments to satisfy the PTO's foreseen need for new links for one or two years. Guidelines on the parameters of links to be operated in the band should be provided by the Administration, to place limits on the distance at which interference to or from radio users in the same frequency band in other areas occurs. There should be provision for consultation with the Administration before selecting a frequency that could involve interference to a foreign station. When a frequency has been selected by the operator for a link, the Administration would be asked to make a formal assignment, to add the assignment to the national register and if appropriate to notify the ITU for registration of the assignment in the MIFR. Periodically, and in particular if the operator asks for more bandwidth, the Administration should audit the efficiency with which the operator is managing the allotment. Having regard to the need for fair competition between operators and to the special value of radio for urgent new requirements, a point may be reached when an application for a new allotment will have to be refused, established routes being transferred to cable.

There will no doubt be a fee to be paid by PTOs for access to spectrum, whether for an allotment or for a single assignment. The fee would probably be made proportional to the allotted bandwidth or the assigned bandwidth, as appropriate. It may also be desirable to include a factor reflecting the relative scarcity of spectrum at the frequency in question, to encourage the use of other bands under less demand, or cabled transmission media instead of radio. The use of auctions in allotting FS spectrum has been implemented in some countries, but it may well be difficult to combine auctions with the maintenance of competition on equal terms between PTOs.

Licensing public land mobile systems

Some providers of land mobile services operate on a small, local scale, providing for example a trunked public access mobile radio (PAMR) network with a few channels from a single base station. These systems are potentially a spectrum-efficient alternative to private mobile systems and their access to spectrum can be treated in much the same way as a private system. However, other public LMS facilities provide dozens or hundreds of channels over wide areas through many base stations. For these big systems, including cellular systems, the Administration and the Regulator may well collaborate to auction licences to competing service providers; see Section 8.5.4.

Licensing broadcasting systems

Broadcasting uses emission characteristics which are nationally standardised and are relatively uniform world-wide. The standards have changed little in several decades, although a phase of extensive technical change, springing from internationally coordinated development of digital systems, has recently begun. The more important frequency assignments are usually planned at government level, nationally and internationally. Thus, the main function of the licensing process is to identify the organisations that are to be authorised to broadcast. The auction mechanism is being used increasingly to resolve the choice between applicants of broadly equal merit; see Sections 7.1.2 and 10.2.

4.4 Identification of stations and emissions

For some kinds of radio system it is very important to be able to identify emissions, so that links can be established between stations as required. For most kinds of system, identification is also required so that sources of interference can be contacted. Accordingly it is a general requirement that all emissions should be identifiable, by the transmission of identification signals at intervals or by some other means, and all transmissions with false or misleading identification signals are prohibited; see RR S19.1 and S19.2. It is recognised, however, that the provision of identification raises technical problems for some kinds of emission, and not all of these problems have been satisfactorily solved. Nevertheless, RR S19.4 to S19.11 and S19.26 insist on the provision of some form of identification for a number of services and specific kinds of emission.

Identification signals usually take the same form as the signals that are the traffic of the emission. A telegraphic emission will be assigned a conventional alpha-numeric call sign. A television emission may give its identity on its test card. An airliner may use its flight number. A ship may use a selective call identifier. A call sign may be given by an Administration to the ITU in notifying a frequency assignment for registration in the MIFR. Ships' call signs and maritime identifiers are notified by Administrations to the ITU and a list of ships' call signs and identifiers is published by the ITU [36].

Call signs and the two kinds of maritime identifier consist of an alpha-numerical group of characters to identify the country responsible followed by a numerical group assigned by the Administration to identify a particular station. There are tables of the national codes for the two kinds of maritime identifier in the preface of post-1995 revisions of the list of ships' call signs [36]. A table of the national codes used in call signs is in RR Appendix S42. A procedure for assigning new and additional codes is set out in RR Article 19 Section II.

4.5 Interference

Radio systems should be designed to minimise the entry of interference to the extent that this is technically and economically feasible, and to tolerate low level interference without failing to attain the performance objectives. Nevertheless, higher levels of interference may occur. This may happen because the prospective level of an interfering emission has been miscalculated in the course of choosing a frequency for a new assignment. Interference may be occasional, arising from sub-normal propagation conditions for the wanted emission or abnormally favourable propagation conditions for the interfering emission. Interference may also arise from maladjusted or faulty equipment and from transmissions made without authority, without due care or mischievously. Action may be necessary to reduce or eliminate such interference; the process followed depends on various factors which are reviewed below.

1 Interfering station within the same country
When unacceptable interference occurs, the receiving station may try to identify the source of the interference. If identification is possible, an informal complaint may be made directly to the transmitting station. If identification is not possible, or a complaint produces no useful result, a report is made to the Administration. With a well-equipped monitoring station and with mobile monitoring stations to locate the source of short range interference, the Administration should be able to locate interfering transmitters, and especially any that are within its own jurisdiction.

If the interfering transmitter is found to be within the jurisdiction of the same Administration and operating without authority, the Administration will order the cessation of transmissions and may prosecute the owner. If the interfering transmission is licensed, the Administration will see to it that any faults and maladjustments at either station and any abuse of the terms of licences, both common causes of interference, are put right. If the interference is found to have arisen from an uncommon abnormality of radio propagation conditions, the Administration may rule that such occasional interference must be endured. If the problem is found to arise from under-achievement of predicted equipment parameters or miscalculation of normal propagation conditions and the severity of the interference justifies action, the Administration will often resolve the problem by assigning a different frequency to one station or the other.

2 Interfering station in a foreign country

If the interfering station is within the jurisdiction of another Administration, it may be more difficult to identify it, but Administrations help one another in combating interference. See also the reference to international monitoring in Section 4.1.1, Stage 6, above. Then, with the source identified, it may be possible to make use of the provisions of the Radio Regulations to eliminate the interference, provided that the interference is severe enough to be classified as 'harmful interference', and the emission suffering interference and the receiving location are registered in the MIFR. Harmful interference is defined in RR S1.169 as

> interference which endangers the functioning of a radionavigation service or of other safety services or seriously degrades, obstructs, or repeatedly interrupts a radiocommunication service operating in accordance with these Regulations.

See also RR Article S15 and RR Appendix S10.

The feasibility of clearing cross-border interference depends on the circumstances, thus:

1   If the assignment suffering interference is registered in the MIFR, if it is conforming with the RR etc. and an emission causing it harmful interference is also conforming but not registered, or is not conforming whether registered or not, the Administration having jurisdiction over the interfering emission must ensure that the interference is reduced to a level that is not harmful.

2   If both assignments are conforming and registered, the outcome in shared frequency bands may depend on the services to which the stations belong and the category of their allocations. If the allocation for one assignment is primary and for the other it is secondary at the location where the interference is occurring, then the assignment with the secondary allocation must cease causing harmful interference to the assignment with the primary allocation and cannot claim protection from whatever interference the former may cause it.

3   However, if the services to which the assignments relate have the same allocation category, whether primary or secondary, then one of three situations arises.

   a)   An emission which is involved in the safeguarding of human life and property (see RR S1.59) is called a 'safety service'; the term is applied to radionavigation systems on which the safety of human life may depend and essential maritime and aeronautical communications services, above all to frequencies reserved for distress and safety communication. RR S4.10 and S15.8 stress the special need to avoid making assignments that will cause interference to safety services and strict mandatory disciplines, designed to ensure that safety messages are not impeded, are applied to operation on distress and safety frequencies. Administrations cooperate and act vigorously to suppress interference to safety services.

   b)   Section 5.4 below reviews the use of internationally agreed plans for assignments in some frequency bands. In drawing up these plans, care is taken to ensure that significant interference should not occur even when all of the assignments have been implemented, and the Administrations which are signatories to such agreements undertake to respect all of the planned assignments, without regard to the date of their being taken into service. Thus:

   i) If interference does, nevertheless, occur between planned assignments, it may be that the parameters of the emissions, as implemented, are not in full accord with the plan or else that assumptions made in planning (for example, as to radio propagation) were incorrect. If so, the emission parameters should be brought into alignment with the plan or the plan should be amended and agreed again by the parties to the agreement.

   ii) It may be that an Administration which is a party to the agreement makes an assignment which is not included in the plan; if this emission caused interference to a planned emission, the Administration must withdraw the assignment.

   iii) If an assignment made by an Administration which is not a party to the agreement causes or suffers interference involving a planned assignment, the problem may have to be resolved as in sub-paragraph c) below if the terms of the plan agreement do not provide a resolution of it.

c) A more typical case of harmful interference is one that arises between assignments of the same allocation category, both registered in the MIFR. Now RR S8.1 reads

> The international rights and obligations of Administrations in respect of their own and other Administrations' frequency assignments shall be derived from the recording of those assignments in the Master International Frequency Register... Such rights shall be conditioned by the provisions of these Regulations...

and RR S4.3 reads

> Any new assignment or any change of frequency or other basic characteristic of an existing assignment (see Appendix S4) shall be made in such a way as to avoid causing harmful interference to services rendered by stations using frequencies in accordance with the Table of Frequency Allocations in this Chapter and the other provisions of these Regulations, the characteristics of which assignments are recorded in the Master International Frequency Register.

See also RR S8.3. There emerges from these paragraphs a strong presumption, when interference occurs between assignments that otherwise have equal status, that the right to continue operating without harmful interference will be conceded to the assignment with the earlier registered date of entry into service. This is indeed how the problem is usually resolved unless some modification of the parameters of one emission or the other, such as a minor change of frequency or the adjustment of antennas, reduces the interference below the threshold of harmfulness. Nevertheless, no such obligation is explicit in the Radio Regulations, perhaps because many Administrations would consider that such an obligation would be inequitable, placing developing countries at a disadvantage relative to developed countries. Thus the resolution of interference problems is sometimes a matter for negotiation between Administrations.

4. Finally it may sometimes happen that harmful interference occurs between emissions, both of which are registered in the MIFR but neither of which is conforming. No provision of the Radio Regulations can help to resolve this problem, but the Administrations responsible may negotiate a resolution, probably acknowledging the claim of the assignment with the earlier registered date of entry into service.

4.6 References

1 'International frequency list (IFL)'. Published every March and September on CD-ROM (ITU, Geneva)

2 'Weekly Circular and Special Sections'. Published by the ITU Radiocommunication Bureau every week. The Weekly Circular is published on paper, as a microfiche and also on diskette. It is also to be published once every two weeks in the form of a CD-ROM starting in 1999, and this is expected to become the main medium for this publication in due course (ITU, Geneva)

3 'Handbook on spectrum monitoring' (ITU, Geneva, 1995)

4 'Tasks of a monitoring service'. Recommendation ITU-R SM.1050, ITU-R Recommendations 1997 SM Series (ITU, Geneva, 1997)

5 'International monitoring system'. Recommendation ITU-R SM.1139, *ibid.*

6 'List VIII'. List of international monitoring stations (ITU, Geneva)

7 'Handbook of national spectrum management' (ITU, Geneva, 1995)

8 'Planning methods for 625-line terrestrial television in VHF/UHF bands'. Recommendation ITU-R BT.1123, ITU-R Recommendations 1997 BT Series (ITU, Geneva, 1997)

9 'A method of spectrum management to be used for aiding frequency assignment for terrestrial services in border areas'. Recommendation ITU-R SM.1049-1, ITU-R Recommendations 1997 SM Series (ITU, Geneva, 1997)

10 'A procedure for determining coordination area between radio-relay stations of the fixed service'. Recommendation ITU-R F.1095, ITU-R Recommendations 1997 F Series Part 1 (ITU, Geneva, 1997)

11 ' Definition of spectrum use and efficiency of a radio system'. Recommendation ITU-R SM.1046-1, ITU-R Recommendations 1997 SM Series (ITU, Geneva, 1997)

12 'National spectrum management'. Recommendation ITU-R SM.1047, *ibid.*

13 'Design guidelines for a basic automated spectrum management system (BASMS)'. Recommendation ITU-R SM.1048, *ibid.*

14 'Design guidelines for developing advanced automated spectrum management systems (ASMS)'. Recommendation ITU-R SM.1370, ITU-R Recommendations 1997 SM Series Supplement 1 (ITU, Geneva, 1997)

15 'Determination of necessary bandwidths including examples for their calculation and associated examples for the designation of emissions'. Recommendation ITU-R SM.1138, ITU-R Recommendations 1997 SM Series (ITU, Geneva, 1997)

16 'Bandwidths and unwanted emissions of digital radio-relay systems'. Recommendation ITU-R F.1191-1, ITU-R Recommendations 1997 F Series Part 1 (ITU, Geneva, 1997)

17 'Frequency tolerance of transmitters'. Recommendation ITU-R SM.1045-1, ITU-R Recommendations 1997 SM Series (ITU, Geneva, 1997)

18 'Technical characteristics of single-sideband transmitters used in the maritime mobile service for radiotelephony in the bands between 1606.5 kHz (1605 kHz Region 2) and 4000 kHz and between 4000 kHz and 27500 kHz'. Recommendation ITU-R M.1173, ITU-R Recommendations 1997 M Series Part 3 (ITU, Geneva, 1997)

19 'Spurious emissions'. Recommendation ITU-R SM.329-7, ITU-R Recommendations 1997 SM Series (ITU, Geneva, 1997)

20 'Techniques for measurement of spurious emissions of maritime radar systems'. Recommendation ITU-R M.1177-1, ITU-R Recommendations 1997 M Series Part 4 (ITU, Geneva, 1997)
21 'Maximum permissible level of spurious emissions from very small aperture terminals (VSATs)'. Recommendation ITU-R S.726-1, ITU-R Recommendations 1997 S Series (ITU, Geneva, 1997)
22 'Essential technical requirements of mobile earth stations for global non-geostationary mobile satellite service systems in the bands 1–3 GHz'. Recommendation ITU-R M.1343, ITU-R Recommendations 1997 M Series Part 5 Supplement 1 (ITU, Geneva, 1997)
23 'Limitation of radiation from industrial, scientific and medical (ISM) equipment'. Recommendation ITU-R SM.1056, ITU-R Recommendations 1997 SM Series (ITU, Geneva, 1997)
24 'Environmental protection of the geostationary-satellite orbit'. Recommendation ITU-R S.1003, ITU-R Recommendations 1997 S Series (ITU, Geneva, 1997)
25 'Station-keeping in longitude of geostationary satellites in the FSS'. Recommendation ITU-R S.484-3, ITU-R Recommendations 1997 S Series (ITU, Geneva, 1997)
26 'Pointing accuracy as a design objective for earthward antennas on board geostationary satellites in the FSS'. Recommendation ITU-R S.1064-1, *ibid.*
27 LEVIN, H. J.: 'The invisible resource; use and regulation of the radio spectrum' (The Johns Hopkins Press, 1971)
28 COASE, R. H.: 'The Federal Communications Commission', *Journal of Law and Economics*, 1959, **2**, pp. 1–40
29 WEBBINK, D. W.: 'The value of the frequency spectrum allocated to specific uses', *IEEE Trans.*, 1977, **EMC-19**, pp. 343–351
30 RUDD, D.: 'A renting system for radio spectrum', *Proc. IEE Pt A*, 1986, **133**, pp. 58–64
31 LEVIN, H. J.: 'Emergent markets for orbit spectrum assignments; an idea whose time has come', *Telecommunications Policy*, 1988, **12**, pp. 57–76.
32 MCMILLAN, J. 'Why auction the spectrum', *Telecommuncations Policy*, 1995, **19**, (3), pp. 191–199
33 NOAM, E.: 'Beyond spectrum auctions – taking the next step to open spectrum access', *Telecommunications Policy*, 1997, **21**, (5), pp. 461–475
34 'Study into the use of spectrum pricing'. Prepared by National Economic Research Associates and Smith System Engineering Ltd for the Radiocommunications Agency (London, April 1996)
35 'Implementing spectrum pricing'. A consultation document of the Radiocommunications Agency (London, May 1997)
36 'List of call signs and numerical identities of stations used by the maritime mobile and maritime mobile-satellite services (List VII A)'. Published periodically and kept up-to-date by quarterly supplements (ITU, Geneva)

Chapter 5

Frequency band planning and mandatory frequency coordination

5.1 Introduction

The frequency assignment practices reviewed in Chapter 4 place the whole responsibility for choosing frequencies for assignment to radio stations on the assigning Administration. In most cases this raises no problem of principle. Interference, if it happens, usually involves stations which are under the jurisdiction of the same Administration and no other body is better able to foresee the risk of interference or remedy the situation if it arises. Interference may be caused or suffered by a station in another country when a newly assigned frequency is brought into use but even so it can usually be avoided by informal coordination between the Administrations of adjacent countries; see Section 4.1.4 above.

However, there are circumstances in which an Administration is unlikely to be able to determine in advance whether an assignment it wishes to make will result in interference. This is particularly true of the satellite services and the circumstances are often made more complex by the frequency band sharing that is practised in almost all frequency bands allocated for satellite services. A new satellite may cause interference to, or suffer interference from, an earth station many thousands of kilometres from its intended service area. Earth stations and terrestrial stations may have interference problems even if they are separated by distances that would ensure no risk of interference if both stations were conventional terrestrial stations. Furthermore, in aggregate, the large number of satellites operating in a frequency band may cause serious interference to vast numbers of terrestrial stations operating in the same frequency band, and vice versa.

In practice, potential interference problems between satellite networks that are already in service and a proposed new satellite network are identified and resolved by a mandatory coordination procedure. A second mandatory coordination procedure is used to identify and prevent interference between a new earth station and forcign terrestrial stations. The principles involved in this latter procedure are summarised in the context of sharing between the FS and the FSS in Recommendation ITU-R

SF.355-4 [1]. Coordination between large numbers of satellite systems and potentially vast numbers of terrestrial stations to prevent unacceptable aggregate interference levels arising would not be feasible, but a system of sharing constraints has been found to be satisfactory for the purpose. These procedures and sharing constraints are reviewed in Section 5.2 below.

Mandatory frequency coordination has also been made a condition of registration in the MIFR of new frequency assignments in a number of special situations where sharing between services has been a subject of controversy; see Section 5.3.

In addition to the mandatory frequency coordination and the associated sharing constraints outlined above and the regulatory provisions reviewed in Chapter 4, which are broad in their application, there are detailed international agreements on the ways in which some specific frequency bands will be used. Some of these agreements are simply radio channel plans like those mentioned in Section 4.1.3 but others go much further, even including plans for the frequency assignments that are expected to use the band; these agreements are reviewed briefly in Section 5.4. Where such plans are of sufficient general interest, they are reviewed in more detail in later chapters.

5.2 Frequency coordination for satellite and terrestrial services

5.2.1 *Coordination between satellite networks*

Terminology

A few of the terms used in ITU texts can usefully be noted here. Formal definitions for most of these terms are to be found in RR Article S1.

A vehicle that is destined to be launched into an orbit, thereby becoming a 'satellite', is formally termed a 'spacecraft'. Satellite communication involves links between 'earth stations' routed via a 'space station', the latter consisting of the radio equipment which is carried by an artificial satellite. However, the distinctions that are drawn between spacecraft, satellite and space station are seldom of great practical importance from the standpoint of telecommunications, and 'space station' has come to have a different meaning in general usage. In this book the word 'satellite' is used for all three concepts, spacecraft, satellite and space station; where the difference between the three formal terms is significant, the context will make clear what is intended.

'Earth stations' may be at fixed locations or they may be mobile, by land, sea or air. 'Terrestrial stations' are radio stations on Earth which provide links to other terrestrial stations.

A 'geostationary' satellite is one that has a circular orbit in the equatorial plane, moving from west to east, having an orbital period of one sidereal day. This orbit is called the geostationary satellite orbit (GSO); its height is about 35 786 km above the Earth's surface. Such a satellite appears, from the Earth, to be stationary in the sky. If the plane of the orbit is allowed to drift out of the equatorial plane, the orbit becomes geosynchronous, or less precisely, quasi-geostationary. Satellites with much lower orbits are used for various scientific purposes and are coming into use for communication with portable, and other very small, earth stations; these 'low Earth orbit' (LEO) satellites are becoming the most common kind of non-GSO satellite.

A ' satellite network' consists of one satellite plus all the earth stations it serves. Some satellite systems employ several or many non-GSO satellites which are used, together or in turn, to provide a service for earth stations to use. The term 'constellation' will be used in this book for a multi-satellite system. A 'satellite link' is a radio link, carrying a single channel or a number of channels multiplexed together, from one earth station to another via one satellite. A satellite link consists of an 'uplink' and a 'downlink'. Direct links between satellites, called 'inter-satellite links' (ISLs), may also be used.

The interference paths between satellite networks

Figure 5.1 shows interference paths between satellite networks which use the same two frequency bands, X and Y, in the same directions of transmission; this is the most common situation. These interference paths, modes 1 and 2, are capable of causing

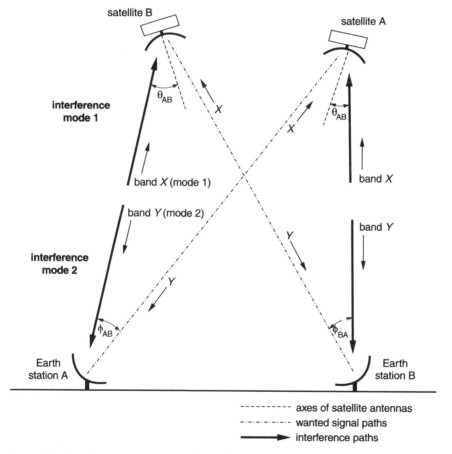

Figure 5.1   Interference paths (modes 1 and 2) between two geostationary satellite networks using the same frequency bands (X and Y) in the same directions of transmission

severe interference over great distances, given an unobstructed propagation path; the level of interference depends largely on the radiation patterns of the satellite and earth station antennas and the angular separation between the satellites.

A number of frequency bands are allocated for uplinks in some satellite networks and downlinks in others. For example, the band 8025–8400 MHz is allocated for uplinks for the FSS and downlinks for the EES. These arrangements allow spectrum to be used more intensively in some circumstances. The interference paths arising with this configuration are illustrated in Figure 5.2 (modes 3 and 4). Interference modes 5A and 5B, illustrated in Figure 5.3, arise when there is a large angular separation between the satellites; they are special cases of mode 3.

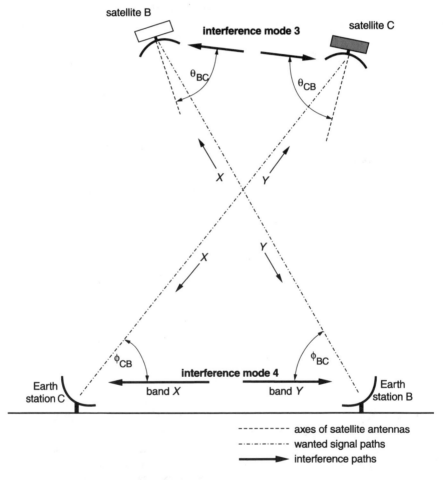

Figure 5.2   Interference paths (modes 3 and 4) between two geostationary satellite networks using the same frequency bands, but in contrary directions of transmission. See also Figure 5.3

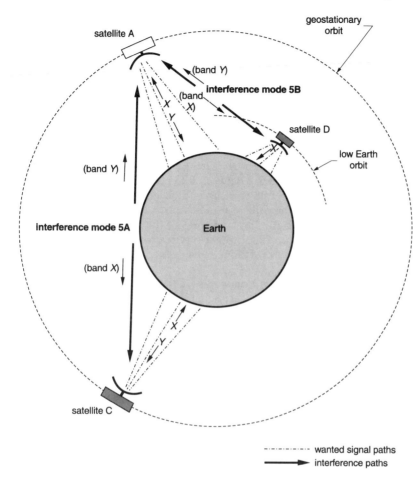

Figure 5.3   *Interference paths between two geostationary satellites over a long orbital arc (mode 5A), and between GSO and non-GSO satellites over arcs of any length (mode 5B), using the same frequency bands but in contrary directions of transmission*

The principles of the coordination process

A satellite network is very expensive to establish, and it usually takes several years to set one up, but once in satisfactory service it may become very important, playing a vital role in the infrastructure of a large area, nation-wide or international. If a major interference problem is not identified until satellite fabrication is well advanced, serious delay to the date of the network's entry into service may arise. This may disrupt the development of major telecommunications facilities and the cost of changing the parameters of the network may be high. The additional cost and the delay of entry into service of the proposed new satellite will be much greater if a need to modify the design of the satellite is not discovered until after the unsatisfactory satellite has been launched and tested in orbit.

Thus, the coordination procedure must be capable of identifying potential interference problems some years before a proposed new satellite network enters service. Once any necessary and sufficient remedies have been identified, the procedure must be robust enough to preserve the new network's opportunity to enter service until any necessary modifications of the system have been implemented. Additionally, the procedure is to ensure that priority of access to spectrum and space is given to real systems in their order of coming into service.

The basic administrative and technical framework that is used for coordination, reinforced to some degree by sharing constraints, was devised originally to deal with interference between GSO satellite networks of the FSS; it is outlined in the present section. There is an extension of the same framework, with a larger admixture of sharing constraints, to control interference between GSO satellite systems and the line-of-sight radio relay systems of the FS that share frequency bands with the FSS; see Section 5.2.3 below. This framework also functions well enough for the small number of non-geostationary (non-GSO) networks of the FSS that operate between 3 and 7 GHz and for the systems of most other services, satellite and terrestrial, that also share frequency bands. The regulations which govern these procedures are set out in RR Articles S9 and S11.

However, new kinds of non-GSO networks of the MSS and FSS are entering service. New forms of sharing constraint are being introduced to facilitate the sharing of frequency bands between non-GSO and GSO networks. Coordination procedures are being adapted to control interference between non-GSO networks, between non-GSO networks and GSO networks, and between non-GSO networks and terrestrial systems. The new procedures, still somewhat tentative, are set out in RR Resolution 46 and in particular in Sections I and II of Annex 1 of the resolution, although they will, no doubt, be transferred into the main body of the Radio Regulations when they have been finalised. The new provisions for controlling interference between satellite networks are also outlined in this section, while the new provisions for controlling interference between non-GSO satellite networks and terrestrial services using the same frequency are outlined in Section 5.2.5.

With a few exceptions, these procedures are used as appropriate in coordinating frequency assignments to all kinds of satellite system and terrestrial system band-sharing with them. The exceptions are the frequency bands allocated for the BSS (see Chapter 10), the AmSS (see Chapter 12) and the FSS allocations that are used for the allotment plan (see Section 9.3). The allocations to which the RR Resolution 46 procedure applies are identified by a reference to 'coordination under No S9.11A' in a footnote to each affected allocation in the international table of frequency allocations; Table S5-1A in RR Appendix S5 also lists all the bands affected. RR Appendix S5 also indicates the coordination thresholds, beneath which coordination is not required.

Constraints, mainly to protect GSO networks from non-GSO networks

There are constraints on the level of radiation to or from non-GSO networks in specific, relatively uncommon, applications that might affect GSO networks. Thus:

- A GSO satellite transmitting an inter-satellite link (ISL) towards a satellite higher than the GSO must direct the beam at least 15° away from the GSO; see RR S22.3.

- More specifically, a GSO EES satellite transmitting an ISL in the 29.95–30 GHz band towards a non-GSO satellite must reduce the interference caused by that emission to a GSO network of the FSS to an acceptable level; see RR S22.4.
- In the frequency band 8025–8400 MHz a non-GSO EES satellite must not illuminate any part of the GSO with a PFD exceeding $-174\,\text{dBW}/\text{m}^2$ in any sampling bandwidth of 4 kHz (that is, $-174\,\text{dBW}\;(\text{W}/\text{m}^2/4\,\text{kHz})$; see RR S22.5. In the band 6700–7075 MHz there is a similar constraint on any non-GSO FSS satellite; see RR S22.5A.

Other constraints, provisionally agreed at WRC-97, arose out of decisions contained in RR Resolutions 130 and 538 to enable non-GSO constellations of the FSS to operate in frequency bands around 12, 19 and 29 GHz which are already in use for GSO FSS networks, and also in the bands near 12 GHz used for the BSS frequency assignment plan. These constraints take three forms:

- A maximum aggregate PFD produced at the GSO by all the satellites of a non-GSO constellation.
- A maximum equivalent PFD produced at any point on the Earth's surface, within the beam of an earth station antenna of a given size, pointing in any direction, by all the satellites of a non-GSO constellation.
- A maximum spectral e.i.r.p. density for the off-beam radiation from the antennas of the earth stations of GSO FSS networks.

See RR S22.5B to S22.5G and RR S22.26 to S22.29. The values of the aggregate PFD and equivalent PFD constraints, which together cover most of the bands intended for non-GSO FSS constellations, are provisional at present. The application of the off-beam e.i.r.p. density constraints, which would cover bands between 12.75 and 14.5 GHz has been suspended, pending a review of the values. All three sets of constraints will be reviewed at WRC-00.

In a number of bands allocated for the FSS (and also in BSS allocations), coordination between GSO and non-GSO networks is simplified by RR S22.2, which states:

> Non-geostationary-satellite systems shall not cause unacceptable interference to geostationary-satellite systems in the fixed-satellite service and the broadcasting-satellite service operating in accordance with these Regulations.

Indeed this regulation applied until recently to all FSS and BSS allocations. Now, for almost every FSS allocation in which the procedure of RR Resolution 46 applies, there is a footnote in the international table of frequency allocations withholding the application of RR S22.2. However, WRC-97 was concerned to enable more non-GSO satellites to get access to spectrum and it considered that the unspecific and unbalanced terms of that paragraph might unnecessarily hinder progress in that respect. The constraints arising from RR Resolutions 130 and 538, outlined below, have been designed to explore other approaches to the problem and RR Resolution 130 foresees that the protection of GSO networks of the FSS from non-GSO networks provided by RR S22.2 will be reformulated at WRC-00.

The coordination procedures

Both coordination procedures have three stages, namely advance publication of information (API), frequency coordination and the notification and registration of frequency assignments. The summary which follows, omitting much detail, is applicable to a proposal for a new satellite network, but a basically similar, though more limited, procedure is employed when any change is made to an existing network which increases any risk of interference.

Stage 1. Advance publication of information

Not more than six years, but preferably not less than two years, before the proposed date of entry into service of a new satellite network (GSO or non-GSO), the proposing Administration informs the ITU of the proposals. Only a few basic details of the proposed network need to be declared at this stage if the RR Resolution 46 procedure does not apply in the frequency bands to be used; these include the orbit, the frequency bands, the service area, the nature of the service to be provided and the expected date of bringing into use. If RR Resolution 46 does apply, rather more information is required for non-GSO networks. The Bureau publishes this information in its 'Weekly Circular' [2].

The publishing of this information serves to inform other Administrations of the new network in prospect. Administrations that perceive a risk that the new network might interfere with a network for which they are responsible, in operation or already coordinated, thus have an opportunity to discuss their concern with the Administration proposing the new network. If a modification of the proposals could eliminate the risk, the implementation of the modification at this early stage will save cost and time. Another significant function of the API is to establish the timeframe within which the process of coordination and entry into service has to be completed, as will be seen below.

Stage 2. Frequency coordination

For proposed new FSS networks that are to operate in bands in which RR S22.2 (quoted above) applies and RR Resolution 46 does not apply, there may be a need for discussions between Administrations if:

- the proposed new network would use the GSO and would overlap in frequency with an existing or planned non-GSO network or
- the proposed new network would not use the GSO and would overlap in frequency with existing or planned GSO FSS networks

to determine how the non-GSO systems could best avoid causing unacceptable interference to the GSO networks. It would probably be desirable to hold these discussions before any mandatory coordination required for GSO networks is finalised. However, such discussions fall outside the mandatory procedures prescribed by RR Article S9. When these informal coordinations have been completed, an Administration responsible for a non-GSO system can proceed to Stage 3 and notify the frequency assignments as soon as it is ready to do so.

Mandatory coordination of a GSO network, and of a non-GSO system for operation in frequency bands to which RR Resolution 46 applies, can begin when the proposing Administration is ready to supply full details of the frequency assignments it proposes to make for transmitting and receiving at the satellite of the new network. Broadly this information, called 'coordination data', will state the radio service to which the network will belong, the proposed satellite orbit and the forecast date of the end of the satellite's operational life. It will also include details of the proposed antenna parameters of the satellite and the frequencies and parameters of the emissions that the satellite is intended to transmit and receive. Information is also required on proposed earth stations but this may be expressed in a generalised form, indicating the characteristics of typical earth stations and the geographical areas (service areas) in which they would be found.

The proposing Administration sends the coordination data to the ITU Radiocommunication Bureau and requests coordination with the other Administrations responsible for satellite networks that might cause interference to, or suffer interference from, the new satellite. If an Administration proposing a new GSO network fails to supply the coordination data within 24 months of the date of receipt by the Bureau of the API, the latter will be cancelled. The proposer must then start the process again from the beginning of Stage 1 if coordination is still required.

The Bureau verifies that the coordination data appear to be complete and in accordance with the international table of frequency allocations and other regulatory provisions. Then it publishes all the information in its 'Weekly Circular' [2]. The Bureau will identify other satellite networks, not the responsibility of the proposing Administration, which might cause or suffer significant interference to or from the proposed new network. This review will take into account all networks which are already in service or, being planned, have already reached the frequency coordination stage. If any of the frequency bands to be used is shared by other satellite services, the review will include any satellites of a service having an allocation category which is not inferior to that of the proposed new network. The Administrations that are responsible for these networks will be informed of the need to coordinate with the proposing Administration and the list is published in the 'Weekly Circular'.

Except for certain unusual situations identified in RR Appendix S5, the Bureau applies the following criteria in selecting the Administrations for listing for coordination:

a) For a proposed new GSO network, whether intended for frequency bands designated for RR Resolution 46 or not, the criterion for coordination with another GSO network using one or more frequency bands which overlap with those of the new network is defined in RR Appendix S8; see also Section 5.2.2 below.

b) For a proposed new GSO network that would use one or more frequency bands designated for the RR Resolution 46 procedure, coordination with a non-GSO network is considered to be required if there is an overlap of the bandwidths occupied by the two networks.

c) For a proposed new non-GSO network for operation in a band for which the Resolution 46 procedure is prescribed, coordination is required with any other network, GSO or non-GSO, if there is an overlap of the bandwidth they occupy.

Some nominally geostationary satellites have significant inclination of the orbital plane, relative to the plane of the equator; such orbits are more accurately described as quasi-geostationary or inclined geosynchronous orbits. When the angle of inclination of the orbit is small, this has relatively little impact on interference between satellite networks (see ITU-R Recommendation S.743-1 [3]) although it may have a significant effect on frequency coordination with terrestrial stations; see Section 5.2.4 below.

The list of Administrations selected for coordination is important. Participation in coordination is the only way that a good estimate of prospective interference levels can be obtained. The proposing Administration needs the participation of all Administrations responsible for systems that could cause or suffer unacceptable interference, so that ways of reducing excessive levels of interference can be found and agreed before satellite design and construction is far advanced. Other Administrations need to establish a well-founded case for objecting to proposed assignments to the new network if indeed their interest is threatened. There is provision for appeal to the Bureau if any Administration with interests likely to be affected has been omitted from the list. Constructive refusal to participate in coordination by a listed Administration may be taken as agreement that unacceptable interference is not foreseen in either direction between the proposed network and those for which the refusing Administration is responsible.

The necessary participants in coordination having been identified, the proposing Administration discusses prospective interference with each of the listed Administrations. At this stage, for the first time, relatively exact calculations can be made to determine whether reception of the worst-affected emissions will suffer interference reaching a level recognised as being unacceptable if implementation of the new network goes ahead as proposed. Acceptable levels of interference are discussed in Section 5.2.6. See also Recommendations ITU-R S.1325 [4] and S.1328 [5]. If prospective unacceptable interference is found, ways of eliminating the excess are sought.

Reduction of interference levels may involve changes in the proposed location of geostationary satellites or the orbital arcs over which non-geostationary satellites are active, changes in the assigned frequencies, the modulation techniques and the parameters of emissions, changes in the characteristics of satellite or earth station antennas etc. These changes are not necessarily limited to the proposed new satellite; while a change to the antennas of satellites that are already in orbit is usually impossible, agreement may be obtained, for example, to changes in the characteristics of emissions in already operating networks. Sometimes agreement can be reached to tolerate a small excess of interference relative to the level that is usually considered to be the maximum that is acceptable.

For interference modes 1, 2, 3 and 5 (see Figures 5.1 to 5.3) the propagation path is almost entirely in free space and, given reliable antenna gain characteristics, the calculation of transmission loss presents few problems. However, interference mode 4 (see Figure 5.2) may have a long propagation path in the troposphere; estimation of the transmission loss by this mode may involve considerable effort, and the best way to identify earth stations of other networks that may cause or suffer interference is to use a process similar to that used to identify terrestrial stations with which an earth station should be coordinated; see Section 5.2.4. Recommendation ITU-R IS.848-1 [6] provides the technical basis for this process; this recommendation is incorporated into the Radio Regulations by reference in RR Appendix S5, Table S5-1.

The terms of RR Article S9 and RR Resolution 46 seek to ensure that the process of coordination proceeds efficiently. The Administrations involved keep the Radiocommunication Bureau informed of the progress being made. The process is complete when all of the Administrations involved have given agreement to the proposals for the new network, with whatever changes have been found necessary. Inevitably there are occasions when it does not prove possible to reach a satisfactory conclusion. The regulations give the Bureau some limited powers to arbitrate if cooperation is unreasonably withheld, but in the absence of unreasonable behaviour the Bureau will not register frequency assignments for the proposed new network unless a satisfactory outcome from coordination has been reached.

Concurrently with the frequency coordination process, the proposer of a new satellite network of the FSS, the MSS or the BSS is expected to be procuring the satellite itself and the means for launching it. The proposing Administration is required to provide information to the Radiocommunication Bureau demonstrating that the prospective user has shown due diligence in these respects by giving, for example, the name of the satellite manufacturer, the contractual delivery window, the name of the launch vehicle and the launcher supplier, the anticipated launch date and the name and location of the launch facility. This 'administrative due diligence' information must be received by the Bureau within five years of the receipt of the API. See RR Resolution 49.

Stage 3. Notification and registration of frequency assignments

Not more than three years before the coordinated assignments are forecast to be brought into use, the proposing Administration notifies all the necessary details of the assignments to the Radiocommunication Bureau. This information includes an up-dated forecast of the date of bringing the assignments into use. All of this information is published in the 'Weekly Circular' [2].

If the Bureau finds the notifications to be timely (see next paragraph) and in order and if the Administrations listed for coordination have consented, the assignments are added to the MIFR. This entry in the MIFR in advance of the entry of the assignments into use is provisional, pending notification of their actual entry into use. If that notification has not been received within 30 days of the up-dated forecast date, the provisional entry will be cancelled.

The forecast date of bringing the assignments into use should be not more than five years after the date of receipt by the Bureau of the API. This deadline may be extended by the Bureau by not more than two years, subject to three conditions, namely that satisfactory due diligence information has been received, that the coordination process has at least begun, and that the notifying Administration certifies that delay has been due to one or more of the following reasons:

- launch failure,
- launch delays beyond the control of the Administration or operator,
- delays due to modification of the satellite design found to be necessary to reach coordination agreements,
- problems in meeting the satellite design specifications,
- delays in effecting coordination despite a request for the assistance of the Bureau,
- financial circumstances outside the control of the Administration or the operator, or
- *force majeure.*

If coordination and bringing the new network into use has not been completed by the five-year deadline with any approved extension, the whole process of coordination is cancelled; see RR S11.48. If the new network is still required, the procedure must be started again from the API submission stage.

5.2.2   The criterion for coordination between GSO networks

Frequency coordination of satellite networks demands expertise and it is costly; no Administration would wish to be involved in it unless there was a significant risk of interference affecting a satellite network for which it was responsible. On the other hand it is vitally important that a new network should not be set up and then be found to be unusable because of interference, to it or from it, which is unacceptable and not reducible at reasonable cost.

RR Appendix S8 provides a criterion that indicates where no significant risk of interference arises in all situations but one (see below), enabling many networks to be released from the need to coordinate. See also Recommendation ITU-R S.738 [7]. This method was developed for the FSS but it has been found to be applicable to geostationary networks in general, and it is used by the ITU Radiocommunication Bureau when listing Adminstrations that should be invited to coordinate for any newly proposed GSO network. The method is briefly as follows.

In the usual arrangement for two potentially interfering geostationary satellite networks, illustrated in Figure 5.1, both networks use the same two frequency bands in the same directions of transmission and both networks use transparent transponders. Then:

a)   By interference mode 1, emissions in frequency band X transmitted from earth station A enter the receiver of satellite B. The highest value of spectral power density at the output of the satellite receiving antenna due to these emissions at any frequency common to the two networks is identified and converted to an equivalent noise temperature ΔT_s referred to the output of the satellite receiving antenna. This noise-equivalent will be passed on to the receiver at earth station B, after frequency changing to band Y, amplification in the transponder and attenuation in the downlink propagation path. The symbol γ is used for the transmission gain (expressed as a numerical power ratio, usually less than unity) between the output of the satellite receiving antenna and the output of the earth station receiving antenna.

b)   By interference mode 2, emissions in frequency band Y transmitted from satellite A enter the receiver at earth station B. Again the highest value of interfering spectral power density is identified and it is converted to an equivalent noise temperature ΔT_e referred to the output of the earth station receiving antenna.

c)   In the worst case, which is also (subject to frequency tolerances) the usual case, these two entries of interference will coincide in frequency at the earth station receiver. Then ΔT, the sum of the two equivalent noise entries at that point, given by

$$\Delta T = \Delta T_e + \gamma \Delta T_s \qquad (5.1)$$

is taken as a measure of the worst interference entry from network A to network B. If ΔT, expressed as a percentage of T, the system noise temperature of earth station B, exceeds 6%, it is concluded that coordination should be carried out, 6% of the system noise at the earth station receiver being of the same order as widely accepted maximum values of interference.

In making these calculations it will be necessary to know, or assume, various data relating to the antennas involved. The gain of satellite antennas in the direction of the earth stations can be obtained from the antenna data and service area data supplied by the Administrations with their proposed or notified frequency assignments; the worst case gains within the relevant service areas would be used in each case. The off-beam angle of earth station antennas in the direction of the unwanted satellite can be calculated from the expression:

$$\phi \approx \text{arc}\cos\frac{d_1^2 + d_2^2 - (84\,332\sin\phi_g/2)^2}{2d_1 \cdot d_2} \tag{5.2}$$

where d_1 and d_2 are the distances in kilometres between the earth station and each satellite (see equation A.5) and ϕ_g is the minimum geocentric angle between the two satellites, allowance being made for their east–west station-keeping tolerances. In the absence of data on the radiation patterns of the earth station antennas, RR Appendix S8 Annex III provides the equations to be assumed.

It may be that interference will be diminished by polarisation discrimination, and Appendix S8 provides guidance as to the degree of protection that may realistically be assumed from it; see also Recommendation ITU-R S.736-3 [8]. However, the polarisation characteristics of earth station antennas vary unpredictably with off-beam radiation angle, making such calculations precarious, and the Radiocommunication Bureau will not take account of such an allowance in identifying Administrations for coordination unless both Administrations have consented to it.

RR Appendix S8 has been drafted primarily for the most common configuration, namely of satellites with transparent transponders using the same two frequency bands in the same directions of transmission. However, other configurations arise. Frequency bands may be used in opposite directions of transmission, having interference modes 3, 4 and 5 as shown in Figures 5.2 and 5.3. Perhaps only one frequency band is used in common. Signal processing transponders, perhaps including a regenerator for digital signals, may be used. RR Appendix S8 also shows how the basic method can be applied to these other configurations.

In practice the criterion provided by RR Appendix S8 successfully identifies situations where interference may be significant, but with one exception. It may fail where a wide-band emission, provided with carrier energy dispersal by frequency modulation of the carrier by a triangular waveform or a similar method, interferes with a narrow-band signal. This is largely a problem for the FSS; see Section 9.2.1. However, the criterion is also unsatisfactory in the contrary sense, since it is often found that $\Delta T/T$ exceeds 6% when, on coordinating, the interference levels found are much less than the level that is usually considered acceptable. Possible ways of improving the discrimination of the method are discussed in Recommendation ITU-R S 739 [9].

5.2.3 *Band sharing by GSO satellite networks and terrestrial stations*

Virtually all frequency allocations for satellite services, which are almost always primary, are shared with terrestrial services, at least one of which usually has a primary allocation also. Methods for limiting interference between stations of the FS and networks of the FSS using geostationary satellites are well established; see RR Articles S9 and S11 and RR Appendix S7. The same methods are also applied in other bands where stations of other terrestrial services and GSO networks of other satellite services share spectrum with primary allocations. These methods are summarised below.

However, the methods used for non-GSO satellite systems in frequency bands designated for the procedure set out in RR Resolution 46 differ in major respects. In addition, it may be necessary for an earth station of a non-GSO system to be coordinated with an earth station of a GSO network using the same frequency bands in the reverse directions of transmission, or vice versa. The RR Resolution 46 methods are outlined in Section 5.2.5.

The methods reviewed in this section are based on the assumption that the satellites involved have ideal geostationary orbits. The parameters of the orbits of most satellites that are treated as geostationary are indeed close to the ideal parameters, especially early in their operational life. However, some satellites that are nominally geostationary develop significant inclination of the orbital plane after some years in operation and this may have to be taken into account; see Section 5.2.4.

Figure 5.4 uses the example of the FSS sharing with the FS in both uplink and downlink bands to illustrate the four interference modes that arise between terrestrial and satellite services, namely:

- Mode 6; interference from a terrestrial transmitter to an earth station receiver in the satellite downlink band.
- Mode 7; interference from an earth station transmitter to a terrestrial receiver in the satellite uplink band.
- Mode 8; interference from a satellite transmitter to a terrestrial receiver in the satellite downlink band.
- Mode 9; interference from a terrestrial transmitter to a satellite receiver in the satellite uplink band.

The spectral e.i.r.p. density of a typical FSS earth station emission is high compared with that of a typical FS emission and interference can arise via mode 7 at considerably greater distances than when two FS stations interfere. Similarly, earth station receivers are usually very sensitive and they may experience interference from very distant FS transmitters via mode 6. Frequency coordination is mandatory where earth stations and terrestrial stations may cause or suffer cross-border interference.

Coordination would be too laborious a way of controlling interference between GSO satellites and terrestrial stations (modes 8 and 9), because of the vast numbers of terrestrial stations that may be located within the footprint of a satellite's antenna beams. (The footprint of a satellite antenna beam is the area on the ground within which the gain of the antenna exceeds a given value, typically 3 dB less than the maximum gain of the antenna.) Fortunately it has been found that sufficient control over interference can be provided by the application of sharing constraints to the power of the emissions of both services, without serious impact on the scope of either service.

Interference modes 6 and 7: frequency coordination around earth stations

Broad measures are taken to limit interference between earth stations and known terrestrial stations. Care is taken in choosing station sites so as to make the best use of distance and site screening by natural and man-made obstructions as a means of reducing the power of interfering signals. Since high sites are often essential for terrestrial stations, the main responsibility for prudent siting usually rests with the earth station owner.

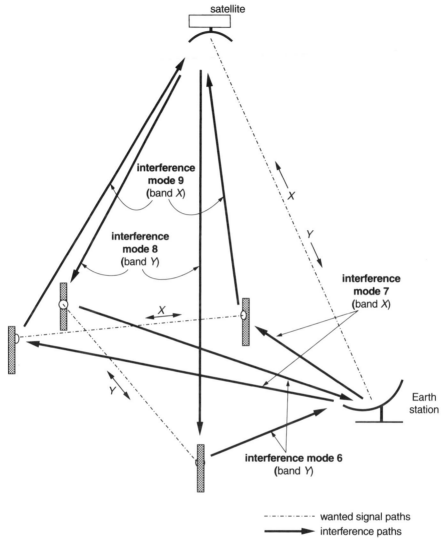

Figure 5.4   Interference paths (modes 6–9) between a satellite network and terrestrial systems having assignments in both the uplink and the downlink frequency bands, X and Y

When a new earth station is being planned, the planners may not have complete knowledge of the location of terrestrial stations in neighbouring countries that might interfere with, or suffer interference from, the earth station. It would be wise to have informal discussions with the Administrations of neighbouring countries at this early stage, to obtain guidance on the best use of whatever site screening might be available from natural features. However, before assigning frequencies to the earth station, the mandatory frequency coordination process must be carried out.

This can be a costly, laborious task, like coordination between satellite networks as described in Section 5.2.1. Accordingly, a procedure has been designed for this case also to ensure that terrestrial stations will not be brought into the process if there is no chance that interference involving them will be unacceptable. The basic procedure is set out in RR Article S9 and RR Appendix S7; it also has three stages, publication, negotiation and registration.

Stage 1 of coordination for modes 6 and 7: publication

When an Administration that has responsibility for a new earth station is ready to announce full details of the proposed frequency assignments to that station, transmitting and receiving, two maps are prepared showing the location of the earth station and the coordination areas around it. One map, making worst-case assumptions where firm knowledge is not available, shows the geographical limit beyond which no terrestrial station with typical parameters would receive significant interference from the earth station's emissions, their power at low angles of elevation being constrained by RR S21.8 or S21.10. The other map shows the limit beyond which a terrestrial station, its power being within sharing constraints imposed by RR S21.2 to S21.7, would not cause significant interference to reception at the earth station. See also Recommendation ITU-R SF.1004 [10].

If any territory within either of these coordination areas lies within the jurisdiction of the Administration of another country where a terrestrial service has the same allocation category, the proposing Administration sends full details of the proposed frequency assignments and copies of the maps to that Administration, with a request for frequency coordination. If the Administration proposing the new earth station finds that no foreign territory lies within the coordination areas, it sends copies of the maps and details of the proposed assignments to the Radiocommunication Bureau. The Bureau checks that the proposals are in order, brings other Administrations into the coordination process if this is seen to be appropriate and publishes the material in its 'Weekly Circular' [2].

The coordination area maps distributed by the proposing Administration must be drawn up in strict accordance with RR Appendix S7. The basis of the method is as follows for earth station emissions; for earth station reception the method is analogous. For an earth station emission with given modulation parameters and a known spectral power density input to the antenna, there will be a value L_c of the transmission loss (that is, the loss from the input to the earth station antenna to the output of the terrestrial station antenna) which would reduce the interference at the output of the terrestrial station antenna to a level that would be acceptable. For this purpose it is considered that two radio propagation modes may be operative, propagation mode 1, consisting of a tropospheric wave or tropospheric scatter, perhaps augmented by defraction, and propagation mode 2, consisting of energy scattered by rain.

For propagation mode 1, the transmission loss L_1 along any chosen azimuth from the earth station to a hypothetical terrestrial station is equal to the loss of the propagation path between the two station sites, offset by the gain of the antennas at the operative angles for interference. It will usually be appropriate to assume that the operative angle at the earth station is the angle of elevation to the local horizon along the azimuth when the antenna is in its normal operating position. An allowance is made for the site screening effect for distant receivers implied by an elevated horizon at the earth station. If the gain of the earth station antenna at the operative angle is not known, the characteristic given by Recommendation ITU-R S.465-5 [11] is available for FSS stations and is at present the best estimate for comparable earth station antennas of other services. It is assumed at this stage that the antenna gain at the terrestrial station is the maximum gain of a standard antenna.

Thus, the coordination distance d along the chosen azimuth is the distance for which the radio path loss is sufficient to make L_1 equal to L_c. Agreed propagation data contained in RR Appendix S7 provides a simple means for determining this distance. The coordination distance is assumed to be 100 km if the distance, calculated as above, has a lower value. Having calculated d for one azimuth, the process is repeated at close angular intervals of azimuth in all directions from the earth station. The locus of the distances at which successive coordination distance values are reached defines the coordination contour for propagation mode 1.

For propagation mode 2, the transmission loss, and the proportion of the time for which rain scatter is active, depend on distance but also on several meteorological factors. The procedure contained in RR Appendix S7 defines the propagation mode 2 coordination contour. It has the form of a circle of radius d_r, its centre being located, not at the earth station, but at a distance Δd from the station along the main beam azimuth.

Thus the coordination area for the earth station for transmissions is the whole area enclosed by either coordination contour; see Figure 5.5. It is helpful to calculate other contours for mode 1, assuming smaller values of required transmission loss;

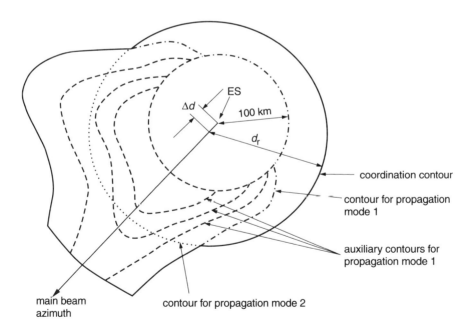

Figure 5.5   An example of the coordination area for an earth station (ES) operating with a geostationary satellite (after Figure 1 of Recommendation ITU-R IS.847-1 [23])

these 'auxiliary contours' simplify the task, arising in Stage 2, of excluding from coordination those terrestrial stations that clearly will not require it, for example because their antennas, not being directed towards the earth station, will have much less gain in the direction of the earth station than the standard value that had been assumed in calculating *d*.

Stage 2 of coordination for modes 6 and 7: negotiation

When an Administration receives the maps of the coordination areas and details of proposed frequency assignments from the proposing Administration, it reviews the assignments that have been made to its terrestrial stations within the coordination area, whether or not those assignments have been registered in the MIFR. Terrestrial stations for which there is no prospective overlap of assigned bandwidths are excluded from further consideration. Where there is an overlap of bandwidth, the Administration may consider whether the basic parameters of the terrestrial station (such as the antenna gain in the direction of the earth station, antenna height, transmitter power and receiver sensitivity) are such as to make the station substantially less likely to cause or suffer interference than the data of RR Appendix S7 assume. The risk of interference may also be reduced by screening by hills in the vicinity of the terrestrial station. These considerations, with the assistance of the auxiliary coordination contours if they have been provided by the proposing Administration, may enable the Administration to exclude many terrestrial stations from further consideration.

If the Administration that has been invited to coordinate finds no risk of interference at all, to or from the earth station, it responds to the request accordingly and advises the Bureau to the same effect. But there may remain a residue of assignments for which the risk of significant interference is too high to be disregarded at this initial stage. If so, the responding Administration provides full details of the frequency assignments to its own stations that give rise to concern. The two Administrations then collaborate in calculating more precisely the levels of interference that can be expected to arise. Actual values of all the significant parameters are used where known. Recommendations ITU-R S.465-5 [11] and ITU-R F.699-4 [12] provide reference antenna characteristics for FSS and FS stations in the absence of measured data. See also Recommendation ITU-R F.1245 [13].

The proposing Administration will be concerned to ensure that the interference entering the earth station receiver from the source in question, added to all the other interference that enters there from other terrestrial stations within range of the earth station, does not cause the acceptable level to be exceeded. The responding Administration will have a more difficult judgement to make if the terrestrial station carries telephone channels with international performance objectives. See Section 5.2.6. The Administrations must make responsible judgements as to which interference liabilities can be accepted, and some may be rejected.

If a prospective interference entry is not acceptable, ways of reducing the interference to an acceptable level will be sought. There are various possible ways of doing this but they depend very much on the circumstances. When solutions have been found to all of these problems, the Bureau is so informed.

Stage 3 of coordination for modes 6 and 7: registration and after

When any necessary coordinations have been successfully completed and the earth station frequency assignments are soon to be brought into use, the proposing Administration notifies full details of the frequency assignments, amended as may have been agreed during coordination, to the Bureau. The details are checked by the Bureau in accordance with RR Article S11, and if they are in order and if the satisfactory completion of all necessary coordination is confirmed, the assignments are provisionally added to the MIFR. When the assignments have been brought into use, the Bureau is notified and the date of bringing into use is added to the MIFR entry.

Any of the responding Administration's assignments to terrestrial stations located within the coordination area that are not already registered in the MIFR will be notified for registration at this stage. All subsequent new assignments to either the earth station or to terrestrial stations within the coordination area must be coordinated before notification for registration.

Interference mode 8: constraints on satellite PFD to protect terrestrial station reception

Interference to FS receivers from transmitters of geostationary satellites of the FSS is limited to an acceptable level by a sharing constraint on the spectral PFD density of satellite emissions as measured at the surface of the Earth. The same constraint is also applied to satellites of other services and to protect other terrestrial services. Exceptionally these limits may be exceeded with the agreement of the Administrations of all the countries that would be affected. Two factors have influenced the form which this sharing constraint takes.

Firstly, almost all stations of the FS above 3 GHz have antennas with high gain in the direction of the local horizon. Consequently, these stations are particularly vulnerable to interference from satellites which are seen, from the FS station, at a low angle of elevation. However, many FSS satellites also have high gain antennas and these antennas are often directed towards areas on the Earth's surface from which the angle of elevation of the satellite is high. Accordingly, it can be to the advantage of both services if the constraint on satellite PFD is related to the angle at which the flux reaches the ground, being more severe at low angles of elevation.

Secondly, almost all FS systems operating above 15 GHz carry either wide-band digital signals or video channels, their emissions having a bandwidth of at least 1 MHz. However, at the time when the constraint was formulated, most radio relay systems operated at lower frequencies and used frequency modulation (FM) for multichannel analogue frequency division multiplex (FDM) telephony systems. The impact of interference on wide-band digital or video systems is a function of the total interference power, integrated over the bandwidth of the emission, but FM/FDM systems are susceptible to peaks of power in the interference spectrum within the channel bandwidth of the FDM system, typically 4 kHz. Accordingly, for most bands below 15 GHz, the PFD constraint is specified for a sampling bandwidth of 4 kHz; above 15 GHz the sampling bandwidth is 1 MHz. With a few exceptions the limit is a constant value of PFD for angles of satellite elevation δ below 5° at any point on the surface of the Earth, a higher constant value where δ is above 25°, with a linear transition between 5° and 25°. For example, for the 4 GHz band, the constraint is $-152\,\text{dB}(\text{W}/\text{m}^2/4\,\text{kHz})$ where δ is $5°$ or less, $-142\,\text{dB}(\text{W}/\text{m}^2/4\,\text{kHz})$ when δ is $25°$ or more and $-152 + 0.5\,(\delta - 5)\,\text{dB}(\text{W}/\text{m}^2/4\,\text{kHz})$ for values of δ between $5°$ and $25°$.

The PFD limits at the Earth's surface are set out in Section V of RR Article S21. Special sharing conditions have given rise to particularly high or low values of the constraint in a few narrow frequency bands; these exceptions are referenced as appropriate in later chapters. The values applying in the major bands are illustrated graphically in Figure 5.6. Some of these have been changed little, if at all, since they were first determined at WARC-71. The technical support for these limits is to be found in CCIR Report 387-2 [14]. In 1971 there was some concern that these constraints might not be sufficiently stringent. However, the potential levels of interference from FSS satellites into FS radio relay chains have recently been reviewed (see Recommendation ITU-R SF.358-5 [15]) and the adequacy of the constraints has been confirmed. See also Recommendation ITU-R F.1107 [16].

Interference mode 9: constraints on terrestrial e.i.r.p. to protect satellite reception

Above 1 GHz, transmissions of the fixed and mobile services are the most likely source of interference to satellite receivers via mode 9. Constraints are applied to the power of stations of the FS and the MS in the frequency bands, listed in RR Table S21.2, which are allocated for satellite uplinks. The constraints, set out in RR S21.2 to S21.7 were implemented retrospectively at a time when many terrestrial systems were already built and consequently some of the constraints have had to be expressed in alternative forms. The constraints are as follows:

a)   Between 1 and 10 GHz; the power delivered to the antenna shall not exceed $+13\,\text{dBW}$. Also, as far as is practicable, the e.i.r.p. radiated within $\pm 2°$ of the direction of the GSO (in effect, at the azimuth at which the GSO cuts the horizon at the station) should not exceed $+35\,\text{dBW}$. If this latter limit cannot be implemented, the e.i.r.p. must not exceed $+47\,\text{dBW}$ within $\pm 0.5°$ of the direction of the GSO, the limit rising on a linear decibel scale (8 dB per degree) to a maximum of $+55\,\text{dBW}$ at azimuths more than $1.5°$ from the direction of the GSO.

b)   Between 10 and 15 GHz; the power delivered to the antenna shall not exceed $10\,\text{dBW}$. Also as far as is practicable, the maximum e.i.r.p. radiated within $\pm 1.5°$ of the direction of the GSO should be $+45\,\text{dBW}$. In no case shall the e.i.r.p. exceed $+55\,\text{dBW}$ in any direction.

c)   Between 25.25 and 27.5 GHz; the power delivered to the antenna shall not exceed $+10\,\text{dBW}$. Also as far as is practicable, the maximum e.i.r.p. density radiated within $\pm 1.5°$ in the direction of the GSO should be $+24\,\text{dBW}$ in any 1 GHz band. If this is impracticable, the e.i.r.p. of the whole emission shall not exceed $+55\,\text{dBW}$.

d)   In other frequency ranges above 15 GHz, the power delivered to the antenna shall not exceed $+10\,\text{dBW}$ and the e.i.r.p. radiated in any direction shall not exceed $+55\,\text{dBW}$.

Where the constraint relates to the direction of the GSO, allowance should be made for atmospheric refraction; see Recommendations ITU-R SF.765 [17] and F.1333 [18]. Many of these constraints are unchanged from the values that were agreed at WARC-

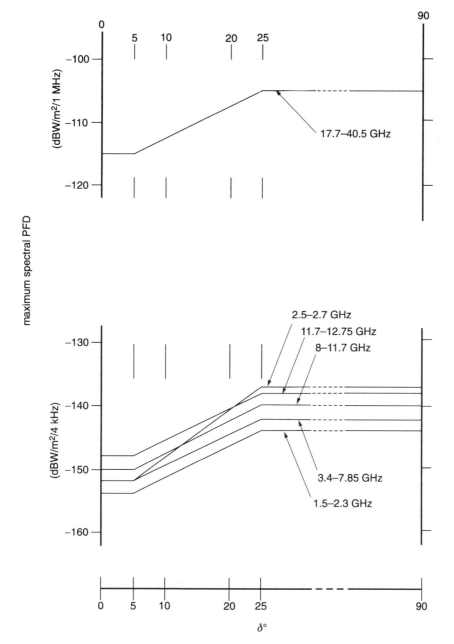

Figure 5.6   The maximum spectral PFD at the Earth's surface of satellite emissions as a function of δ, the angle of elevation on arrival at the Earth, in the major satellite frequency allocations that are shared with terrestrial services, both services having primary allocations. Above 31 GHz the values are provisional

63 and extended at WARC-71. The rationale for these power levels is to be found in the versions of CCIR Report 209 [19] and CCIR Recommendation 450 [20] that were current in 1963 and 1971, respectively. See also Recommendation ITU-R SF.406-8 [21].

5.2.4 Band sharing by quasi-GSO satellite networks and terrestrial stations

While an earth station antenna is directed at a truly geostationary satellite, its gain in the direction of a terrestrial station is constant. Likewise, in the direction of a satellite, the gain of the antenna of a terrestrial station is also constant. The methods used for calculating interference levels that are incorporated in RR Appendix S7 and the constraints on satellite PFD and terrestrial station e.i.r.p. that are reviewed in Section 5.2.3 assume that these conditions are valid.

However, natural forces acting on a satellite that was initially geostationary cause perturbations in the parameters of the orbit. Departures of the orbit from circularity and from the ideal period cause the position of the satellite as seen from the Earth to move to east and/or west of its nominal longitude. These perturbations can be corrected by the operation of thrusters on board the satellite. The close mandatory constraint on the east–west station-keeping of nominally geostationary satellites (see Section 4.2.4) which has been adopted to minimise inefficiency in the use of the orbit ensures that this movement is not allowed to become big enough to affect the coordination process significantly.

Natural forces also cause changes in the inclination of the satellite orbit, producing an apparent figure-of-eight movement of the satellite in the sky as seen from the Earth, mainly north–south in orientation. This orbital perturbation, within limits, has relatively little effect on satellite-to-satellite interference, especially if the gain of the satellite antennas is low. Nonetheless, it may have a significant effect on the mutual antenna gains, and therefore the interference levels, between satellites and terrestrial stations. Furthermore, if the earth station varies the pointing direction of its antenna to track the apparent movement of the satellite, the gain of the antenna varies in the direction of terrestrial stations. There is at present no mandatory constraint upon this perturbation. However, a substantial mass of thruster fuel is required, with present-day technology, to counteract the perturbation, tending to limit a satellite's operational lifetime. In consequence, the orbits of some satellites, nominally geostationary, acquire an inclination of several degrees in the course of their operational lives.

The effect of such quasi-geostationary orbits on interference between satellite and terrestrial systems has been studied; see Recommendation ITU-R SF.1008-1 [22]. The conclusions drawn include the recommendation that account needs to be taken of foreseen orbital inclination at the coordination stage, especially if the angle of inclination is expected to exceed 5°. The 1998 version of RR Appendix S7 does not provide directions on how this should be done but Recommendation ITU-R IS.847-1 [23], which has been recommended as a draft for the up-dating of the appendix, does so. This recommendation has been incorporated into the Radio Regulations by a reference, for example, in RR Appendix S5 Table S5-1. The revision of the appendix will probably be considered at WRC-00.

5.2.5 Band sharing for non-GSO satellite systems and terrestrial stations under RR Resolution 46

The coordination procedures for limiting interference between non-GSO systems of the MSS and the FSS on the one hand, and terrestrial services on the other hand, in frequency bands where RR Resolution 46 applies, are to be found in Annex 1, Sections III and IV, to the resolution. The circumstances differ considerably from those arising when the stations to be coordinated are the earth stations of GSO FSS networks and radio relay stations of the FS and consequently the procedures differ from those reviewed in Section 5.2.3 above. The basic texts containing the criteria for coordination are RR Article S21 and RR Appendix S5 Annex 1, although RR Resolution 46 Annex 2 may also be found helpful. These criteria vary from band to band and, at present, they are incomplete. Many of the criteria are subject to review at WRC-00.

MSS systems below 1 GHz

In the 401 MHz (downlink) band the threshold for coordination is a downlink PFD of $-125\,dB(W/m^2/4\,kHz)$. In the 137 GHz (downlink) band the same criterion is generally applicable but a PFD of $-140\,dB(W/m^2/4\,kHz)$ is appropriate in some circumstances; see RR Appendix S5, Annex 1, Section 1.1.

The coordination distances for coordination between mobile earth stations and terrestrial stations are given in RR Appendix S5, Annex 1, Table 1. These give co-ordination areas consisting of the service area extended in all directions by a distance ranging from 500 km to 1080 km, depending on the nature (airborne or ground-based) of the earth stations and the terrestrial stations. Special conditions apply in the 148–149.9 MHz band; see Table 1 and Recommendation ITU-R M.1185-1 [24]; the recommendation has been incorporated into the Radio Regulations by reference in RR Appendix S5.

MSS systems between 1 and 3 GHz

Three sets of criteria have been drawn up for coordination in the MSS bands between 1 and 3 GHz, namely:

- Thresholds of downlink PFD at the Earth's surface from non-GSO satellites; if these thresholds are not exceeded there is no requirement for coordination with terrestrial stations.
- If these coordination thresholds are exceeded, some limited coordination will be required at least but additional criteria are provided which may make detailed coordination unnecessary, depending on the nature of the terrestrial stations within the coordination area.
- There are criteria for determining coordination areas for mobile earth stations.

The thresholds for downlink coordination in MSS are given, in terms of the PFD per satellite at the Earth's surface, in Table S5-2 of RR Appendix S5, Annex 1. The threshold takes the form $\Phi\,dB(W/m^2/F) + \delta$ where Φ is a value between -154 and -128, F is the sampling bandwidth (4 kHz or 1 MHz) and δ is a function of the angle of elevation of the wave at the ground. Φ and F vary from band to band, depending mainly on the nature of the terrestrial services with which the MSS shares the band.

The same table also gives coordination threshold values for use with GSO MSS satellites.

The concept of fractional degradation of performance has been developed as a simple method of enabling detailed coordination to be dispensed with in appropriate cases when the terrestrial service consists of digital systems of the FS, even when the PFD of the satellite emission exceeds somewhat the values given in RR Table S5-2. See Section 1.2.2 of RR Appendix S5, Annex 1 and Recommendation ITU-R M.1141-1 [25]. Recommendation ITU-R M.1142-1 [26] may also be noted.

For use where the method of fractional degradation of performance cannot be applied, an alternative method for avoiding detailed coordination in marginal cases is being developed in the form of a system specific methodology. The present state of development of this method is to be seen in Recommendation ITU-R M.1143-1 [27]. If neither of these methods provides assurance of acceptable sharing, it will be necessary to carry out a detailed coordination. See also Recommendations ITU-R S.1257 [28], F.1108-2 [29] and S.1323 [30].

The coordination distances for mobile earth stations and terrestrial stations between 1 and 3 GHz are given in RR Appendix S5, Annex 1, Table 2. For non-GSO systems the coordination distances for ground-based earth stations and ground-based terrestrial stations are to be determined in accordance with Recommendations ITU-R IS.849-1 [31] and IS.847-1 [23]; see also Recommendation ITU-R M.1187 [32]. All of these recommendations have been incorporated into the Radio Regulations by reference in RR Appendix S5. For situations where airborne earth stations and/or airborne terrestrial stations, including radiosondes, are involved, Table 2 provides coordination distances between 500 and 1080 km by which the service area is to be extended in all directions to obtain the coordination area.

Non-GSO FSS systems above 3 GHz

There is a primary allocation for non-GSO systems of the FSS in the bands 18.8–19.3 GHz (downlinks) and 28.6–29.1 GHz (uplinks), sharing with GSO FSS networks operating in the same directions of transmission. In addition, however, there is provision for non-GSO FSS systems to operate (with some variation from Region to Region), under the terms of RR Resolution 130 or 538 in the bands 10.7–12.75 GHz, 17.8–18.6 GHz and 19.7–20.2 GHz (all for downlinks) and the bands 12.5–13.25 GHz, 13.75–14.5 GHz, 27.5–28.6 GHz and 29.5–30.0 GHz (all for uplinks).

All of these frequency bands are shared with terrestrial services in some or all countries but RR S5.524 withholds protection of terrestrial services from satellite downlink signals at 19.7–20.2 GHz, and the allocation for terrestrial services by RR S5.542 at 29.5–30.0 GHz has secondary status. In the other downlink bands listed in the previous paragraph the constraints on downlink PFD given in RR Article S21, discussed in Section 5.2.3 above (interference mode 8) in connection with GSO networks and illustrated in Figure 5.6, apply in general to non-GSO systems also. However, RR Resolution 131 expresses concern that a large increase in the number of satellites setting up such a substantial downward power flux may make difficult the sharing of these bands with the FS. The resolution urges Administrations to consider reducing either the number of satellites in non-GSO FSS constellations or the power flux that each satellite sets up. The resolution also reduces by 10 dB the maximum PFD per satellite that constellations of more than 100 satellites operating in the 18.8–19.3 GHz may set up relative to the value given by RR Article S21.

In general it will be necessary to coordinate the earth stations of non-GSO FSS systems with terrestrial stations. RR Appendix S5, Annex 1, Table 4 directs the use of Recommendations ITU-R IS.847 [23] and IS.849 [31] in the bands 18.9–19.3 GHz and 28.7–29.1 GHz in calculating coordination distances and the same texts would be found useful elsewhere in this range of frequencies.

Feeder links for non-GSO MSS systems

At 5091–5250 MHz, used for feeder uplinks and in part for feeder downlinks, the PFD of a feeder downlink must not exceed $-164\,dB(W/m^2/4\,kHz)$ at any angle of elevation. See also RR S5.444A. The principal sharing service in this band is the ARNS and the coordination area about a feeder link earth station for aircraft stations is circular, with a radius of 500 km.

At 6700–7075 MHz, used for feeder downlinks and shared with uplinks for GSO FSS networks, the FS and the MS, the maximum PFD at the Earth's surface may not exceed the following limits at or below 5° of elevation:

$$6700\text{–}6825\,MHz; \;-137\,dB(W/m^2/1\,MHz),$$
$$6825\text{–}7075\,MHz; \; both \;-154\,dB(W/m^2/4\,kHz) \; and$$
$$-134\,dB(W/m^2/1\,MHz).$$

In all of these cases the limit at or above 25° is 10 dB greater, with a linear transition between 5° and 25°. See also Recommendation ITU-R SF.1320 [33]. The coordination area about a feeder link earth station for ground-based terrestrial stations is circular with a radius of 300 km; where the terrestrial stations are on aircraft, the coordination radius is 500 km.

The band 15.43–15.63 GHz, used for feeder uplinks and downlinks, is shared with the ARNS. There is a constraint on the angle of elevation of earth station antennas to protect the ARNS; see Recommendation ITU-R S.1341 [34], which also recommends on minimum coordination distances, and see also RR S5.511A. There is also a constraint on the downlink PFD at the Earth's surface but there is a discrepancy between RR Appendix S5, Annex 1, Section 2.1 and RR Resolution 46, Annex 2, Section A2.2.1 as to the values of this constraint. Precautions are also required to prevent interference from satellite emissions to the RAS in the band 15.35–15.40 GHz; see Recommendation RA.769-1 [35] and also RR S5.511D. This recommendation has been incorporated into the Radio Regulations by the reference in RR S5.511A.

The bands 19.3–19.7 GHz (feeder downlinks) and 29.1–29.5 GHz (feeder uplinks) are shared with links for GSO FSS networks operating in the same directions of transmission and also the FS and the MS. The constraint on the downlink PFD for the feeder links is the same as for the GSO network downlinks; see Figure 5.6. The coordination distances for ground-based terrestrial stations are to be determined in accordance with Recommendations ITU-R IS.847-1 [23] and IS.849-1 [31].

5.2.6 Acceptable levels of interference

When coordinating frequency assignments, it is necessary for the participants to have as an objective a level beyond which interference from a single other system is unacceptable. This level is important. If the level is set too high, circuit performance may fall short of end-to-end standards, depending on the level of interference and other degradations entering from other sources. On the other hand, if the accepted level is very low it will be more difficult, and ultimately impossible, to coordinate new systems. Standard values have been agreed for some situations, based on the objectives set for all degradations for standard connections. For some other situations an agreed basis exists from which to develop criteria. But in some situations, *ad hoc* criteria have to be agreed.

The most complete set of criteria relates to telephone channels used for international connections transmitted by cable, by line-of-sight radio relay systems or by GSO satellites. A less complete basis is available for television channels within FS or GSO FSS networks. A corresponding set of basic standards is currently emerging for telephone channels in MSS networks, GSO and non-GSO; see, for example, Recommendations ITU-R M.1183 [36], M.1228 [37], M.1229 [38], M.1234 [39] and M.1232 [40]. For other services and other types of signal, it is usually necessary, for the present, to adapt from the criteria for telephone channels.

There are performance objectives, set in ITU-T Recommendations, for international telephone connections of defined length and make-up, which recommend maximum levels of the total channel noise power plus equivalent interference power for analogue transmission and the total BER for digital transmission. The analogue telephony case is used here to illustrate the principles involved, since it is simpler to express than the digital case. Thus:

a) Recommendations ITU-T G.222 [41] and ITU-R F.392 [42], F.393-4 [43], ITU-R S.352-4 [44] and S.353-8 [45], taken together, recommend that the total noise and equivalent other degradations in an analogue telephony channel of the FS and of the FSS using a GSO satellite should have the following limits.

  i) In a hypothetical reference circuit (HRC) 2500 km long, made up of 50 line-of-sight radio relay links of the FS, the noise and interference should not exceed 10 000 picowatts (pW), psophometrically weighted, at a point of zero reference level (that is, 10 000 pW0p), averaged over one minute, for more than 20% of any month. Similarly, the noise and interference should not exceed 50 000 pW0p for more than 0.1% of any month. Both limits include an allowance of 2500 pW0p for degradations arising in frequency division multiplex (FDM) equipment.

  ii) For a single satellite link the same long term objective as is given in the previous sub-paragraph (that is, for 20% of any month) also applies. The same short term level objective also applies, but in this case it is applicable to not more than 0.3% of any month.

b) Between geostationary networks of the FSS, Recommendation ITU-R S.466-6 [46] recommends that the total interference in any analogue telephony channel should not exceed the equivalent of 2500 pW0p of noise, nor should the interference from any single source exceed the equivalent of 800 pW0p of noise, for more than 20% of any month. A lower level of total interference from other GSO networks is recommended where the satellite network employs frequency re-use. Interference into GSO networks from non-GSO networks has not been factored into these budgets yet.

c)   For analogue systems of the FS, Recommendation ITU-R SF.357-4 [47] recommends that the sum of all the interference entries into any one channel in an HRC from all networks of the FSS, averaged over one minute, should not exceed 1000 pW0p for more than 20% of any month, nor should it exceed 50 000 pW0p for more than 0.01% of any month.

d)   For geostationary networks of the FSS, Recommendation ITU-R SF.356-4 [48] recommends the same long term objective as is given in (c) for the aggregate interference entering any analogue channel from all systems of the FS. The same short term level objective also applies, but in this case for not more than 0.03% of any month.

Some frequency bands allocated for the FSS are used for uplinks, some are used for downlinks, but almost all FSS allocations are shared with FS systems. Thus, disregarding for the moment the FSS allocations which are for both uplinks and downlinks, interference from FSS networks may enter FS systems operating in an FSS uplink band (via mode 7; see Figure 5.4) or it may enter a different population of FS systems operating in a downlink band (via mode 8). However, FS systems will not suffer interference from FSS networks via both modes. On the other hand, FSS networks may suffer interference from FS systems operating in the uplink band (via mode 9) and also from other FS systems operating in the downlink band (via mode 6).

Consequently, the constraint on the downlink PFD of geostationary satellites (see Figure 5.6) is designed to ensure that the total entry of interference from all satellite transmitters into any channel in any radio relay chain equivalent to the HRC (mode 8) operating in a downlink band of the FSS will not exceed 1000 pW0p for more than 20% of any month. For shorter radio relay chains the interference entry should be proportionately less. Correspondingly, the frequency coordination process is used to prevent the total entry of interference from the transmitters of all earth stations to any channel in a radio relay chain equivalent to the HRC (mode 7) from exceeding 1000 pW0p for more than 20% of any month via stable propagation modes, nor 50 000 pW0p for more than 0.01% of any month via sporadic tropospheric propagation modes. Thus, the level of interference entering a single terrestrial station from any single earth station that would be regarded as acceptable will be much smaller than the values quoted in Recommendation ITU-R SF.357-3 [47]. The ITU does not make general recommendations on what that level should be; this is a matter for coordinating Administrations to decide, and much will depend on the circumstances.

The allotment of interference from FS systems to FSS networks must be divided between modes 6 and 9. An equal division of the long term allotment, with the whole of the short term allotment being available for mode 6, may be as appropriate as any other ratio. The constraints on the power and the e.i.r.p. of terrestrial stations are designed to prevent the total entry of interference from all terrestrial transmitters into a channel of an FSS uplink (mode 9) from exceeding 500 pW0p. The frequency coordination process (mode 6) is used to prevent the total entry of interference from all terrestrial transmitters within the coordination area of an earth station into any one channel of a satellite downlink from exceeding 500 pW0p for more than 20% of any month via stable propagation modes, nor 50 000 pW0p for more than 0.03% of any month via sporadic tropospheric propagation modes. As before, the ITU offers no guidance as to the level that should be accepted from a single source.

Some of the bandwidth allocated for the FSS has recently been taken into use for both uplinks and downlinks. Consequently, some interference may enter FS systems via mode 7 in frequency bands where previously only Mode 8 was active, and vice versa. This usage is likely to increase. Entries of mode 7 and mode 8 interference in these new ways from single sources can be kept within limits by frequency coordination and by constraints on satellite PFD at the earth's surface, as usual. However, there is concern that the aggregate of the entries into radio relay systems may eventually become unacceptably high. Recommendation ITU-R SF.1005 [49] proposes that this risk should be eliminated by making the PFD constraint on satellite emissions somewhat more severe.

The computation of interference levels

During the coordination process, the Administrations involved and their advisors representing the service providers, existing and proposing, will have the best information available on the characteristics of the antennas and the parameters of the emissions that may cause or suffer interference. This will enable computations to be made of all relevant wanted carrier-to-interference ratios. Recommendation ITU-R SF.1006 [50] provides assistance regarding the more difficult case of interference paths involving tropospheric propagation. It will usually be necessary to compute the degree of degradation which this interference will cause to wanted signals, to ascertain whether the interference will be acceptable. Recommendations ITU-R SF.766 [51] and S.741-2 [52] provide guidance on this latter process, of particular relevance to the kinds of signal used in the FS and the FSS. See also the Appendix below, Section A9.

5.2.7 Geostationary network coordination in practice

The frequency coordination of satellite networks, summarised in Section 5.2.1, is surely the most complex, administratively and technically, of all the international procedures involved in spectrum management. Coordination under the provisions of RR Resolution 46 has only just begun, but there is long experience of coordination between GSO networks in accordance with RR Article 9 and it is timely to consider whether the GSO procedure is working well. Resolution 18 [53] of PP-94 revealed the concern of the Plenipotentiary Conference about the effectiveness of the procedures and instructed the Director of the Radiocommunication Bureau to review the situation and to propose ways forward to WRC-97.

Document 8 [54], presented to WRC-97, is that report. It shows grounds for concern. As at 31 August 1997:

- notifications to the Bureau of assignments to satellites indicated that a total of 365 geostationary networks were in service,
- proposals for new geostationary networks were being received at a rate of about 500 per year, up from about 100 per year before 1993,
- proposals for 672 new networks were at some stage of coordination,
- coordination data for 403 other proposed new networks had been received but had not been published in the Weekly Circular, there being an average delay in publishing coordination data of about 18 months.
- API for 419 additional proposed new networks had been delivered to the Bureau. Of these, 246 had been published but coordination data had not yet been submitted. The remaining 173 sets of API were awaiting publication.

The Director's report does not indicate the rate at which proposed new networks were reaching a successful conclusion to coordination, nor how long coordination was typically taking, but it seems likely that the rate was low and progress was slow.

Clearly the procedure is not working well. The facts given in the report do not allow informed conclusions to be drawn as to why this should be, but it may be supposed that at least two factors are involved. A significant proportion of the new proposals are, no doubt, for satellites to be located in arcs of the GSO which are already in heavy use; interference levels may already be high enough in these arcs to make coordination of additional networks difficult to achieve. Secondly, a scrutiny of the list of proposed new networks suggests strongly that Administrations with one new satellite to coordinate, aware of the risk of delays and uncertainties in securing the coordination at their first choice of orbital location, are initiating the simultaneous coordination of satellites at several different locations with the intention of implementing the first to achieve satisfactory coordination. This practice is probably the main source of the 'paper satellites' which are said to be clogging up the procedure.

Insufficient manpower at the Radiocommunication Bureau, delaying the publication of coordination data, may be a third significant factor. It might, however, be argued that expedited handling of data in the Bureau, by adding more proposals to the large number already in active coordination, might create more congestion and increase delays at that later stage.

Any indication that the coordination of new GSO networks is failing for procedural reasons would be of great concern. There would be even greater concern if exhaustion of the capacity of the GSO was the reason. Since the mid-1970s it has been recognised that satellite communication can be of vital importance, not only for international communication but also for many services involving national coverage. It is a basic principle of the ITU that all nations should have equitable access to spectrum and space, especially the GSO. This principle is asserted in Article 44 of the ITU Constitution; see also paragraph S0.3 of the preamble to the Radio Regulations. The allotment plan for the FSS and the frequency assignment plan for the BSS (see Section 5.4 below) go some way towards meeting this commitment to equity but some Administrations may consider them insufficient.

The Director's report suggested a number of measures for improving the situation. WRC-97 decided to implement several proposed changes to the procedure immediately to hasten progress in coordination. A requirement for due diligence to be demonstrated in the procurement of satellites and launch facilities was approved. The workload falling on the Bureau was reduced somewhat by eliminating much of the technical detail previously included in the API for proposed satellites. Consideration of the implementation of other measures suggested by the Director was delayed until WRC-00 to allow time for the effectiveness of the measures approved by WRC-97 to be assessed.

The Director's report had suggested, subject to the adoption by WRC-97 of the requirement for demonstrating due diligence in the procurement of satellites and launching services, that consideration should be given at the subsequent WRC to the imposition of fees for the recovery from the paying Administration of the cost of the coordination and registration procedures falling on the ITU. Another suggestion was that an Administration proposing a new satellite should be required to pay a financial deposit, to be returned if and when the satellite was launched. The Plenipotentiary Conference meeting in 1998 gave broad approval to the concept of cost recovery by the ITU for its products and services. The ITU Council is to consider how cost recovery is to be implemented.

It will be noted that all of these new measures address procedural issues. Difficulties in coordination arising from the density of satellites already in operation in some arcs of the GSO are mainly a problem of the FSS, and are considered further in Chapter 9.

5.3 Miscellaneous allocations requiring coordination

Proposals for new frequency allocations, to share a frequency band with existing allocations, often meet resistance at WRCs from Administrations having strong interest in the existing allocations and little or none in the proposed new one. Sometimes agreement can be reached, for example, by making the allocation secondary, by placing geographical limitations on it, by applying a sharing constraint to systems of the new service or by some combination of such measures; see Section 3.2 above. Finally, when concern about the impact of a new allocation is strong but the Administration proposing it foresees relatively few assignments being made, it has sometimes been found possible to agree to the new allocation provided that every assignment made under it is first coordinated with other Administrations with a recognised concern, to ensure that such concerns would not be damaged by stations of the new service. About 80 allocations have been treated in this way to date.

Such allocations are identified in the international table of frequency allocations by a footnote referring to RR S9.21. If an Administration notifies to the Radiocommunication Bureau an assignment relating to one of these allocations, the Bureau will identify and inform the Administrations that might be affected by the assignment and publish the details in its 'Weekly Circular' [2]. After any necessary coordination has been satisfactorily completed, the Bureau will consider registering the assignment in the MIFR.

5.4 International planning of frequency band utilisation

Frequency band planning takes several forms; for example:

a)   The spectrum between 1606.5 and 3800 kHz is allocated in a large number of narrow bands, many of them for the MS. The frequency band 2173.5–2190.5 kHz is allocated world-wide specifically for distress and related traffic using radio-telephony and for calling (that is, for ship and shore stations to make initial contact, preparatory to passing traffic on other frequencies). In Region 1 the frequency bands used for maritime traffic in other parts of this frequency range are listed in RR S52.9 and S52.10, sub-divided according to whether they are to be used for ship-to-shore, shore-to-ship or ship-to-ship links, for telephony or for direct printing telegraphy.

b)   The eight bands between 4 and 27 MHz that are allocated exclusively for the MMS are sub-divided world-wide according to their use, as in the previous example. In addition, radio channel plans like the one illustrated in Figure 4.1 have been drawn up for all of the sub-bands identified for transmission from ships, and for some of the sub-bands for coast station transmitters. Details of these plans are to be found in RR Appendix S17; see also Section 8.2 below.

c)   The planning of the MMS allocations in the HF range (see the previous example) has been taken a stage further in the sub-bands used for telephony. A frequency allotment plan has been drawn up, each of the channels in the sub-bands used for the transmission of telephony from coast stations being allotted to a number of countries, well-separated geographically, so that each country can use its channels independently with an acceptably low incidence of interference from a coast station in another country which has been allotted the same channel; see RR Appendix S25. Rules for the use of these channels and limits on transmitter power, set out in RR Articles S51 and S52, assist in the satisfactory operation of the plan. An Administration assigns the various channels that it has been allotted to whichever it chooses of its coast stations, and these assignments are notified to the Radiocommunication Bureau in the usual way. There is a procedure for adding new allotments to the plan as new needs arise.

d)   For the principal frequency bands around 12 GHz that are allocated for the BSS, an agreement has been drawn up for the management of a frequency plan which defines geostationary satellite locations, national coverage areas and much else for television broadcasting channels in these bands. There is a corresponding agreement and frequency plan for feeder links to the satellites. See RR Appendices S30 and S30A and Section 10.3 below. These are, in effect, frequency assignment plans. Comprehensive sets of regulations for the use of these channels and for the management of assignments for systems in other services sharing these frequency bands are included in the agreements. When an Administration wishes to bring one of these channels into use, it assigns the frequency etc. and notifies the assignment to the Radiocommunication Bureau.

The various plans mentioned above, and others of the same kind, have been developed within the ITU and approved by WARCs and this is usual for plans that have global or Regional applicability. Plans have also been drawn up, where appropriate, for smaller geographical areas and in less formal circumstances, as is authorised by Article 42 of the ITU Constitution [55]. RR S6.2 recognises that Administrations may decide to meet, plan frequency assignments, and conclude 'special agreements', subject to various conditions. This option for special agreements is not available in the frequency range 5.06–27.5 MHz, where ionospheric propagation would make it impracticable. These agreements often take the form of frequency assignment plans. At present, frequency allotment and frequency assignment agreements are used for spectrum management in many broadcasting, maritime mobile and aeronautical mobile bands, and in some radionavigation bands, in addition to the broadcasting-satellite and fixed-satellite plans mentioned above.

The frequency assignment/allotment planning process

The production of a frequency assignment or frequency allotment agreement is complex and varies in detail according to the nature of the service to be planned, but the essence of it is as follows. One or more frequency bands are identified for planning. Administrations state the systems that they require the plan to accommodate, typically identifying the location of stations or service areas, the emission bandwidths, transmitter powers if stations are already in existence etc. Performance and interference objectives and data for radio propagation predictions are agreed. Then a plan of frequency assignments is devised, nowadays typically using computer methods, providing if possible for all the systems required with the agreed standard of performance and appropriately low levels of interference.

If, as may happen, no way can be found to produce a plan that satisfies all the coverage requirements and meets all the performance objectives, a reiteration of the process with less stringent, though acceptable, performance objectives or more stringent equipment requirements or with the omission of some systems may enable an acceptable conclusion to be reached. The agreement will be completed with regulations for transition to the new regime, for managing the application of the plan in operation, for amending it as needs change and so on. Such agreements are ratified by governments, acquiring treaty status.

A comparison between ad hoc *assignment and planning methods*

The various *ad hoc* methods, reviewed in Sections 4.1, 5.2 and 5.3, that are used for choosing frequencies and orbits for assignment, have major advantages. They are flexible and pragmatic, usable for mixtures of diverse kinds of system, and capable, within limits, of dealing in an evolutionary way with changing circumstances and unforeseen needs. Plan agreements, on the other hand, can promote efficient spectrum utilisation when providing for large numbers of essentially similar systems which are already in operation or clearly foreseeable. On the other hand, forecasts of long term requirements tend to be unrealistic, providing an unreliable basis for planning and tending to increase needlessly the cost of the systems that are implemented. Each method has fields of application for which it is to be preferred.

However, there is another difference between the two basic methods of assigning frequencies, namely the form taken by the right of the user of an assignment to the international recognition that registration in the MIFR provides. *Ad hoc* selection of frequencies for assignment bases that right to operate without harmful interference on priority of the date, registered in the MIFR, of taking the frequency into operation. This right is not explicit in the Radio Regulations but it is seldom disputed. On the other hand, a frequency allotment or a frequency assignment in a plan agreement that has been ratified by the governments involved is specifically guaranteed by those governments, irrespective of the date of taking the assignment into operation.

This difference is relatively unimportant for short range terrestrial services in frequency bands that do not support the propagation of interference over long distances. On the other hand, for long distance terrestrial services and all satellite services the difference may be vital. *Ad hoc* methods give advantage to the countries that make early use of new transmission media; these are usually the developed countries. Developing countries may subsequently find it difficult to get established in favourable frequency bands or favourable orbital locations. Consequently, *ad hoc* methods may be expedient but they may also be considered inequitable.

Equitable national access to spectrum

In 1947, as part of the process of reconstruction after World War II, a resolution [56] of the International Radio Conference set up the Provisional Frequency Board and charged it with responsibility for drawing up global frequency assignment plans for 14–150 kHz and almost all frequency bands between 2.85 and 27.5 MHz. (Military use of frequencies above 27.5 MHz was declining in 1947; civil use had only just begun and the interference range of its systems was short.) Responsibility for plans for the BS, the MMS and the AMS was entrusted to specialised conferences.

A satisfactory frequency allotment plan was produced for HF telephony in the MMS, together with a radio channel plan for MMS telegraph links. Satisfactory frequency allotment plans were produced for the AMS (R) and the AMS (OR) in HF bands. Indeed all of these plans, modified as necessary in the course of time, are still in use. Satisfactory frequency assignment plans were produced for a few other frequency bands allocated for other services. But in general, and most conspicuously at HF for the FS and the BS, the attempts to plan frequency assignments failed. The total demand for assignments, for systems that were already in existence and systems that were foreseen for the future, exceeded by far the total capacity of the HF spectrum and no acceptable way was found for scaling down the stated requirements to what the bandwidth available could accommodate. A majority of ITU members concluded that management of access to spectrum by priority of use was, for most services, the best available, workable, option although not necessarily fully satisfactory.

However, by the 1960s the accession to the ITU of many newly independent countries considerably increased the proportion of developing countries among the Members of the Union. By 1965 a majority of Members favoured a change of regulatory methods for bands below 30 MHz, seeing planning as a way for new countries to get equitable access to spectrum. A succession of WARCs met with no success in planning for HF broadcasting, although improvements have been made in the handling of *ad hoc* use of frequencies for this service; see Section 7.4 below.

By the mid-1970s, by which time the FSS began to be perceived as the radio medium of the future for international communication and an important potential medium for domestic communication and broadcasting, concern for change for the FS was replaced by concern about the FSS and the BSS; see RR Resolution 2 and RR Recommendation 700, both from WARC-79 and also RR Resolution 4.

WARC-71 set in motion a train of events that led at WARC Orb-88 to a complete frequency assignment plan for satellite television in frequency bands allocated for BSS close to 12 GHz; see Section 10.3 below. A resolution of WARC-79 led, also at WARC Orb-88, to the production of a frequency and orbit allotment plan for the FSS; see Section 9.3 below. However, there has been relatively little take-up of these assignments and allotments so far. The main FSS activity continues in the bands in which it has long been established, despite great coordination problems. Broadcasting by satellite has also developed mainly in bands allocated to the FSS. It seems likely that the constraints adopted in setting the objectives for these two plans, and in particular the limitation of service areas to national coverage and the provision of equal shares for every country, large or small, while they made the production of plans achievable, have led in most cases to provision for systems that are not commercially viable in any but very large countries.

5.5 References

1 'Frequency sharing between systems in the fixed-satellite service and radio-relay systems in the same frequency bands'. Recommendation ITU-R SF.355-4, ITU-R Recommendations 1997 SF Series (ITU, Geneva, 1997)
2 'Weekly Circular and Special Sections', Published by the ITU Radiocommunication Bureau every week

3 'The coordination between satellite networks using slightly inclined geostationary-satellite orbits (GSOs) and between such networks and satellite networks using non-inclined GSO satellites'. Recommendation ITU-R S.743-1, ITU-R Recommendations 1997 S Series. (ITU, Geneva, 1997)

4 'Simulation methodology for assessing short-term interference between co-frequency, codirectional non-GSO FSS networks and other non-GSO FSS or GSO FSS networks'. Recommendation ITU-R S.1325, *ibid*.

5 'Satellite system characteristics to be considered in frequency sharing analyses between GSO and non-GSO satellite systems in the fixed-satellite service including feeder links for the mobile-satellite service'. Recommendation ITU-R S.1328, *ibid*.

6 'Determination of the coordination area of a transmitting earth station using the same frequency band as receiving earth stations in bidirectionally allocated frequency bands'. Recommendation ITU-R IS.848-1, ITU-R Recommendations 1997 IS Series (ITU, Geneva, 1997)

7 'Procedure for determining if coordination is required between geostationary-satellite networks sharing the same frequency bands'. Recommendation ITU-R S.738, ITU-R Recommendations 1997 S Series (ITU, Geneva, 1997)

8 'Estimation of polarization discrimination in the calculations of interference between geostationary-satellite networks in the fixed-satellite service'. Recommendation ITU-R S.736-3, *ibid*.

9 'Additional methods for determining if detailed coordination is necessary between geostationary-satellite networks in the fixed-satellite service sharing the same frequency bands'. Recommendation ITU-R S.739, *ibid*.

10 'Maximum equivalent isotropically radiated power transmitted towards the horizon by earth stations of the fixed-satellite service sharing frequency bands with the fixed service'. Recommendation ITU-R SF.1004, ITU-R Recommendations 1997 SF Series (ITU, Geneva, 1997)

11 'Reference earth station radiation pattern for use in coordination and interference assessment in the frequency range from 2 to about 30 GHz'. Recommendation ITU-R S.465-5, *ibid*.

12 'Reference radiation patterns for line-of-sight radio-relay system antennas for use in coordination studies and interference assessment in the frequency range from 1 to about 40 GHz'. Recommendation ITU-R F.699-4, ITU-R Recommendations 1997 F Series Part 2 (ITU, Geneva, 1997)

13 'Mathematical model of average radiation patterns for line-of-sight point-to-point radio-relay system antennas for use in certain coordination studies and interference assessment in the frequency range from 1 to about 40 GHz'. Recommendation ITU-R F.1245, *ibid*.

14 'Protection of terrestrial line-of-sight radio-relay systems against interference due to emissions from space stations in the fixed-satellite service in shared frequency bands between 1 and 23 GHz'. CCIR Report 387-2. Texts of the CCIR XIIIth Plenary Assembly, Geneva, 1974 (ITU, Geneva, 1974)

15 'Maximum permissible values of power flux-density at the surface of the Earth produced by satellites in the fixed-satellite service using the same frequency bands above 1 GHz as line-of-sight radio relay systems'. Recommendation ITU-R SF.358-5, ITU-R Recommendations 1997 SF Series (ITU, Geneva, 1997)

16 'Probabilistic analysis for calculating interference into the fixed service from satellites occupying the geostationary orbit'. Recommendation ITU-R F.1107, ITU-R Recommendations 1997 F Series Part 2 (ITU, Geneva, 1997)

17 'Intersection of radio-relay antenna beams with orbits used by space stations in the fixed-satellite service'. Recommendation ITU-R SF.765, *ibid*.

18 'Estimation of the actual elevation angle from a station in the fixed service towards a space station taking into account atmospheric refraction'. Recommendation ITU-R F.1333, ITU-R Recommendations 1997 F Series Part 2 (ITU, Geneva, 1997)

19 'Communication-satellite systems – Frequency sharing between communication-satellite systems and terrestrial services'. CCIR Report 209. Documents of the Xth Plenary Assembly, Geneva 1963, Volume IV (ITU, Geneva, 1963)

20 'Frequency sharing between line-of-sight radio-relay systems and communication-satellite systems at frequencies above 10 GHz'. CCIR Recommendation 450, CCIR XIIth Plenary Assembly, New Delhi, 1970, Volume IV, Part 1 (ITU, Geneva, 1970)

21 'Maximum equivalent isotropically radiated power of radio-relay system transmitters operating in the frequency bands shared with the fixed-satellite service'. Recommendation ITU-R SF.406-8, ITU-R Recommendations 1997 SF Series (ITU, Geneva, 1997)

22 'Possible use by space stations in the fixed-satellite service of orbits slightly inclined with respect to the geostationary-satellite orbit in bands shared with the fixed service'. Recommendation ITU-R SF.1008-1, *ibid.*

23 'Determination of the coordination area of an earth station operating with a geostationary space station and using the same frequency band as a system in a terrestrial service'. Recommendation ITU-R IS.847-1, ITU-R Recommendations 1997 IS Series (ITU, Geneva, 1997)

24 'Method for determining coordination distance between ground based mobile earth stations and terrestrial stations operating in the 148.0–149.9 MHz band'. Recommendation ITU-R M.1185-1, ITU-R Recommendations 1997 M Series Part 5 (ITU, Geneva, 1997)

25 'Sharing in the frequency bands in the 1–3 GHz frequency range between the non-geostationary space stations operating in the mobile-satellite service and the fixed service'. Recommendation ITU-R M.1141-1, *ibid.*

26 'Sharing in the frequency bands in the 1–3 GHz frequency range between geostationary space stations operating in the mobile-satellite service and the fixed service'. Recommendation ITU-R M.1142-1, *ibid.*

27 'System specific methodology for coordination of non-geostationary space stations (space-to-Earth) operating in the mobile-satellite service with the fixed service'. Recommendation ITU-R M.1143-1, *ibid.*

28 'Analytical method to calculate visibility statistics for NGSO-satellites as seen from a point on the Earth's surface. Recommendation ITU-R S.1257. ITU-R Recommendations 1997 S Series (ITU, Geneva, 1997)

29 'Determination of the criteria to protect fixed service receivers from the emissions of space stations operating in non-geostationary orbits in shared frequency bands'. Recommendation ITU-R F.1108-2, ITU-R Recommendations 1997 F Series Part 2 (ITU, Geneva, 1997)

30 'Maximum permissible levels of interference in a satellite network (GSO/FSS; non-GSO/FSS; non-GSO/MSS feeder links) for a Hypothetical Reference Digital Path in the fixed-satellite service caused by other codirectional networks below 30 GHz'. Recommendation ITU-R S.1323, ITU-R Recommendations 1997 S Series (ITU, Geneva, 1997)

31 'Determination of the coordination area for earth stations operating with non-geostationary spacecraft in bands shared with terrestrial services'. Recommendation ITU-R IS.849-1, ITU-R Recommendations 1997 M Series (ITU, Geneva, 1997)

32 'A method for the calculation of the potentially affected region for a mobile-satellite service network (MSS) in the 1–3 GHz range using circular orbits'. Recommendation ITU-R M.1187, ITU-R Recommendation 1997 M Series Part 5 (ITU, Geneva, 1997)

33 'Maximum allowable values of power flux density at the surface of the Earth produced by non-geostationary satellites of the fixed-satellite service used in feeder links of the mobile-satellite service and sharing the same frequency bands with radio-relay systems'. Recommendation ITU-R SF.1320, ITU-R Recommendations 1997 SF Series (ITU, Geneva, 1997)

34 'Sharing between feeder links for the mobile-satellite service and the aeronautical radionavigation service in the space-to-earth direction in the band 15.4–15.7 GHz and the protection of the radio astronomy service in the band 15.35–15.4 GHz'. Recommendation ITU-R S.1341, ITU-R Recommendations 1997 S Series (ITU, Geneva, 1997)

35 'Protection criteria used for radioastronomical measurements'. Recommendation ITU-R RA.769-1, ITU-R Recommendations 1997 RA Series (ITU, Geneva, 1997)

36 'Permissible levels of interference in a digital channel of a geostationary network in mobile-satellite service in 1–3 GHz caused by other networks of this service and fixed-satellite service'. Recommendation ITU-R M.1183, ITU-R Recommendations 1997 M Series Part 5 (ITU, Geneva, 1997)

37 'Methodology for determining performance objectives for narrow-band channels in mobile satellite systems using geostationary satellites not forming part of the ISDN'. Recommendation ITU-R M.1228, *ibid.*

38 'Performance objectives for the digital aeronautical mobile-satellite service (AMSS) channels operating in the bands 1525 to 1559 MHz and 1626.5 to 1660.5 MHz not forming part of the ISDN'. Recommendation ITU-R M.1229, *ibid.*

39 'Permissible level of interference in a digital channel of a geostationary satellite network in the aeronautical mobile-satellite (R) service (AMS(R)S) in the bands 1545 to 1555 MHz and 1646.5 to 1656.5 MHz and its associated feeder links caused by other networks of this service and the fixed-satellite service'. Recommendation ITU-R M.1234, *ibid.*

40 'Sharing criteria for space-to-Earth links operating in the mobile-satellite service with non-geostationary satellites in the 137-138 MHz band'. Recommendation ITU-R M.1232, *ibid.*

41 'Noise objectives for design of carrier transmission systems of 2500 km'. Recommendation ITU-T G.222, CCITT Blue Book, Fascicle III, Part 2 (ITU, Geneva, 1988)

42 'Hypothetical reference circuit for radio-relay systems for telephony using frequency-division multiplex with a capacity of more than 60 telephone channels'. Recommendation ITU-R F.392, ITU-R Recommendations 1997, F Series Part 1 (ITU, Geneva, 1997)

43 'Allowable noise power in the hypothetical reference circuit for radio-relay systems for telephony using frequency-division multiplex'. Recommendation ITU-R F.393-4, *ibid.*

44 'Hypothetical reference circuit for systems using analogue transmission in the fixed-satellite service'. Recommendation ITU-R S.352-4, ITU-R Recommendations 1997, S Series (ITU, Geneva, 1997)

45 'Allowable noise power in the hypothetical reference circuit for frequency-division multiplex telephony in the fixed-satellite service'. Recommendation ITU-R S.353-8, *ibid.*

46 'Maximum permissible level of interference in a telephone channel of a geostationary-satellite network of the fixed-satellite service employing frequency modulation with frequency-division multiplex caused by other networks of this service'. Recommendation ITU-R S.466-6, *ibid.*

47 'Maximum allowable values of interference in a telephone channel of an analogue angle-modulated radio-relay system sharing the same frequency bands as systems in the fixed-satellite service'. Recommendation ITU-R SF.357-4, ITU-R Recommendations 1997 SF Series (ITU, Geneva, 1997)

48 'Maximum allowable values of interference from line-of-sight radio-relay systems in a telephone channel of a system in the fixed-satellite service employing frequency modulation when the same frequency bands are shared by both systems'. Recommendation ITU-R SF.356-4, *ibid.*

49 'Sharing between the fixed service and the fixed-satellite service with bidirectional usage in bands above 10 GHz currently unidirectionally allocated'. Recommendation ITU-R SF.1005, *ibid.*

50 'Determination of the interference potential between earth stations of the fixed-satellite service and stations in the fixed service'. Recommendation ITU-R SF.1006, *ibid.*

51 'Methods for determining the effects of interference on the performance and the availability of terrestrial radio-relay systems and systems in the fixed-satellite service'. Recommendation ITU-R SF.766, *ibid.*

52 'Carrier-to-interference calculations between networks in the fixed-satellite service'. Recommendation ITU-R S.741-2, ITU-R Recommendations 1997 S Series (ITU, Geneva, 1997)

53 'Review of the ITU's frequency coordination and planning framework for satellite networks'. Resolution 18, Final Acts of the Plenipotentiary Conference (ITU, Geneva, 1994)
54 Report on Resolution 18 (Kyoto 1994), Document 8 of WRC-97
55 The current text of Article 42 of the Constitution of the International Telecommunication Union is contained in The Final Acts of the Plenipotentiary Conference 1998 (ITU, Geneva, 1998)
56 Resolution relating to the preparation of the New International Frequency List. Final Acts of the International Telecommunication and Radio Conferences (Atlantic City, 1947)

Chapter 6

The fixed service

6.1 The FS operating below 30 MHz

6.1.1 VLF and LF systems

At the lower end of the spectrum, radio waves from terrestrial transmitters are propagated mainly by the ionospheric wave and the ground wave and they can cause interference at terrestrial receivers at great distances. At higher frequencies propagation between terrestrial stations is mainly by the tropospheric wave and the range at which interference arises is much less. Thus, the control of interference, the primary purpose of spectrum management, is mainly an international problem at the lower frequencies and mainly a national problem at the higher frequencies, and different management measures are used for the two propagation regimes. The transition from one mode to the other is gradual and varies with the state of the ionosphere but 30 MHz is found to be an appropriate changeover point for most purposes. Section 6.1 is concerned with the management of the lower frequencies and Section 6.2 with the higher frequencies. There is a similar division in the chapters concerned with the other major terrestrial services.

The fixed service (that is, terrestrial links between radio stations at specified fixed points) began in the lowest part of the usable radio spectrum and did not spread into higher bands until the 1930s. Even now 90% of the spectrum between 9 and 150 kHz is allocated for the FS, plus a further 40 kHz just above 150 kHz in Regions 2 and 3. However, when the international table of frequency allocations was revised in 1947, all of the FS allocations below 150 kHz were shared with the MMS, the RNS or the MRNS. The terms of these sharing arrangements are particularly onerous for the FS, which must avoid causing harmful interference to these safety services, which have area coverage; see also RR Resolution 706. By 1998 the sharing situation had become marginally more disadvantageous for the FS, some of its allocations having been made secondary. Relatively few FS stations operate now in the VLF or LF ranges and their numbers are likely to continue to decline.

6.1.2 MF and HF systems

The international frequency allocations

There are a number of primary FS allocations between 1.6 and 4.0 MHz, used for links up to a few hundred kilometres long. However, all of these bands are shared, mainly with mobile services which also have primary allocations and are the predominant users of this part of the spectrum.

Between 2300 and 5060 kHz four bands allocated for the FS are shared with the BS, used for broadcasting in the Tropical Zone. The allocations to both services are primary, but broadcasting has priority within the Tropical Zone; see Section 7.3 below.

The FS is the main user of spectrum between 4 and 30 MHz. Between 1947 and 1979 the FS had primary allocations totalling about 15 MHz in this frequency range, 58% of the whole, split up into many narrow bands. Some of these bands were shared with other terrestrial services which also had primary allocations but most were exclusively for the FS or shared with a service having only a secondary allocation. Until about 1980 these bands were very heavily loaded with long distance links, mainly providing telephone and telegraph circuits for the international public telecommunications network.

Repeatered coaxial transoceanic cables and satellite systems of the FSS began to come into use in the mid-1950s and mid-1960s respectively, eventually relieving the HF medium of a large part of the international public network traffic. By the time of WARC-79 it had become feasible to plan the transfer of some of the FS HF allocations to other services with more pressing needs. At that conference it was agreed that the FS should withdraw from 17 bands within 10 years (15 years for bands below 10 MHz), in order that seven bands with a total bandwidth of 850 kHz might be transferred to the BS, eight bands totalling 800 kHz to the MMS and two bands totalling 200 kHz to the AmS and AmSS. See RR Resolution 8.

A similar arrangement agreed at WARC-92 will release a further ten bands totalling 790 kHz to the BS by 2007; RR Resolution 29 calls for study to be made of the feasibility of earlier access to these bands for the BS. However, the FS will retain a right to use these bands with minimum transmitter power for links within national borders provided that harmful interference is not caused to broadcasting services; see RR S5.136, S5.143, S5.146 and S5.151. Arrangements for the transition are set out in RR Resolution 21. Allowing for these transfers, accomplished and prospective, the remaining allocations for the FS between 4 and 30 MHz total 12.6 MHz in 30 bands, distributed over the frequency range, all primary, about 40% being shared with other services which also have primary allocations.

The use of two FS allocations, 21 850–21 870 kHz and 23 200–23 350 kHz, is limited to services related to aircraft flight safety by RR S5.155B and S5.156A. This is a remnant of the aeronautical fixed service, for which these bands were previously allocated but which was merged into the FS at WRC-95.

Typical MF and HF systems in use

Despite competition from the submarine cable and satellite media, there are still HF radio systems providing telephone and telegraph channels for the public network. These systems are usually operated continuously or for long daily schedules. However, most long distance HF systems serve other purposes nowadays. Many of these systems are used for government communications, often military. Some links function in an operator-to-operator mode and some operate intermittently, call by call or message by message. To minimise the necessary bandwidth of emissions, the use of FM or phase modulation (PM) for analogue signals, typically speech, is prohibited in bands allocated for the FS below 30 MHz and the use of double sideband (DSB) amplitude modulation (AM) is deprecated; see RR S24.1 and S24.2.

The public network radiotelephone systems use single sideband (SSB) or independent sideband (ISB) AM duplex radio systems with a reduced carrier level carrying up to four analogue channels, usually limited to a bandwidth of 3.0 kHz per channel; see Recommendation ITU-R F.348-4 [1] and CCIR Recommendation 454-1 [2]. Some systems use automatic level control and singing suppressors at the transmitting terminal to maintain a high signal-to-noise ratio at the receiver without loss of circuit stability. Alternatively, a system such as Lincompex using sub-carrier signal level control is often preferred, providing better suppression of noise and interference; see CCIR Recommendation 335-2 [3] and Recommendations ITU-R F.455-2 [4] and F.1111-1 [5].

Public radiotelegraphy systems supply channels for five-unit start–stop apparatus employing International Telegraph Alphabet No. 2 for message traffic, 50-baud telex, leased circuits etc. Automatic error correction by repetition (ARQ) is usual for duplex systems, channels typically being multiplexed by time division multiplex (TDM) in pairs or fours. Single channel duplex and single channel simplex systems may be used where the flow of traffic is light; see Recommendations ITU-R F.342-2 [6], F.518-1 [7] and F.519 [8]. Where one multiplex or one channel is enough, the radio emissions will probably use frequency shift keying (FSK) of the carrier, in which case the necessary bandwidth will be a few hundred hertz, depending mainly on the frequency shift; see CCIR Recommendation 246-3 [9]. Alternatively, and especially if many channels are to be transmitted, the digital signals may modulate voice-frequency (VF) tones carried by an SSB or ISB radio system which may carry telephone channels also; the VF tones are modulated by FSK or PSK; see Recommendation ITU-R F.436-4 [10].

Systems that do not provide channels for the public network are more varied in their configurations and in the types of modulation that they use. Many, especially short distance links operating in the MF range, provide radiotelephone channels or single teleprinter channels, with or without error correction. Others are sophisticated systems, providing large numbers of teleprinter channels, medium capacity data links or speech channels with privacy.

Sky wave propagation at MF and HF often exhibits multipath behaviour, the various paths having transmission time differences of several milliseconds (ms). This can cause a high BER in digital transmission systems with a symbol rate exceeding a few hundred symbols per second. The problem can be diminished by distributing a higher speed binary data stream between a number of parallel channels in FDM, each with a lower bit rate, or by using high order modulation, typically PSK with four or eight significant phase states. Further improvement in the BER may well be necessary, typically by means of ARQ. See, for example, Recommendation ITU-R F.763-3 [11].

If effective privacy is required for speech channels, it is necessary to convert the analogue signals to the digital form and scramble the digital signal. To avoid an unacceptably large necessary bandwidth, Recommendation ITU-R F.1112-1 [12] recommends that vocoder techniques should be used in the digital phase. For short range links using ground wave propagation, vocoders requiring up to 4800 bits per second (bit/s) are recommended but this should be reduced to 2400 bit/s or less for longer links using the sky wave. Transmission techniques such as those mentioned in the previous paragraph will be needed for these relatively high information rates.

Choice of operating frequency

Short distance links, typically 100 or 200 km long, are propagated by the ground wave. It is sometimes convenient to assign two frequencies for such links, one at about 3 or 4 MHz being used by day and another at about 20 MHz being used by night. In this way interference to and from distant stations is avoided, the 3 MHz assignment being protected by ionospheric absorption and the 20 MHz assignment by penetration of the ionosphere.

To communicate in the HF range over a distance exceeding a few hundred kilometres, ionospheric propagation is used, and the operating frequency at any time must be appropriate to the radio propagation conditions at that time. The frequency should be high enough to avoid excessive absorption of the radio wave in the ionosphere, which is most severe over the parts of the world that are illuminated by the Sun, but not so high that the wave breaks through the ionosphere and is lost in space without reaching the receiving station. These mechanisms are outlined in the Appendix, Section A4.3. The lowest usable frequency (LUF) and the maximum usable frequency (MUF) depend on the location of the transmitting and receiving stations and vary with the time of day, the season of the year and with cyclical changes in the strength of the radiation from the Sun. Furthermore, the MUF, in particular the MUF for the F region of the ionosphere, which dominates prediction of propagation conditions over very long distances, varies erratically from hour to hour and from day to day. Occasional events associated with sporadic disturbances of the Sun's outer layers may make sky wave communication impossible on Earth for periods lasting minutes or hours.

Figure A.19 below sketches typical variations of the MUF, LUF and the optimum working frequency (OWF) in the course of a quiet day, that is, a day on which propagation conditions are not disturbed. Clearly no single frequency assignment will provide for communication over this link throughout a single day. Additional assignments will be needed at the extremes of the annual and the solar activity cycles. Typically, a link that is thousands of kilometres long will need the use of five assignments for continuous operation, although four or three may be sufficient for shorter links. The group of frequencies assigned for an HF link is called a complement of frequencies. On any one day it may be necessary to change the working frequency half a dozen times if communication is to be maintained. Furthermore, when a change of ionospheric conditions has caused communication to fail, it is not unusual to retune the system to another frequency, expecting to re-establish communication, but without success.

The long-established method for choosing the frequency to change to when the frequency in use on a link has failed to maintain communication is based on the use of predictions of MUF, OWF and LUF, interpreted by an experienced system controller who may also be aware of the behaviour of propagation on other HF links at the same time. But this method is not satisfactory, in particular because conventional predictions cannot take short term variations of conditions into account. What is needed is information on the radio path in real time.

Several real time channel evaluation (RTCE) systems have been developed. Some enable a transmitting station, when its link to a receiving station is failing, to determine automatically whether propagation conditions at another frequency in its complement would be satisfactory and whether reception is likely to be sufficiently free from interference. If the result is favourable in both respects, the operating frequency would be changed, also automatically, thus constituting a frequency adaptive system. See Recommendation ITU-R F.1110-2 [13]. Corresponding information on the current state of HF propagation can be obtained by using oblique-incidence ionosphere sounding; see Recommendation ITU-R F.1337 [14].

It has been demonstrated that the reliability of communication over MF and HF links is greater using RTCE than when operating frequencies are chosen by traditional methods. The test signals that are transmitted to probe propagation conditions arguably increase the burden on the spectrum. On the other hand, many links managed in the traditional way are maintained continuously in service although the traffic may be sporadic. With an efficient RTCE system and especially if the complement of frequencies were enlarged, it would be feasible to operate a link only when there was traffic to pass; an enlarged complement could then be shared between a number of links having different propagation paths. Operating in this way the burden on the spectrum would probably be less for a given total of useful traffic. See also Recommendation ITU-R SM.1266 [15].

The paragraphs of the RR which deal with the use of frequency assignments, drafted at times when frequency adaptive systems did not exist, do not accommodate the practice and place great stress on the need to use the least number of assignments at HF; see for example RR S4.1. There have been doubts as to whether the RR permit the use of frequency adaptive systems. However, WRC-97 recognised the potential usefulness of frequency adaptive systems and RR Resolution 729, agreed at WRC-97, specifically authorises their use at MF and HF, subject to safeguards as to the frequency bands that may be used in this way. Studies in the relevant ITU-R study groups are to continue.

Efficient use of spectrum

Efficient use by the FS of the HF range of frequencies has two main aspects. Systems should be designed to provide circuits that are of a quality that can be used for the purpose intended throughout the desired schedule. Secondly, systems should not avoidably deny use of the spectrum to other potential users. However, these two aspects are not independent. On the one hand, the factors that make a system function well (for example, the use of ARQ) may also reduce its vulnerability to interference. On the other hand, measures taken to avoid interference when it does occur (for example, the inclusion of assignments in the complement which are not justified solely by propagation considerations) may make worse the problem of access for other users, especially in the absence of an efficient automatic frequency adaptive system.

The ITU provides guidance in this matter. RR S4.1 calls for the number of frequencies used to be minimised. RR S3.2 urges the use of transmitting and receiving equipment which is technically up-to-date. RR S15.2 requires that no more power be transmitted than is necessary for satisfactory service. Recommendation ITU-R F.162-3

[16] stresses the value of using directional antennas for transmitting and receiving. For links that will be served by the sky wave it is important for the gain characteristics of the antennas in the vertical plane to be well matched to the dominant propagation modes. Spaced-antenna diversity reception and ARQ are usually important for sky wave radiotelegraphy and should be used where feasible. Similarly, Lincompex or a similar device substantially improves the performance of radiotelephony systems.

6.2 The FS operating above 30 MHz

6.2.1 Frequency allocations and sharing

It would be tedious to list the FS frequency allocations above 30 MHz; they are many and complex, and for the present purpose it is unnecessary. Summarising, between 30 and 1000 MHz around 45% of the spectrum is allocated for the FS with primary status in Regions 1 and 2; in Region 3 the corresponding figure is 90%. Between 1 and 3 GHz the primary FS allocations include about 55% of the spectrum world-wide and in the frequency ranges 3–30 GHz and 30–60 GHz the corresponding figures are 64% and 57%. Almost all of these allocations are primary. In addition, much of the bandwidth that is not allocated for the FS world-wide or Region-wide is allocated for this service in specified countries in footnotes to the international table of frequency allocations. There are further large allocations for the FS at frequencies above 60 GHz but these are little used at present.

Thus, the FS allocations are very extensive. However, they are all shared in the inter-national table with other services which also have primary allocations. Below 1 GHz the FS shares almost all of its allocations with the MS although often with the exclusion of the AMS to limit the range at which interference from mobile stations can occur. In addition, much spectrum allocated for the FS below 1 GHz is shared with the BS, especially in Region 3 and some is shared with other services. Finally, virtually all FS allocations above 1 GHz are shared with primary allocations for one or more of such services as the BS, the BSS, the FSS, the MS, the MSS, the RLS, the SR and the ISS.

To some degree, systems of the FS and geostationary systems of some satellite services, especially the FSS, can operate at the same frequency at the same broad loca-tion, making use of antenna directivity, sharing constraints and mandatory frequency coordination to limit interference; see Sections 5.2.2 and 5.2.3 above. With other combinations of allocations, sharing provides Administrations with choice, but not with an opportunity for double spectrum use. For some services, in particular the MS, the MSS and the BS, the lower frequency end of this great range of spectrum is much to be preferred, whereas for the other services the choice between VHF and SHF carries much less weight.

In general, Administrations use the measures discussed in Chapters 4 and 5 to reduce interference between stations to an acceptable level. However, various addi-tional management devices have been developed, typically sharing constraints, to allow the best use to be made of specific frequency bands by stations of the various services for which that band has been allocated. Thus, various factors influence the ways in which different FS frequency allocations are used and managed, as follows.

1 Sharing with the MS (general)

a)   The frequency range between 100 MHz and 1000 MHz is technically the best for vehicle-borne and hand-portable radiotelephone and related systems of the

LMS. Spectrum between 1 and 3 GHz is also usable for these purposes but it becomes less satisfactory as the frequency increases. There has been strong growth in the use of these mobile systems since the mid-1980s and in many countries the FS is being squeezed out of the bands that are allocated for both services below 1 or 2 GHz. Recommendation ITU-R F.1334 [17] provides protection criteria for use when bands are used for both services.

b) Above 3 GHz the principal use of MS allocations is for transportable stations providing temporary wide-band links, typically for video signals, using antennas of relatively high gain; see Section 8.5.6 below. The parameters of such links are similar to those of typical line-of-sight FS links operating in the same frequency bands. Efficient use of spectrum for this mixture of technically similar transportable and fixed links requires careful management but raises no technical problems.

2 Sharing with geostationary satellite networks (general)

The sharing constraints and mandatory frequency coordination procedures used for the management of band sharing between FS and GSO satellite networks are reviewed in some detail in Section 5.2. Exceptionally, special regulations apply to planned assignments for satellite broadcasting, including the associated feeder links; see item 11 below.

3 Sharing with the MSS and the BSS at 1350–1530 MHz.

a) New frequency allocations have been made to the MSS and the BSS in this frequency range in recent years. In consequence the frequency assignments made to multichannel FS systems in this part of the spectrum may no longer provide the best way of using this frequency range by the various sharing services. Recommendation ITU-R F.1242 [18] recommends new plans for frequency assignments for multichannel digital FS systems in this part of the spectrum which are designed to harmonise as well as may be with the new situation. See also Recommendation ITU-R F.1246 [19], which recommends the reference bandwidth to be assumed for FS digital receivers when coordinating with satellite transmitters of the MSS in this frequency range.

b) Until a frequency plan has been agreed for the BSS in the band 1452–1492 MHz, assignments to the FS and the BSS will be coordinated in accordance with RR Resolution 33. Recommendation ITU-R F.1338 [20] provides a criterion for determining whether coordination is necessary.

4 Sharing with the MSS, the BSS and the science services between 1900 and 2670 MHz

This frequency range is allocated for the FS, with sharing arrangements of long-standing with various other services. Several sharing allocations have been added or upgraded in recent years, mainly for the MSS and various scientific services. Intensive current use and the pressure of new services has made it urgent to do what can be done to maximise the use that can be made of this part of the spectrum by the various services sharing it; see for example RR Resolution 716 and RR Recommendation 622. Various measures have been adopted to contribute to the solution of these problems and the following involve the FS.

a) RR Resolution 716 forbids the installation of new tropospheric scatter stations of the FS in the band 1980–2010 MHz and, in Region 2, at 2010–2025 MHz also, to reduce interference to MSS uplinks in those bands. The Resolution also urges

Administrations to phase out existing tropospheric scatter systems in these bands by the year 2000.

b)   Recommendation ITU-R F.1248 [21] recommends severe limits on the power of new tropospheric scatter systems operating in the bands 2025–2110 MHz and 2200–2290 MHz. These constraints are to protect allocations for the SO, SR and EES which are used for uplinks and downlinks and also for ISLs between low-orbiting satellites and geostationary data relay satellites. For the same reasons, Recommendation ITU-R F.1247 [22] recommends constraints on the power of FS line-of-sight stations operating in these frequency bands. Both of these recommendations apply especially severe constraints to stations with antennas directed towards a number of locations in the GSO, identified in Recommendation ITU-R SA.1275 [23] as preferred locations for data relay satellites.

c)   Recommendations ITU-R F.382-7 [24], F.1243 [25] and F.1098-1 [26] offer new radio channel plans for line-of-sight FS systems designed to harmonise with MSS and science service allocations between 1980 and 2300 MHz.

d)   Recognising that it may be necessary to transfer some FS stations from frequency bands which have recently been allocated for the MSS, Recommendation ITU-R F.1335 [27] advises on the transition process.

e)   The use of the band 2520–2670 MHz by the BSS is limited to national and regional systems for community reception by RR S5.416. This protection of the FS, and other terrestrial services with allocations in this band, is quantified by RR Article S21, which imposes a constraint on the PFD delivered at the Earth's surface by broadcasting satellites. However, it may be noted that RR S5.418 allocates 2535–2655 MHz for digital audio broadcasting by satellite (S-DAB) in several countries in Regions 1 and 3, under the terms of RR Resolution 528 and that this constraint will not apply to such systems.

5   *Sharing with EES downlinks at 8025–8400 MHz*

At WRC-97 a secondary allocation for the EES (downlinks) in Regions 1 and 3 in the band 8025–8400 MHz was up-graded to primary status, subject to the severe sharing constraint set out in RR S5.462A to protect the FS which also has a primary allocation in this band. This constraint is, however, provisional. RR Resolution 124 sets out the circumstances of the imposition of this constraint in more detail and calls for studies to be made to ascertain whether a milder constraint would be satisfactory.

6   *Sharing with the RLS around 10 GHz*

There are primary allocations for the RLS sharing a few frequency bands with the FS, also with primary status; see in particular at 10–10.5 GHz. Digital communication systems are particularly susceptible to interference from pulse radars. If radars used in a shared band are mobile it will usually be desirable for an Administration to divide the band between the two services, preferably with a scheme that has been agreed with the Administrations of neighbouring countries. If, however, the radars are few and fixed, various measures for reducing their impact on FS systems are available, see Recommendation ITU-R F.1097 [28]. Recommendation ITU-R F.1190 [29] shows the level of interference power from a radar that can be expected to be tolerable at a digital FS receiver.

7 Sharing with the RAS at 10.6–10.68 GHz

In general, interference from FS stations to radio astronomy stations in bands shared by the FS and the RAS is controlled by prudent siting of observatories and by with-holding FS assignments that would lead to interference. Exceptionally, for the band 10.6–10.68 GHz, RR S5.482 assists in the limitation of interference to the RAS by apply-ing a severe constraint to the power of FS stations in most countries unless a relaxation of this constraint has been agreed by the mechanism of RR S9.21 with the countries concerned; see also Section 5.3 above.

8 Sharing with satellite-borne passive sensors at 10.6 and 18.6 GHz

a)   Where an FS allocation is shared with allocations for passive sensors of the SR and EES there may be no agreed means of ameliorating the interference. However, in the band 10.6–10.68 GHz the constraint applied to the power of FS transmitters by RR S5.482 (see item 7 above) gives some protection to the passive sensors of the SR and the EES which also have allocations in this band.

b)   There is a similar sharing situation in the band 18.6–18.8 GHz. Section 13.2.2 refers to concerns about interference from FS transmitters to EES passive sensors in the band 18.6–18.8 GHz in Regions 1 and 3, where the EES allocation is secondary. RR S5.522 invites Administrations to limit the transmitter power and the e.i.r.p. of FS stations in this band as much as possible, to eliminate avoidable interference. See also Recommendation ITU-R F.761 [30] and CCIR Report 850 [31].

9 Sharing with non-GSO FSS systems at 10.7–12.75 GHz and 17.8–18.6 GHz

WRC-97 made tentative allocations for downlinks of non-GSO FSS systems in various frequency bands already allocated for the FSS and used for GSO systems, sharing with the FS; see RR Resolution 130 and RR S5.484A. RR Resolution 131 applies to these non-GSO downlinks the same constraints on PFD at the Earth's surface as are given in RR Article S21 Section V to limit interference to terrestrial services from GSO satellites. However, WRC-97 thought that such a constraint may be unneces-sarily severe, and Resolution 131 calls for the matter to be studied further. See also RR Resolution 538.

10 Sharing with GSO FSS systems at 11.7–12.2 GHz in Region 2

The FS and downlinks of GSO systems of the FSS share 11.7–12.2 GHz in Region 2. RR S5.488 limits FSS use to national and sub-Regional systems. RR Article S21 applies no sharing constraints to either service but RR S5.488 requires prior coordination (see Section 5.3 above) before any use is made of this band by the FSS. Recommenda-tion ITU-R SF.674-1 [32] provides a value of PFD at the Earth's surface for use as a criterion of a need to coordinate.

11 Sharing in frequency bands planned for the BSS

The BSS frequency assignment plan, involving downlinks around 12 GHz and feeder uplinks around 18 and 15 GHz, includes special provisions to protect broadcasting systems from interference from other services; see Section 10.3.3 below. These provi-sions have the effect of making the FS allocations inferior in status to those of the BSS and subject to a special coordination process. When similar plans are agreed for other BSS allocations, the same consequences can be expected for the sharing services, including services sharing the feeder link bands.

12 Sharing with unplanned BSS bands at 17.7 and 21.4 GHz

Part of the Region 2 BSS allocation at 17 GHz, namely 17.7–17.8 GHz, is shared with the FS. The Regions 1 and 3 BSS allocation at 21.4–22.0 GHz is also shared with the FS. Interference between these services would be controlled by coordination in accordance with RR Resolution 33, prior to the agreement of a frequency plan for the BSS. At present there is no constraint on the PFD of the BSS emissions at the Earth's surface in these bands but RR Recommendation ITU-R F.760-1 [33] provides recommended values, closely aligned with the values applicable to the FSS in adjacent bands.

13 Sharing with the ISS at 25.25 27.5 GHz

This ISS allocation, shared with equal primary status with the FS and the MS, is to be used for inter-satellite links involving geostationary data relay satellites. Recommendation ITU-R F.1249 [34] recommends constraints on the power of FS stations in this band and more severe constraints on the power of the FS stations which have antennas that are directed towards the locations on the GSO which are identified in Recommendation ITU-R SA.1276 [35] as preferred locations for data relay satellites. RR S21.2 to S21.5 set limits on the power of FS stations in this band and urge that a severe limit (e.i.r.p. not to exceed 24 dBW in any 1 MHz sampling bandwidth) be implemented within ± 1.5° of the GSO as far as is practicable.

14 Sharing with the FSS at 37.5–40.0 GHz

The FS frequency allocation at 37–40 GHz is identified in RR Resolution 133 for possible future designation for high-density applications; see Section 6.2.3. Of this band, 37.5–40.0 GHz is shared with equal primary status with FSS downlinks, subject to a PFD constraint at the Earth's surface which is applied to the FSS by RR Article S21. The resolution calls for studies to be made of means that would facilitate the use of this band by the various services for which it is allocated and specifically to ascertain whether the constraint applied by RR Article S21 will be sufficient to enable the FS allocation to be used for high-density applications.

15 Sharing with the FSS at 40.5–42.5 GHz

At WRC-97 an allocation for the FS, formerly secondary, was made primary. This band is shared with primary allocations for the FSS (downlinks), the BSS and the BS. RR Resolution 129 calls for studies to be done to develop means for the use of this band by the services for which it is allocated.

6.2.2 Line-of-sight systems, 30–1000 MHz

At frequencies below 1 GHz there is rather more diffraction of radio waves at the edges of obstacles, natural or man-made, than occurs at higher frequencies. This property is of value in the FS, because it may allow communication over transmission paths which are somewhat obstructed. Also the cost of narrow-band radio equipment for use in this part of the spectrum is relatively low. However, almost all of the spectrum allocated internationally for the FS in this frequency range is shared with the MS or the BS or both. These factors of propagation and cost have even greater importance for the BS and the LMS. These latter services are large and growing and they have secured the use of a large proportion of the available spectrum in this frequency range.

Very little of this spectrum is left in some countries for the FS. What is available may be used for multichannel systems, connected in tandem to form radio-relay chains providing channels for the PSTN; see for example Recommendation ITU-R F.754 [36]. More usually, however, it is used for the specific applications for which the particular qualities of this part of the spectrum, technical and economic, make its use particularly attractive. Such uses include, for example, long-hop links to offshore islands, links serving sparsely developed territory where the cost of relay stations is particularly high and single channel multipoint networks providing telephone or telemetry connections to isolated stations.

6.2.3 Line-of-sight systems above 1 GHz

The circumstances which limit access for FS systems to spectrum below 1 GHz extend, although with less force, to higher frequencies. In recent years the allocations in the range 1–2 GHz that are shared between the FS and the MS are being taken over by the LMS and, since 1992, by new allocations for the MSS. Many FS systems that operated between 1 and 2 GHz have had to be replaced by new systems using higher frequencies. This trend continues. Substantial bandwidth below 3 GHz remains available for the FS for the present and is used for line-of-sight links as well as tropospheric scatter links. However, much more FS spectrum remains available above 3 GHz and massive growth in the use of radio for fixed links is taking place in wide allocations between 20 and 50 GHz.

As the frequency rises, radio propagation becomes less favourable for long distance links because the transmission loss in the troposphere, mainly owing to rain, rises. Between 3 and about 12 GHz conditions are suitable, in temperate climates, for links up to about 50 km long; in rainy climates maximum link lengths are considerably shorter. Above 12 GHz typical maximum link lengths decline to 10–20 km at 20 GHz and to just a few kilometres at 50 GHz, the link lengths achievable at any given location depending in detail on such factors as climate, system parameters and type of modulation.

The FS includes a number of different applications using different kinds of system. Which frequency allocations are used for each of the various applications varies to a considerable extent from country to country depending, for example, on the propagation factors mentioned in the previous paragraph and the history of the development of the service in each country. A larger degree of uniformity is probably emerging in the bands above about 25 GHz which are now coming into use. The following notes draw attention to the impact of international agreements on these various applications.

Wide-band radio relay systems, mainly for the PSTN

In the FS allocations between 3 and about 15 GHz, the predominant application is for wide-band links carrying multiplexed telephony channels. These links are often connected in tandem to form radio relay chains which may be hundreds, sometimes thousands, of kilometres long. The telephone channels are used mainly for trunk telephone connections, national or international, between switching centres of the PSTN. The multiplex systems for telephony channels, FDM for analogue transmission and TDM for digital transmission, have been standardised by ITU-T; for details, reference should be made to recommendations in the ITU-T G series. These systems often carry video channels which are used, for example, for connecting locations where broadcasters assemble programmes to terrestrial broadcasting stations, cable network head ends and feeder link earth stations.

Radio channel plans designed to fit these wide-band radio systems efficiently into the principal FS allocations above 1 GHz have been standardised by recommendations of the ITU-R. These radio channel standards have been defined to provide efficient use of spectrum and convenient interconnection of radio relay systems at national frontiers, although their use is not mandatory. They presuppose that two adjacent stations in a radio relay chain will often be linked by several multiplexed carriers, forming a large trunk route totalling perhaps thousands or tens of thousands of channels or their equivalent. This is a situation that commonly arises in the trunk networks of PTOs. Recommendation ITU-R F.746-3 [37] lists these plans and explains some of the concepts that they embody. Recommendation ITU-R F.635-4 [38] may be taken as an example of these plans, suitable for radio relay systems operating in the 2 and 4 GHz bands. Some of these plans have been devised to harmonise the use of a band with systems of other services which share some part or all of that band; see for example Recommendations ITU-R F.1242 [18] and F.1098-1 [26], which are noted in Section 6.2.1 above.

Other wide-band systems: 'high density services'

An increasing proportion of line-of-sight systems are short, consisting of a single hop carrying a single multiplexed carrier or one video channel. The system described in Section 4.1.1 above is of this type. Others provide video connections to broadcasting stations. Many are used to link the base stations of LMS cellular systems together and into the public network. Others give wide-band access from office buildings to the public network for high speed data or for multichannel private telephone links. Sometimes radio links of this type are used to make quick temporary provision of facilities, pending permanent provision by cable. Often a number of these links radiate from a central station or a concentration point, and at such locations the use of a point-to-multipoint terminal may be found economic; see Recommendation ITU-R F.755-1 [39]. Simple radio channel plans like those illustrated in Figures 4.3 and 4.4 are likely to be more convenient for such applications than the plans reviewed in Recommendation ITU-R F.746-3 [37].

The longer links may require assignments in frequency bands below 20 GHz. However, satisfactory performance using low transmitter power is being obtained, for example around 38 and 55 GHz, over distances of a few kilometres. With careful management spectrum can be used very intensively for links of this kind; these are the so-called high-density FS applications. However, much of the band at 59–64 GHz is unusable for this purpose because of the very high absorption in the atmosphere due to molecular resonance in oxygen. Much bandwidth elsewhere already allocated to the FS would be difficult to use efficiently in this way because of current or planned use by sharing services using high power transmissions.

WRC-97 perceived an urgent need to make new provision for growth in the use of this part of the spectrum for FS high-density applications. New primary FS allocations were made at 31.8–33.4 GHz, 51.4–52.6 GHz and 64–66 GHz. The previously existing 54.25–58.2 GHz FS allocation was amended, becoming 55.78–59 GHz, to avoid a difficult sharing situation that arose with the previous allocation. These four bands are identified by RR S5.547 for high-density applications. WRC-97 called on ITU-R to propose criteria to WRC-00 for sharing these bands between high-density applications of the FS and the other services to which they are allocated; see RR Resolutions 126 and 726. It was also foreseen that the FS allocation at 37–40 GHz will need to be used with high density in the future; see the reference to RR Resolution 133 in Section 6.2.1, item 14.

High altitude platforms

Recent developments enable lighter-than-air platforms to be maintained semi-permanently in chosen locations in the stratosphere, between 20 and 50 km above the ground (see the definition of 'high altitude platform station' in RR S1.66A) to be used for relaying signals between terrestrial stations of the FS within line-of-sight. WRC-97 designated the sub-bands 47.2–47.5 GHz and 47.9–48.2 GHz, within the allocation 47.2–50.4 GHz, for this purpose; see RR S5.552A. For the present, such systems may not operate in any other bands; see RR S4.15A. However, 47.2–50.4 GHz is shared with various services and technical studies had not, at the time of WRC-97, been made to determine the basis on which the band could be shared between them. RR Resolution 122 calls on the ITU-R to brief WRC-00 on this matter.

The wireless local loop

It may be economic, in particular where the population is sparse, to connect telephone subscribers to their local exchange by radio instead of by wire. Various radio link designs are used for 'radio fixed access', typically line-of-sight systems operating between 1 and 10 GHz, incorporating some form of circuit concentration. See for example Recommendations ITU-R F.701-2 [40] and F.756 [41].

6.2.4 Trans-horizon systems

Meteor-burst systems

The physical basis for this mode of propagation is outlined in the Appendix, in Section A6. The frequency range most often used for meteor-burst systems is 30–50 MHz. A data stream of several kilobits per second can be sustained for the short period, typically a fraction of a second, that a meteor trail is active and in sight of the terrestrial stations involved in a link, which may be separated by up to 2000 km. For much of the time there may be no meteor burst simultaneously in sight at both stations needing to communicate, but with a suitable packet protocol an average data rate of a few hundred bits per second can usually be maintained. The deployment of meteor-burst systems has recently begun although a major role for them has not emerged yet. For information on some existing networks, see Recommendation ITU-R F.1113 [42].

These systems, operating in the preferred frequency range, may also be expected to cause intermittent interference at great distances by ionospheric reflection via the regular F2 mode and by sporadic E layer propagation, but the transmitters employed are not powerful enough to cause significant interference by the ionospheric forward scatter mode. At the time of writing no special international measures have been adopted to harmonise these systems with line-of-sight systems.

Tropospheric scatter systems

The physical basis for tropospheric scatter propagation is outlined in the Appendix, in Section A5. This propagation mechanism can be observed over a wide range of frequency, starting well below 1 GHz and extending beyond 10 GHz. The optimum frequency range is around 2 GHz, where wide-band links, 200–400 km long, capable of carrying multichannel telephony systems or video channels, are feasible. Lower frequencies, given high transmitter power, favourable station sites and complex diversity systems, provide links up to about 700 km long, although with narrower usable bandwidth. Frequencies up to at least 5 GHz may be used for shorter distances, up to 200 km. See Recommendations ITU-R F.698-2 [43] and F.1106 [44]. Such links are quite widely used, sometimes connected in tandem to form long connections.

Tropospheric scatter is a high loss propagation mode which uses high transmitter power, very sensitive receivers and hill-top station sites. These characteristics make it necessary to provide wide geographical separation between the stations of different tropospheric scatter systems and between such stations and the stations of line-of-sight FS and MS systems.

It is also difficult to share frequency bands between tropospheric scatter systems and satellite systems. The spectral PFD which RR Article S21 Section V permits satellite downlinks in frequency bands shared with the FS to deliver to the Earth's surface is often more than tropospheric scatter receivers can accept; see footnote 9 to RR Table S21-4 (RR S21.16.3). Furthermore, whereas it may sometimes be feasible to site tropospheric scatter stations so that their receiving antennas are directed so that the interference from geostationary satellites is acceptably low this strategy is not effective for the non-GSO MSS systems that are planned for frequency bands near 2 GHz. Also, the power of most tropospheric scatter transmitters and also the e.i.r.p. from their antennas exceed the constraints imposed by RR Article S21 Section II upon FS stations in frequency bands shared with satellite uplinks. Finally, the current vigorous growth of terrestrial LMS and MSS satellite systems is putting great pressure on the spectrum around 2 GHz, the optimum frequency for tropospheric scatter systems.

RR S21.7 withholds the application of power constraints from tropospheric scatter transmitters operating in the bands 1700–1710 MHz, 1970–2010 MHz, 2025–2110 MHz and 2200–2290 MHz except for radiation in the direction of the geostationary orbit. However, these concessions have recently been negated, to a large degree, by RR Resolution 716 and Recommendation ITU-R F.1248 [21]. The resolution forbids the installation of new stations in the band 1980–2010 MHz and urges Administrations to phase out existing stations in this band by the year 2000 to protect MSS uplinks. The recommendation recommends severe limits on the power of new tropospheric scatter stations operating in the bands 2025–2110 MHz and 2200–2290 MHz. See also Section 6.2.1, item 3, above.

Furthermore, RR S5.409 and S5.411 urge that new tropospheric scatter systems should not be set up in the band 2500–2690 MHz. If such installations cannot be avoided, it is required that all possible measures should be taken to avoid new systems with antennas directed towards the GSO. RR S5.410 requires that any new systems set up in this band in Region 1 should first be coordinated with other Administrations with an interest in accordance with the mechanism of RR S9.21; see also Recommendation ITU-R F.302-3 [45] and Section 5.3 above.

The frequency range 2300–2487.5 MHz seems likely to be the first choice in the future in many countries for new tropospheric scatter systems and for existing systems that have been displaced from other bands. However, satisfactory new frequency assignments will be increasingly difficult to find, and they will need to be coordinated with care with the Administrations of neighbouring countries. RR Recommendation 100 foresees a need for a future WRC to give attention to this problem.

6.3 References

1 'Arrangement of channels in multi-channel single-sideband and independent-sideband transmitters for long-range circuits operating at frequencies below about 30 MHz'. Recommendation ITU-R F.348-4, ITU-R Recommendations 1997 F Series Part 3 (ITU, Geneva, 1997)
2 'Pilot carrier levels for HF single sideband and independent-sideband reduced carrier systems'. CCIR Recommendation 454-1. Volume III of the CCIR XVIIth Plenary Assembly (ITU, Geneva, 1990)
3 'Use of radio links in international telephone circuits'. CCIR Recommendation 335-2, *ibid*.
4 'Improved transmission system for HF radiotelephone circuits'. Recommendation ITU-R F.455-2, ITU-R Recommendations 1992 RF Series (ITU, Geneva, 1992)
5 'Improved Lincompex system for HF radiotelephone circuits'. Recommendation ITU-R F.1111-1, ITU-R Recommendations 1997 F Series Part 3 (ITU, Geneva, 1997)
6 'Automatic error-correcting system for telegraph signals transmitted over radio circuits'. Recommendation ITU-R F.342-2, *ibid*.
7 'Single-channel simplex ARQ telegraph system'. Recommendation ITU-R F.518-1, *ibid*.
8 'Single-channel duplex ARQ telegraph system'. Recommendation ITU-R F.519, *ibid*.
9 'Frequency-shift keying'. CCIR Recommendation 246-3. Volume III of the CCIR XVIIth Plenary Assembly (ITU, Geneva, 1990)
10 'Arrangement for voice-frequency, frequency-shift telegraph channels over HF radio circuits'. Recommendation ITU-R F.436-4, ITU-R Recommendations 1997 F Series Part 3 (ITU, Geneva, 1997)
11 'Data transmission over HF circuits using phase-shift keying'. Recommendation ITU-R F.763-3, *ibid*.
12 'Digitized speech transmissions for systems operating below about 30 MHz'. Recommendation ITU-R F.1112-1, *ibid*.
13 'Adaptive radio systems for frequencies below about 30 MHz'. Recommendation ITU-R F.1110-2, *ibid*.
14 'Frequency management of adaptive HF radio systems and networks using FMCW oblique-incidence sounding'. Recommendation ITU-R F.1337, *ibid*.
15 'Adaptive MF/HF systems'. Recommendation ITU-R SM.1266, ITU-R Recommendations 1997 SM Series (ITU, Geneva, 1997)
16 'Use of directional transmitting antennas in the fixed service operating in bands below about 30 MHz'. Recommendation ITU-R F.162-3, ITU-R Recommendations 1997 F Series Part 3 (ITU, Geneva, 1997)
17 'Protection criteria for systems in the fixed service sharing the same frequency bands in the 1 to 3 GHz range with the land mobile service'. Recommendation ITU-R F.1334, ITU-R Recommendations 1997 F Series Part 2 (ITU, Geneva, 1997)
18 'Radio-frequency channel arrangements for digital radio systems operating in the range 1350–1530 MHz'. Recommendation ITU-R F.1242, ITU-R Recommendations 1997 F Series Part 1 (ITU, Geneva, 1997)
19 'Reference bandwidth for receiving stations in the fixed service to be used in coordination of frequency assignments with transmitting space stations in the mobile-satellite service in the 1–3 GHz range'. Recommendation ITU-R F.1246, ITU-R Recommendations 1997 F Series Part 2 (ITU, Geneva, 1997)
20 'Threshold levels to determine the need to coordinate between particular systems of the broadcasting-satellite service (sound) in the geostationary-satellite orbit for space-to-

Earth transmissions and the fixed service in the band 1452–1492 MHz'. Recommendation ITU-R F.1338, *ibid*.

21 'Limiting interference to satellites in the space science services from the emissions of trans-horizon radio relay systems in the bands 2025–2110 MHz and 2200–2290 MHz'. Recommendation ITU-R F.1248, *ibid*.

22 'Technical and operational characteristics of systems in the fixed service to facilitate sharing with the space research, space operation and Earth-exploration services operating in the bands 2025–2110 MHz and 2200–2290 MHz'. Recommendation ITU-R F.1247, *ibid*.

23 'Orbital locations of data relay satellites to be protected from the emissions of fixed service systems operating in the band 2200–2290 MHz'. Recommendation ITU-R SA.1275, ITU-R Recommendations 1997 SA Series (ITU, Geneva, 1997)

24 'Radio-frequency channel arrangements for radio-relay systems operating in the 2 and 4 GHz bands'. Recommendation ITU-R F.382-7, ITU-R Recommendations 1997 F Series Part 1 (ITU, Geneva, 1997)

25 'Radio-frequency channel arrangements for digital radio systems operating in the range 2290–2670 MHz'. Recommendation ITU-R F.1243, *ibid*.

26 'Radio-frequency channel arrangements for radio-relay systems in the 1900–2300 MHz band'. Recommendation ITU-R F.1098-1, *ibid*.

27 'Technical and operational considerations in the phased transitional approach for bands shared between the mobile-satellite service and the fixed service at 2 GHz'. Recommendation ITU-R F.1335, ITU-R Recommendations 1997 F Series Part 2 (ITU, Geneva, 1997)

28 'Interference mitigation options to enhance compatibility between radar systems and digital radio relay systems'. Recommendation ITU-R F.1097, ITU-R Recommendations 1997 F Series Part 1 (ITU, Geneva, 1997)

29 'Protection criteria for digital radio-relay systems to ensure compatibility with radar systems in the radiodetermination service'. Recommendation ITU-R F.1190, *ibid*.

30 'Frequency sharing between the fixed service and passive sensors in the band 18.6–18.8 GHz'. Recommendation ITU-R F.761, ITU-R Recommendations 1997 F Series Part 2 (ITU, Geneva, 1997)

31 'Frequency sharing by passive sensors with the fixed, mobile except aeronautical mobile and fixed-satellite services in the band 18.6–18.8 GHz'. CCIR Report 850-1. Recommendations and Reports of the XVIIth Plenary Assembly of the CCIR, Annex to Volume II, (1990)

32 'Power flux-density values to facilitate the application of Article 14 of the Radio Regulations for FSS in relation to the fixed-satellite service in the 11.7–12.2 GHz band in Region 2'. Recommendation ITU-R SF.674-1, ITU-R Recommendations 1997 SF Series (ITU, Geneva, 1997)

33 'Protection of terrestrial line-of-sight radio-relay systems against interference from the broadcasting-satellite service in the bands near 20 GHz'. Recommendation ITU-R F.760-1, ITU-R Recommendations 1997 F Series Part 2 (ITU, Geneva, 1997)

34 'Maximum equivalent isotropically radiated power of transmitting stations in the fixed service operating in the band 25.25–27.5 GHz shared with the inter-satellite service'. Recommendation ITU-R F.1249, *ibid*.

35 'Orbital locations of data relay satellites to be protected from the emissions of fixed service systems operating in the band 25.25–27.5 MHz'. Recommendation ITU-R SA.1276, ITU-R Recommendations 1997 SA Series (ITU, Geneva, 1997)

36 'Radio-relay systems in bands 8 and 9 for the provision of telephone trunk connections in rural areas'. Recommendation ITU-R F.754, ITU-R Recommendations 1997 F Series Part 1 (ITU, Geneva, 1997)

37 'Radio-frequency channel arrangements for radio-relay systems'. Recommendation ITU-R F.746-3, *ibid*.

38 'Radio-frequency channel arrangements based on a homogeneous pattern for radio-relay systems operating in the 4 GHz band'. Recommendation ITU-R F.635-4, *ibid*.

39 'Point-to-multipoint systems used in the fixed service'. Recommendation ITU-R F.755-1, *ibid*.

40 'Radio-frequency channel arrangements for analogue and digital point-to-multipoint radio systems operating in frequency bands in the range 1.427 to 2.690 GHz (1.5, 1.8, 2.0, 2.2, 2.4 and 2.6 GHz)'. Recommendation ITU-R F.701-2, *ibid.*

41 'TDMA point-to-multipoint systems used as radio concentrators'. Recommendation ITU-R F.756, *ibid.*

42 'Radio systems employing meteor-burst propagation'. Recommendation ITU-R F.1113, ITU-R Recommendations 1997 F Series Part 3 (ITU, Geneva, 1997)

43 'Preferred frequency bands for trans-horizon radio-relay systems'. Recommendation ITU-R F.698-2, ITU-R Recommendations 1997 F Series Part 1 (ITU, Geneva, 1997)

44 'Effects of propagation on the design and operation of trans-horizon radio-relay systems'. Recommendation ITU-R F.1106, *ibid.*

45 'Limitation of interference from trans-horizon radio-relay systems'. Recommendation ITU-R F.302-3, *ibid.*

Chapter 7

The broadcasting service

7.1 Introduction

7.1.1 The international regulatory background

The broadcasting service comprises transmissions that are intended to be received directly by the general public. It does not include radio links that are used to deliver sound or vision programme signals to the broadcasting transmitter; these are typically part of the FS. On the other hand it does not exclude radio transmissions that, in addition to direct reception by the public, are received for onward local distribution, for example by cable.

Bands between 5.95 and 26.1 MHz that are allocated exclusively for the BS are used for broadcasting to both domestic and foreign audiences. Nowhere do the RR forbid the use of other frequency bands for broadcasting to audiences in foreign countries, but there is an implication that any coverage of foreign territory is an unintentional overspill of a domestic service. RR S23.3 requires the power of transmitters to be limited to what will provide a national service, except in the HF exclusive bands and the band 3.9–4.0 MHz. RR S23.2 prohibits broadcasting from ships or aircraft outside national territory.

The expectation that broadcasting signals will be received by the general public is unique among major radio services and it leads to extreme stability in system design, with special care taken to avoid interference, and high transmitter power to minimise the cost of satisfactory reception. The systems used for sound radio in the LF/MF/HF and VHF bands have evolved since their introduction 80 and 40 years ago respectively, enhancing facilities and improving performance by refinement but without basic system changes. A few countries have changed their television (TV) picture standards, but with that exception the same is almost true of 40 years of terrestrial TV, households being equipped to make use of colour, teletext and the ancillary signals facilitating video recording etc. at the householder's individually chosen pace. However, this period of tranquility is being brought to an end by the introduction of digital broadcasting.

7.1.2 National regulation of broadcasting

Spectrum management

In many parts of the world, frequency assignment plan agreements have been drawn up by conferences attended by delegates of the Administrations of all or most of the countries in an ITU Region or some large sub-Regional area. These agreements include provision for all the foreseen broadcasting stations within the area covered, together with a procedure for amending the plan when necessary to provide for changed or new requirements. Planning for broadcasting in the LF/MF and VHF/UHF frequency ranges is discussed more specifically in Sections 7.2 and 7.5.2 below. When an Administration decides to bring one of these planned assignments into use, the frequency is assigned to the station operator and the assignment is notified to the ITU for registration in accordance with the procedure outlined in Section 5.4 above.

Where there is no plan, the Administration will, in most cases, choose a frequency for assignment that seems unlikely to suffer interference from, or cause interference to, other stations that have an equal or superior right to operate in the frequency band in question. The Administration will usually discuss its choice with the Administrations of other countries that are likely to be affected by interference if the new assignment is taken into use. It will then make the assignment to the station operator and notify it to the ITU for registration. The process will be along the lines of the treatment of an assignment to an FS station, described in Section 4.1.1 above and in particular Stage 5. If interference arises, it will be cleared as discussed in Section 4.5 above.

The treatment of frequency assignments for HF broadcasting differs from both of these procedures. This special procedure has been found to be necessary because of the intense pressure on the spectrum available for HF broadcasting, the long reach of interference at HF and the vagaries of HF radio propagation. See Section 7.4 below.

The Administration can be expected to require payment in some form, ultimately if not directly from the user of the assignment, in compensation for the cost of these spectrum management functions. This fee may, however, be a small part of the total charge for a licence to broadcast.

The choice of broadcaster

Some broadcasting is done as a public service, usually by bodies under a measure of public or government oversight. Public service broadcasting may be financed through a charge, levied on members of the public who receive the programmes, as part of a licence fee paid to the government and disbursed to the broadcasting organisation. The receiving licence fee may also cover the cost of suppressing interference to the broadcasting service.

However, in recent years there has been a great increase in the number of programmes of sound and television that are broadcast. This growth in the supply of broadcasting to the public is financed mainly through advertising fees and charges for access to programmes, paid by advertisers, listeners and viewers directly to the broadcaster. Commercial broadcasting is potentially very profitable and there may be strong competition for a licence to broadcast. Often the Regulator must choose between applicants.

Regulators are concerned to ensure that the organisations chosen for licensing from among a number applying for licences will provide programmes of an appropriate quality. The choice may be made arbitrarily by the Regulator, taking into account, for example, an appraisal of the perceived merits of applicants as broadcasters, their past performance and their statements of intentions; beauty competitions, as they are called. However, it is difficult to assess prospective quality from past performance or statements of intentions, and beauty competitions may lead to appeals, litigation and delay.

Arguably, the revenue that a commercial broadcaster draws from advertisers and from the public is a valid measure of the quality of the programmes. Arguably too, an applicant's confidence in the profitability of the programmes that he would provide can be measured by the fee that he is prepared to pay for the licence. It may also be argued that competition between broadcasters for the support of advertisers will lead to the provision of programmes that please the public. For these reasons there is a growing tendency for auctioning to play a major part in the selection of licensees, subject to the maintenance of a sufficient degree of competition within the broadcasting medium and within the larger field of all media of mass information and entertainment.

However, to sustain programme quality it is important that commercial broadcasters should be required to reapply for their licences at intervals, in competition with other applicants, and that the competition between established licensees and potential new licensees should be on terms that are, as far as possible, equal. An established broadcaster has advantages relative to an applicant not already broadcasting, in the form of experienced staff and established programme production facilities. An established broadcaster owning the means for transmitting programmes to the public is doubly entrenched. It is important that the regulation of the industry should minimise these advantages. In the United Kingdom the Broadcasting Acts of 1990 and 1996 show one approach to this problem; the following is a much-simplified account of the principal features of these Acts.

Commercial broadcasting regulation in the United Kingdom: (a) Commercial television licensing

The Broadcasting Act of 1990 established the Independent Television Commission as the Regulator of commercial television in the United Kingdom, except for the public service programmes of the British Broadcasting Corporation and the Welsh Authority. The Act divides terrestrial television broadcasting into two functions, 'local delivery service' and 'programme provision', and makes the Commission the source of licences for these functions, in collaboration with the Radiocommunications Agency, the British Administration.

Licensing a TV local delivery service

When the Commission proposes to grant a local delivery service licence, it announces its intention, stating the area to be served and the frequency assignments that would be made available for it by the Administration if terrestrial radio is the medium favoured by an applicant. Interested applicants respond, giving their technical proposals, stating a timetable, offering an annual sum as a bid for the licence and providing information that will enable the Commission to judge the adequacy of the resources of the applicant to carry out its proposal if licensed. Unacceptable proposals having been discarded, a licence is granted for a period of 15 years to the highest bidder. The licensee then applies to the Administration for the necessary frequency assignment(s). At the end of the period of the local delivery service licence, the licence may be renewed by the Commission, subject to conditions.

Licensing digital TV multiplexes
The Act of 1996 makes provisions, similar in principle to those summarised in the previous paragraph, for the licensing by the Commission of local delivery service providers to distribute multiplexed groups of digital TV channels and the various other services which digital TV systems can supply.

Licensing a TV programme provider
In 1990 TV programmes numbers three and four were being operated commercially in the United Kingdom, each being assigned a number of radio channels in UHF Bands IV and V (see Table 7.2 below) and providing substantially nation-wide coverage. Plans had been made for licensing a fifth programme, also commercial, with nation-wide coverage in as far as the availability of unassigned radio channels permitted; programme five is now in operation. The government's objectives for the services that these three programmes were to provide were not the same, and the provisions in the Act for licensing programme providers for them differ accordingly. Programme number three is the most basically commercial in concept. A number of companies, financed by advertising, are licensed to provide programming in different regions of the United Kingdom; in the London area different companies operate on weekdays and on weekend days. Licensing procedure for programme three is as follows.

When the Commission proposes to grant a TV programme provider's licence, it announces its intention, stating the area to be served and giving guidelines (with the help of its own market research) as to the kinds of programming that it would expect to be provided. One of the government's objectives is to stimulate a thriving industry in the production of programme material by independent producers, and a percentage (usually 25%) of the licensee's programmes will be required to be from independent producers. There are special provisions as to the source of news programmes.

Interested applicants respond, giving their proposals as to programming, explaining for example what would be done to take into account the special needs of the deaf, the hard of hearing, the blind and the partially sighted, giving extensive information as to their business and financial status and offering an annual sum as a bid for the licence. After detailed scrutiny of proposals and the elimination of any that fail to respond adequately to the Commission's objectives, an award of the licence is made. It will usually be made to the highest bidder, although there is provision for the award to be made to a particularly meritorious under-bidder instead.

The usual period of a programme three licence is 10 years, during which time the fee offered in the bid, indexed in line with changes in the retail prices index, is payable annually. There is an additional charge, related to the turnover of advertising revenue.

Commercial broadcasting regulation in the United Kingdom: (b) Sound radio licensing

The Act of 1990 established the Radio Authority as the Regulator of sound radio, other than the services provided by the British Broadcasting Corporation.

Licensing a sound radio broadcaster

Sound radio programmes are classified by the Radio Authority, as 'national' (which includes coverage areas that would more accurately be described as regional in extent) and 'local'. In awarding national licences, the Authority follows practices similar to those of the Commission in awarding TV programme provision licences. However, it is recognised that local broadcasting is not always a profit-making business; indeed it may not be commercially viable. Often local radio is to be seen as a kind of public service broadcasting for local minority interests. Thus, while care is taken to ensure that licences for local sound broadcasting are granted only to fit persons, the views of people living within the service area are taken into account in assessing proposals and the charge for the licence is based on a tariff, not an auction. Sound broadcasting licences have a maximum duration of eight years.

The Broadcasting Act of 1996 authorised the Radio Authority to license service providers to assemble digital sound programmes supplied by broadcasting service providers into multiplexes for transmission.

Commercial broadcasting regulation in the United Kingdom: (c) Enforcement of licence conditions

The Independent Television Commission and the Radio Authority state objectives for the quality, scope, coverage, acceptability etc. of broadcasting programmes and grant licences against credible undertakings that the objectives will be achieved. There is provision for surveillance to ensure that licensees fulfil their commitments.

The Act of 1990 maintained the already-existing Broadcasting Complaints Commission to investigate reports from the public of offensive handling of personal issues, infringements of privacy etc. The Act also set up the Broadcasting Standards Council to establish codes of practice in the treatment of such matters as violence, sex, taste and decency. The Council was to monitor programmes, assess observance of these codes and consider complaints from the public. Both bodies referred complaints that they considered, on examination, to be matters of substance to the broadcasters responsible and, if not satisfied by the response, had the power to require their findings to be published by the offender. The Act of 1996 amalgamated these two bodies to form the Broadcasting Standards Commission.

The two Regulators also have a duty to monitor programmes, in order to verify that the terms of programme provision licences are implemented and they are, of course, made aware of the findings of the Broadcasting Standards Commission. If they are not satisfied with the programmes that are provided, the Regulators have various powers to enforce the terms of licences, ranging from a requirement that the licensee broadcasts a correction to revocation of the licence.

7.2 Sound broadcasting below 2 MHz

Frequency allocations

The band 148.5–283.5 kHz is allocated for broadcasting in Region 1. However, the high level of atmospheric noise makes this part of the spectrum unsatisfactory for broadcasting in tropical regions and footnotes to the international table of frequency allocations allocate some or all of this band for other services in many African countries. There is no LF broadcasting allocation in Regions 2 and 3.

The band 526.5–1606.5 kHz is allocated for MF broadcasting in Regions 1 and 3. In Region 2 the corresponding allocation is 525–1705 kHz; a constraint is applied to the power of Region 2 BS transmitters below 535 kHz to facilitate sharing with the ARNS; see RR S5.86.

At the time of writing all broadcasting below 2 MHz uses double sideband (DSB) amplitude modulation (AM), and Section 7.2 is based entirely on this present practice. However, the performance of AM systems is much criticised and it is now thought that better performance may be obtainable using digital modulation systems which have an occupied bandwidth not greater, and interference characteristics which are no worse, than DSB AM. Recommendations ITU-R BS.1348 [1] and BS.1349 [2] set objectives for the development of such systems. See also Report ITU-R BS.2004 [3].

Transmitting antennas and service areas

Most transmitting antennas used for LF broadcasting are vertical mast radiators, fed at the base, the height being small compared with the wavelength. The gain is greatest in a horizontal direction but it remains substantial up to high angles of elevation. The wave radiated is almost entirely vertically polarised. At higher frequencies, in the MF broadcasting band, vertical antennas up to a quarter of the wavelength in height are practicable and usual, providing rather higher gain in the horizontal plane and rather less gain at high angles of elevation. Some MF antennas have an effective height exceeding a quarter of the wavelength, further increasing the low angle gain and reducing the high angle gain; these are called anti-fading antennas.

Figure 7.1 sketches the distribution of field strength set up at ground level by a vertical transmitting antenna at LF. Ground wave propagation (see Appendix, Section A3 below) dominates in the vicinity of the transmitting station. The rate at which the ground wave field strength declines with increasing distance from the transmitting antenna increases with the frequency and is also affected by the nature of the ground, mainly its conductivity. There may also be a significant sky wave, reflected from the D or E region of the ionosphere (see Appendix Section A4.2), much weaker than the ground wave near the transmitter but rising, perhaps to equal it in strength, at greater distances. Farther from the transmitter the sky wave may become dominant before both waves become too weak for satisfactory reception in the presence of atmospheric noise and interference.

Thus, surrounding an LF transmitting station there will be the primary service area, served by the ground wave, relatively constant in amplitude, typically circular, with a radius of several hundreds of kilometres. Surrounding the primary service area there is an annular area where the field strengths of the ground wave and the sky wave are comparable. The two waves interfere in this zone, causing fading which degrades to the point of unacceptability the quality of the signal. Beyond this fading zone there may be a secondary service area where the ground wave is virtually absent and the sky wave is able to provide a quality of service that may be acceptable, though inferior to that of the primary service area. At a still greater distance from the transmitter the sky wave becomes too weak to overcome the background of atmospheric noise and low-level interference and becomes itself a source of interference to distant stations broadcasting at the same frequency and in the adjacent channels.

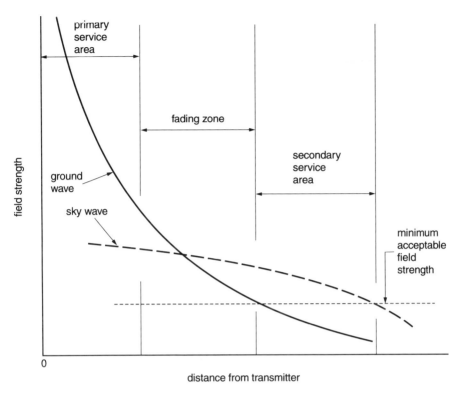

Figure 7.1   A sketch of the field strength from an LF transmitter with a short vertical antenna as a function of distance, showing the radius of the primary service area and the inner and outer radii of the secondary service area

There is no significant sky wave component by day in MF broadcasting and the ground wave provides a service area out to the distance at which the declining field strength is insufficient to overcome noise and low-level interference. See Figure 7.2. However, a sky wave becomes established at night, setting up concentric primary and secondary service areas, separated by a fading zone, as at LF. By using an anti-fading transmitting antenna the sky wave field strength close to the transmitter is made weaker, extending the radius of the primary service area, as may be seen from the figure. An anti-fading antenna may also extend the secondary service area and marginally increase the emission's potential for causing distant interference.

With a single vertical radiating element the radiation pattern in the horizontal plane is omnidirectional. However, radiation patterns corresponding more closely to a wanted geographical power distribution, improving the coverage of the service area or reducing interference to another station, can be obtained by using two or more radiating elements, appropriately spaced and with suitably phased feeds.

If horizontal radiators form the antenna at MF, the signal radiated will contain a large horizontally polarised component but only a small vertically polarised component. The horizontally polarised component in the ground wave is strongly attenuated,

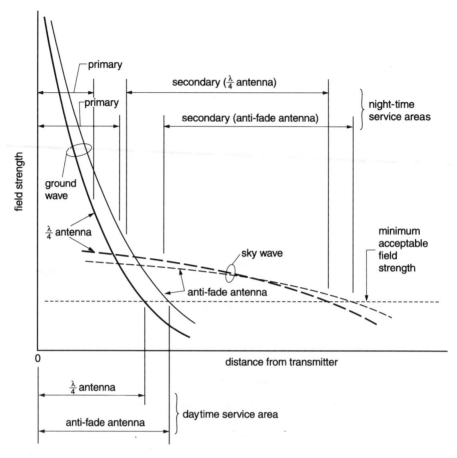

Figure 7.2   A sketch of the field strength from MF transmitters with vertical antennas as a function of distance, showing the radius of the daytime service areas and the limits of the primary and secondary service areas by night. One antenna has a height of a quarter of the wavelength (λ), the other has anti-fading characteristics

giving a small service area by day. At night, however, the sky wave due to high-angle radiation of the horizontally polarised component is predominant. This produces an area of severe fading close to the transmitter due to interference between the sky wave and the ground wave, surrounded by a large secondary service area served by the sky wave, which may be what is required in some circumstances.

There is a discussion of many aspects of transmitting antenna design that bear on efficient use of spectrum in Recommendation ITU-R BS.598-1 [4].

Broadcasting with multiple transmitters

Many LF and MF transmitters are powerful, delivering hundreds of kilowatts to their antennas and serving large areas. Other configurations, using several or many less powerful transmitters distributed over a region and using a common assigned frequency, are found preferable in some circumstances. For example, with carrier frequencies made very precisely equal and preferably phase-locked, with common programmes and synchronisation of the modulating signals, it becomes possible to overlap the ground wave service areas of several transmitting stations to serve a single extensive coverage area with little fading due to wave interference.

Alternatively, each of a number of towns within a province may be served by its own low-power transmitter. The ground wave field strength will be low between the towns and there will be fading at night in such locations due to wave interference between the carriers of the various ground waves and sky waves but the ground wave will be strong enough inside the towns to prevent significant fading. By day it may be feasible to modulate the transmitters with different programme material without significant loss of coverage but at night the service areas may be significantly reduced in size if a common programme is not used.

Frequency assignment plans

The amplitude modulated emissions currently used in broadcasting are susceptible to interference and the radio propagation conditions operative in the lower frequency bands support the transmission of interference over great distances, particularly at night. Consequently, large geographical separations are necessary between stations assigned the same or adjacent frequencies. Thus it is necessary to plan with care, nationally and internationally, the distribution of frequency assignments in order to control interference and to maximise the number of programmes that can be transmitted in finite frequency bands. Fortunately, the high degree of uniformity of system parameters makes planning relatively straightforward.

A plan for assignments in the LF band (Region 1) and the MF band (Regions 1 and 3) was agreed at an RARC in Geneva in 1975. A MF plan for Region 2 covering the band 525–1605 kHz was agreed at an RARC in Rio de Janeiro, in 1981. A further RARC was held in Rio de Janeiro in 1988 and it agreed a plan covering 1605–1705 kHz, an extension of the broadcasting allocation in Region 2 that had been agreed at WARC-79. These plans, incorporating the various configurations outlined above, are to be found in the Final Acts of the three conferences [5–7]. The plans will remain in force, modified as may be agreed from time to time, until they are replaced by other agreements produced by competent conferences. Some of the technical and regulatory aspects of these plans are reviewed below.

Necessary bandwidth and channel bandwidth

The frequency separation required between the carriers of adjacent DSB AM emissions carrying speech, music etc., depends on such factors as:

- the audio frequency bandwidth of the modulating signal,
- non-linear distortion in the transmitter,
- the intermediate frequency passband of representative receivers,
- the peak demand for assignments over areas continental in extent, and
- subjective appraisal of received signals.

See Recommendation ITU-R BS.639 [8]. The frequency separation adopted in planning for Regions 1 and 3 was 9 kHz, providing 15 LF radio channels for Region 1 and

120 MF channels for both Regions. For Region 2, a separation of 10 kHz was used, the whole allocated band providing 118 radio channels.

Planning criteria

The planning process is an iterative one, the objective being a distribution of frequencies for assignment to specified stations which would provide an acceptable quality of service throughout the area that each station was required to cover. The criteria which defined acceptability were:

- a field strength from the wanted transmitter sufficient to provide an adequate carrier-to-noise ratio in the absence of interference, this is called the minimum usable field strength;
- the achievement of a ratio between the powers of the wanted and interfering carriers, measured at the input to a receiver, that is sufficient to ensure an acceptable quality of reception, subjectively defined. This ratio is called the radio frequency protection ratio.

These terms are defined formally in Recommendation ITU-R BS.638 [9]. It is considered that a radio frequency protection ratio of 40 dB for co-channel interference provides an excellent quality of reception. It is, however, recognised that it may not be feasible to produce a plan of assignments for all the stations that Administrations propose to licence which achieves these criteria throughout the whole of the desired coverage areas.

There is a review of these criteria in Recommendation ITU-R BS.560-4 [10]. Values of the minimum usable field strength used in drawing up the current plans range from 100 µV/m to 10 mV/m depending on the level of atmospheric noise. The co-channel protection ratios adopted ranged between 30 and 26 dB.

In the 1975 plan for LF broadcasting, assignments are planned for an average of about 12 stations in each of the 15 channels. The 1975 MF plan for Regions 1 and 3 set aside three channels for low power stations and assignments were planned for about 2500 stations in these. The remaining 117 MF channels are planned for stations of medium or high power, and on average about 60 assignments were planned for each channel. The 1981 MF plan for Region 2 provided on average about 90 assignments for each of the 107 channels.

Realignment of LF assigned frequencies

Part of the interference perceived in LF reception arises from non-linear amplification in early stages of the receiver of combinations of the wanted and interfering carriers. Such interference is considerably less if the carrier frequencies are multiples of the carrier separation. In 1975, when the Region 1 LF plan was drawn up, the allocation was 150–285 kHz and so this desirable relationship would have involved a waste of spectrum; instead, increased liability to intermodulation products was accepted. However, WARC-79 changed the limits of the allocated band to its present values and, by stages between 1986 and 1990, the carrier frequencies assigned to broadcasting stations in the plan were reduced by 2 kHz, becoming multiples of 9 kHz.

7.3 Tropical broadcasting, 2–5 MHz

The atmospheric noise level, too high for LF broadcasting in tropical regions, also causes service areas to be restricted in the MF band. There is less noise above 2 MHz, but most of the broadcasting bands in the HF range are unsatisfactory for domestic broadcasting for three reasons:

- the primary service area provided by the ground wave is small,
- energy radiated from the transmitting antenna at a high angle of elevation penetrates the ionosphere at night and consequently there is no equivalent of the MF secondary service area, and
- these bands are heavily used for long distance international broadcasting.

To provide for broadcasting in tropical regions in a frequency range where the sky wave provides service at night and the atmospheric noise is not too severe, four frequency bands have been allocated for the BS, for use for broadcasting within a defined Tropical Zone to a service area within national borders; see RR S5.113 and S23.4–S23.10. The bands are as follows:

>2300–2495 kHz (extended to 2498 kHz in Region 1)
>3200–3400 kHz
>4750–4995 kHz
>5005–5060 kHz.

The Tropical Zone is defined formally in RR S5.16–S5.21. In Regions 1 and 3 the Zone is bounded to the south by latitude 35°S and to the North by latitude 30°N, being extended northwards to latitude 40°N between longitudes 40°E and 80°E and including the whole of Libya. In Region 2 the Zone is bounded to the south by the Tropic of Capricorn; to the north it is bounded by the Tropic of Cancer, although it may be extended to latitude 33°N, subject to the agreement of the countries concerned.

These bands are shared with the FS world-wide and in some of the bands with the MS also, all allocations being primary. Within the Tropical Zone (except for the parts of Libya north of latitude 30°N; see RR S23.9) the BS has priority over the other services for which the bands are allocated (see RR S23.8), although the benefits that flow from this priority are not defined with precision in the RR. See also Recommendation ITU-R BS.48-2 [11], which makes the point that, damaging as interference from FS stations may be to broadcasting, interference from MS stations is likely to be even more damaging; the recommendation asks Administrations outside the Tropical Zone to use these bands for the FS in preference to the MS.

The power of broadcasting transmitters in the tropical broadcasting bands is limited to 50 kW by RR S23.7. Recommendation ITU-R BS.215-2 [12] recommends that the power should be not more than 10 kW where the maximum distance to be covered does not exceed 400 km. There is advice on the design of transmitting antennas in Recommendation ITU-R BS.139-3 [13]. At the time of writing, broadcasting in the tropical broadcasting bands uses DSB AM. However, the development of digital modulation systems having an occupied bandwidth not greater than 10 kHz for use in the LF and MF broadcasting bands, referred to in Section 7.2 above might be equally relevant to tropical broadcasting.

7.4 HF sound broadcasting, 5.9–26.1 MHz

The frequency bands allocated for the BS between 5.9 and 26.1 MHz (see Table 7.1 below) are used for two purposes. Firstly they are used for domestic broadcasting, typically by day, in countries where the atmospheric noise level is so high that the MF band cannot be used economically; used in this way, the HF allocations complement the tropical broadcasting allocations which are used mainly at night. Secondly the HF bands are used for international broadcasting.

With negligible exceptions DSB AM is used at present for both of these applications. As mentioned in Section 7.2 above, the feasibility of designing a digital modulation system to replace this analogue system, giving better audio quality, interference characteristics that are not worse and requiring no more bandwidth, is under study. Such a system might be used in the HF bands also and this too is under study. However, the main thrust of the development of modulation methods for HF broadcasting at present is towards the minimisation of the occupied bandwidth per channel, and the use of single sideband (SSB) AM is more likely to meet this requirement; see below.

Domestic broadcasting at short distances at HF uses ground wave propagation, which is relatively stable. However, at long range both domestic and international broadcasting are dependent on sky wave propagation and the variability of ionospheric propagation that complicates the operation of long distance HF FS systems (see Section 6.1.2 above and Appendix Section A4.3 below) affects HF broadcasting also. Unfortunately, a fairly effective solution to this problem that is coming into use in the FS, namely automatic frequency adaptive systems, is unlikely to be feasible in broadcasting.

Frequency assignments for HF broadcasting

Under the radio regulations agreed at the International Radio Conference at Atlantic City in 1947, frequency assignments for HF broadcasting were to be treated like any other frequency assignment to a transmitting station at a fixed location. Frequency assignment plans were to be drawn up meeting all foreseen requirements; as assignments were taken into use they would be notified to the ITU and, subject to checks, they would be entered in the MIFR; see Section 5.4 above. A conference met in Mexico City in 1948 to draw up the frequency assignment plan but it could find no way of fitting all the assignments that would be necessary for all the broadcasting that was said to be required into the available allocations. Lacking an agreed plan, Administrations notified their assignments, which were registered in the MIFR and gained status thereby.

Many developing countries saw as inequitable the advantage that this practice gave to developed countries that had had earlier opportunities to register assignments, and the demand for the opportunity to broadcast at HF has continued to grow. Consequently, three themes have dominated the regulation of HF broadcasting since 1947, namely:

- the allocation of additional bandwidth for HF broadcasting;
- the establishment of means, acceptable in a service which is particularly sensitive to cost at the receiving station, that will enable more broadcasting to be done with the available spectrum;

- a search for an international spectrum management procedure that would provide an equitable distribution between the interested nations of the available allocated spectrum and the elimination of the benefit of priority derived from the registration of assignments in the MIFR.

Additional frequency allocations for broadcasting

The eight frequency bands allocated for HF broadcasting in 1947 are listed in the second column of Table 7.1. The total bandwidth was 2350 kHz in Regions 1 and 3 and 2150 kHz in Region 2. All was allocated exclusively for the BS except for part of the 7 MHz band, which was shared with the AmS until 1959. These bands are commonly identified by the order of magnitude of the wavelength, given in the first column of the table.

By the time of WARC-79 it was evident that the availability of wide-band, repeatered submarine cables and of satellite systems was reducing the pressure on the HF allocations for the FS. WARC-79 arranged for FS assignments that were still in use to be changed to enable seven frequency bands to be transferred to the BS, the bands above 10 MHz becoming available for the BS by 1989 and those below 10 MHz by 1994. These bands, with a total bandwidth of 850 kHz, are allocated almost exclusively for the BS; they are listed in the third column of Table 7.1. Exceptionally, three of these bands remain available for low-power FS links wholly within one country, provided that harmful interference is not caused to broadcasting; see RR S5.147.

WARC-92 agreed to a second rearrangement of FS assignments, and also some MS assignments, to provide for additional new BS allocations with effect from 1 April 2007; see RR Resolution 21. The bands to be transferred are listed in the fourth

Table 7.1 Frequency allocations for HF broadcasting

Waveband	Allocated pre-1979 (kHz)	Additions, available for BS by 1989 or 1994 (kHz)	Planned additions, available for BS by 2007 or sooner* (kHz)
49 m	5950–6200		5900–5950
41 m	7100–7300**		7300–7350
31 m	9500–9775	9775–9900	9400–9500
25 m	11700–11975	11650–11700	11600–11650
		11975–12050	12050–12100
22 m		13600–13800	13570–13600
			13800–13870
19 m	15100–15450	15450–15600	15600–15800
17 m	17700–17900	17550–17700	17480–17550
16 m			18900–19020
13 m	21450–21750	21750–21850	
11 m	25600–26100		

*To be reserved for SSB emissions.
**Regions 1 and 3 only.

column of Table 7.1; their bandwidth totals 790 kHz. These bands are to be allocated virtually exclusively for the BS, but with the same concession as before for domestic links of the FS and the MS; see RR S5.136, S5.143, S5.146 and S5.151. The use of these new BS allocations is limited to SSB emissions or another spectrum-efficient modulation technique that may be recommended by the ITU-R; see RR S5.134. RR Resolution 29 calls for information on the extent of the use of these bands by systems of the FS and the MS to be gathered, with a view to their possible transfer to broadcasting use, wholly or in part, before 2007.

More effective frequency band utilisation

Almost all HF broadcasting at present uses DSB AM but it is expected that the use of SSB will grow and it is intended that DSB will be completely superseded by SSB, or perhaps some other bandwidth-efficient modulation method, by 2015; see below. RR S23.12 requires DSB and SSB systems to comply with the system specifications in RR Appendix S11; see also Recommendation ITU-R BS.640-3 [14]. Among the provisions of these specifications is the requirement that the necessary bandwidths of DSB and SSB emissions do not exceed 9 and 4.5 kHz respectively. Recommendation ITU-R BS.597-1 [15] recommends that HF DSB carriers should be spaced by 10 kHz world-wide, on nominal frequencies that are multiples of 10 kHz, and that HF SSB carriers should be spaced by 5 kHz on multiples of 5 kHz.

The frequency tolerance for broadcasting transmitters at HF, DSB and SSB, is 10 Hz. However, footnote 21 to the table of tolerances in RR Appendix S2 suggests that frequency errors of a few hertz should be avoided when using DSB emissions; frequency errors of this order produce an effect on received signals when interference is present similar to fading, even when the interference is well below the level that causes the unwanted modulating signal to be audible. A carrier frequency error approaching 10 Hz is less objectionable but an error of the order of 0.1 Hz is even better.

Recommendation ITU-R BS.80-3 [16] stresses the importance of using directional transmitting antennas wherever the relative disposition of transmitting site and service area permits. This recommendation provides guidance on the radiation characteristics that are desirable and on what is achievable with typical antenna systems. There is also a very extensive study of HF antenna design principles in Recommendation ITU-R BS.705-1 [17].

Some of the unreliability of broadcasting based on predictions of OWF could be offset by transmitting simultaneously on more than one frequency. This practice is deprecated in general as unacceptably costly of spectrum, although Recommendation ITU-R BS.702-1 [18] accepts that it may be justifiable where radio propagation is especially difficult. However, this same recommendation and also RR Recommendation 503 give firm support to the use of a common frequency, in appropriate circumstances, for transmitting a common programme simultaneously to two different service areas.

Finally, RR Recommendations 503 and 518 urge Administrations to draw the attention of receiver manufacturers to the need to ensure that the receivers that they make will function well in present-day conditions. In particular, low-cost receivers should be equipped to cover all HF broadcasting bands and should, if possible, provide digital display of the frequency to which they are tuned. Also the selectivity characteristics of receivers should take Recommendation ITU-R SM.332-4 [19] into account.

The introduction of SSB

The most important visible prospect of increasing the efficiency with which the BS uses HF spectrum is the replacement of DSB modulation with SSB. When fully implemented this could more than double the number of channels that would be available and improve the audio quality obtained. Discussion of the replacement of DSB with SSB in HF broadcasting has been pursued vigorously since 1987. RR Resolution 517 sets out a plan of campaign including a target for all DSB emissions to cease by end-2015. Considerable progress has been made already. SSB transmissions can be accommodated in the 10 kHz channel plan that is used, primarily for DSB. As indicated above, the additional bands that will become available not later than 2007, listed in the third column of Table 7.1, are reserved entirely for SSB emissions.

The SSB system specification in RR Appendix S11 and Recommendation ITU-R BS.640-3 [14] uses initially a reduced level carrier, 6 dB below the peak envelope power of the emission. Such emissions will be receivable with DSB receivers using envelope detection, typical of present-day low-cost receivers, without significant reduction of quality. After DSB transmissions have ceased, the SSB carrier level is to be reduced to 12 dB below the peak envelope power, but satisfactory reception of these emissions, and the realisation of the full benefits that will eventually flow from reduced necessary bandwidth, may require DSB receivers using envelope demodulators to be replaced by SSB receivers using a synchronous demodulator. However, many transmitters now in use are not suitable for operation in the SSB mode. Thus a lack of SSB transmitters and receivers may cause the target date of 2015 to be missed by some broadcasters.

A Recommendation [20] of WARC HFBC-87 urged Administrations to ensure that new transmitters installed post-1990 were capable of working in the SSB mode. The same recommendation invited Administrations to encourage manufacturers to produce low-cost SSB receivers. RR Recommendation 515 (as revised by WARC-97) renews this pressure on manufacturers via Administrations. RR Resolution 537 calls for a survey of SSB transmitter and receiver availability to be made and the results submitted to WRC-00.

Equitable international spectrum management

Following on from the failure of the Mexico City conference in 1948, band planning was discussed again at WARC-59 with no greater success. Accordingly WARC-59 changed the way in which ITU was to deal with notices of frequency assignments for HF broadcasting. Thenceforth assignments would not be registered in the MIFR and the assignments that had already been registered were deleted, the then-established procedure being replaced by spectrum management measures intended to harmonise national plans for the use of frequencies in the near term.

With the prospect in view of increases in 1989 and 1992 of the bandwidth allocated for the BS, a two-session WARC was arranged to try once more to plan assignments for the service. These sessions took place in 1984 and 1987 and the IFRB was directed by the second session to continue the development of means for planning, reporting back to a future WARC. However, while WARC-92 provided for a further, albeit smaller, transfer of allocated bandwidth from the FS to the BS, to take effect by 2007, the conclusion was reached once again that frequency assignment planning is not feasible for this service.

Meanwhile, the procedure adopted by WARC-59 has been in use, with minor modifications from time to time. The latest revision, adopted at WRC-97, is to be found at RR Article S12. The basis of the revised procedure is briefly as follows:

a)   Years are divided into two seasons, approximately April to October and November to March. Administrations will inform the Radiocommunication Bureau of the ITU of the assignments they propose to make for broadcasting during a season, at a time some months before the start of that season. Alternative frequencies may be proposed. It is recommended that Administrations coordinate their proposals with other Administrations before submitting them and Administrations are encouraged to set up regional coordination groups to facilitate this process.

b)   The Bureau combines all the proposed assignments into a single schedule after analysing the probabilities of interference and selecting the best option where Administrations have offered alternatives. This Tentative Schedule is published not less than two months before the start of a season.

c)   Administrations study the Tentative Schedule. If prospects of interference are perceived, the Administrations, or their broadcasters acting as their delegates (see RR Recommendation 522) meet bilaterally or multilaterally to find ways of avoiding the interference. The Bureau is informed of any changes in the Administrations' intentions as set out in the Tentative Schedule at least two weeks before the start of the season.

d)   The Bureau prepares a new consolidated high frequency broadcasting schedule and publishes it, together with comments on the compatibility of the proposals as revised.

e)   In the light of the Bureau's compatibility analysis and of reports that the broadcasters may receive from listeners, there may be changes of assignments during the course of the season. These changes are notified promptly to the Bureau, which publishes a revised Schedule and a revised compatibility analysis monthly.

f)   One month after the end of each season the Bureau publishes a Final Schedule.

This procedure involves the handling of a great amount of data within a stringent timetable, and RR Resolution 535 addresses the organisation of this process.

7.5 Sound and television broadcasting above 30 MHz

7.5.1 *The frequency allocations*

Broadcasting allocations and sharing between 30 MHz and 1 GHz

The European VHF/UHF Broadcasting Conference at Stockholm in 1961 designated the five frequency bands that it was planning as Bands I to V. The frequency limits are shown in the second column of Table 7.2. These frequency limits differed in detail from the BS allocations then in effect in Regions 2 and 3 and they no longer represent exactly the allocations for broadcasting in any Region. Nevertheless, these band designations are still found to be a convenient way of identifying in broad terms the broadcasting bands in this part of the spectrum and they are widely used. The present band limits for the three Regions are also shown in the table.

Table 7.2   The VHF/UHF bands allocated for broadcasting

Band	Frequency (MHz)			
	Stockholm Conference 1961	Region 1 1998	Region 2 1998	Region 3 1998
I	41–68	47–68	54–72	47–50 and 54–68
II	87.5–100	87.5–108	76–108	87–108
III	162–230	174–230	174–216	174–230
IV	470–582	470–582	470–582	470–582
V	582–960	582–960	582–608 and 614–890	582–960

With the systems for TV and sound radio which are now in general use above 30 MHz, quite large amounts of bandwidth are required to broadcast a single programme with continuous coverage to an area indefinite in extent. The requirement is, perhaps, 80 MHz per programme for TV and about 2 MHz per programme for sound radio. This is because these analogue systems are very susceptible to interference, even when the interfering emission is modulated with the same signal as the wanted emission. Furthermore, in recent years the number of programmes that are broadcast in many countries has been increasing. The VHF and UHF allocations for the BS, which total about 60% of the whole spectrum below 960 MHz, include a large proportion of all the spectrum with the most satisfactory propagation characteristics for short distance services. This is probably more than any country can afford to devote entirely to broadcasting and much more than many countries need for that purpose. Consequently there is extensive sharing of the five Bands with other services.

For example, in Region 1 the upper part of Band V, from 790 to 960 MHz, is shared with the FS, and also with the MS (except AMS) over much of that sub-band; these allocations are all primary. There is similar sharing in Region 2 between 806 and 890 MHz, and in Region 3 almost all of the five Bands are shared on equal terms with the FS and the MS. In addition to these allocations in the framed part of the international table of frequency allocations, a large number of footnotes to the table make allocations of substantial bandwidth for the FS and/or the MS in many countries. There are also footnote allocations for various other services, including the RAS at 606–614 MHz (see RR S5.304 and S5.306), the AMS at 849–851 MHz and 894–896 MHz (see RR S5.318) and various radiodetermination services.

In Region 1, broadcasting is protected from interference from these sharing services in various ways, typically by the provisions of frequency assignment planning agreements, by the terms of the footnotes that make the sharing allocations or by a requirement for frequency coordination as described in Section 5.3. Specific sharing criteria have been established in a few specific cases. However, in Regions 2 and 3 broadcasting does not in general enjoy an internationally protected status and problems of interference if they should arise would be resolved in the ways outlined in Section 4.5.

Nevertheless, mention should be made of some of the more significant allocations to other services, sharing with the BS at UHF, as follows:

1   WARC-92, WRC-95 and WRC-97 had made a number of new frequency allocations for systems of the MSS using non-GSO satellites but WRC-97 perceived a need for still more bandwidth to be allocated for this purpose, especially below 1 GHz. The report of the CPM to WRC-97 had referred to possibilities, tentatively identified in ITU-R, of sharing between such systems and the services for which the band 470–862 MHz is now allocated. Accordingly, RR Resolution 728 calls for further studies to be made to determine whether such sharing would, in fact, be feasible, the conclusion to be reported to a future WRC.

2   The band 585–610 MHz is allocated for the RNS in Region 3, sharing with the BS, the FS and the MS on equal terms. There are also footnote allocations for the RNS or the ARNS in various frequency bands within Bands IV and V in various countries in other Regions. Some of these footnote allocations are primary and for some there is a requirement to coordinate new RNS assignments by the mechanism of RR S9.21 (see also Section 5.3 above) or more specifically by the terms of a footnote such as RR S5.302. Recommendation ITU-R BT.565 [21] gives protection ratios to protect BS emissions from radionavigation transmitters.

3   RR S5.311 permits assignments to FM TV transmitters in the BSS in the band 620–790 MHz, sharing with the BS and other terrestrial services, subject to the coordination procedure of RR Resolution 33 and a sharing constraint on the satellite emissions which is set out in RR Recommendation 705.

4   RR S5.317, S5.319 and S5.320 allocate certain frequency bands above 806 MHz for the MSS within various geographical limits. There are provisions in these footnotes to ensure that broadcasting will not suffer harmful interference from these systems.

Broadcasting allocations and sharing above 1 GHz

There are four sets of frequency allocations for the BS above 1 GHz, as follows.

a)   1452–1492 MHz; exceptionally the allocation is at 2310–2360 MHz in the USA and both bands are allocated for the BS in India and Mexico; see RR S5.344 and S5.393. There is an additional allocation at 2535–2655 MHz in the countries listed in RR S5.418.

b)   Near 12 GHz there are allocations for the BS, at 11.7–12.5 GHz (Regions 1 and 3) and 12.2–12.7 GHz (Region 2).

c)   40.5–42.5 GHz is allocated world-wide for the BS.

d)   84–86 GHz is allocated world-wide for the BS.

With the exception of the band 12.2–12.5 GHz in Region 3, all of these bands are also allocated for the BSS and their use for terrestrial broadcasting is seen as complementary to broadcasting by satellite, typically to be used for filling gaps in satellite coverage. The allocations in sub-paragraph (a) are limited to digital audio broadcasting (DAB); see RR S5.345, S5.393 and S5.418. The plans for the use of the BSS allocations at 12 GHz (see Section 10.3 below) strongly constrain the use that is made of the terrestrial allocations in these bands.

All of these bands are also shared with primary allocations for other services. Note in particular two decisions of WRC-97.

i)   New allocations were made for downlinks of non-GSO systems of the BSS in the frequency bands given in (b) above. The terms on which these allocations may be used initially, and provisional limits on the PFD from these satellites, are given in RR Resolution 538; see also RR Resolution 131.

ii) A new allocation was made for FSS downlinks in the band 40.5–42.5 GHz at WRC-97 and the FS allocation, previously secondary, was made primary. However, WRC-97 was not able to decide how this band could best be shared between these various services and RR Resolution 129 calls for the necessary studies to be done.

7.5.2 Television broadcasting

Many advanced TV systems have been developed in recent years. Some give high definition and are intended for use in the programme production phase or for the international exchange of programme material. Some are designed to make use of the technical opportunities that satellite broadcasting provides, in particular greater bandwidth per channel, for providing signals of better picture quality, including higher definition, for service to the public. Some are prototypes of new systems, analogue or digital, that may be used for terrestrial broadcasting in the future. Information on a number of these systems is to be found in recommendations and reports of the ITU-R BT and BO Series.

But the systems in general current use for terrestrial broadcasting to the public are analogue colour systems that have evolved by gradual refinement and elaboration from monochrome systems that came into use almost half a century ago, although maintaining backward compatibility. The digital systems that are coming into use at the time of writing are bit-rate-reduced systems that combine acceptable picture quality with a requirement for substantially less bandwidth per channel than analogue emissions need. New systems, both analogue and digital, often include not only a television channel (or several digital TV channels multiplexed onto a single carrier) but they have in addition the means to transmit a substantial amount of digital data which may be used to provide information services to the general public, or to a more limited clientele via a conditional access system.

Analogue television broadcasting systems

Almost all terrestrial TV broadcasting at present uses Band I, III, IV or V, one of two sets of picture standards and one of three systems for conveying colour information. The picture standards are 625-lines/25-frames-per-second and 525-line/30-frames-per-second, with a picture aspect ratio of 4:3. In each case the frames are transmitted as two interlaced fields. Chrominance information is transmitted by the phase alternate line (PAL), séquentiel couleur avec mémoire (SECAM) or the National Television Standards Committee (NTSC) systems. The video carrier is amplitude modulated by an analogue luminance signal and by a sub-carrier which is itself amplitude modulated by the analogue chrominance signal. This video signal is filtered asymmetrically, eliminating much of one sideband (almost always the lower sideband) and reducing the carrier level, so producing a vestigial sideband (VSB) emission. See Figure 7.3.

One or more carriers are transmitted offset in frequency from the VSB video emission, modulated by the sound signal(s), usually by FM but sometimes by AM or by a digital modulation system. Depending mainly on the picture standards, the necessary bandwidth for these various TV emissions ranges from less than 6 MHz to 8 MHz.

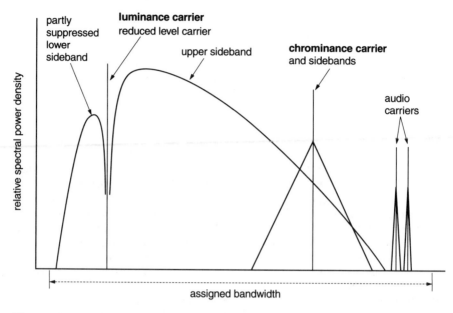

Figure 7.3   A sketch of the spectrum of a VSB colour television emission with two sound carriers

There are, however, many differences of technical detail between the systems that are used for broadcasting in the various countries; Recommendation ITU-R BT.470-5 [22] gives extensive data on these differences.

In addition there is terrestrial TV broadcasting in Japan at 12 GHz and experimental broadcasting has been carried out in that frequency band in Germany, The Netherlands and Switzerland. Analogue VSB systems such as those in Bands I to V and also analogue FM systems have been used. There is a brief report on these transmissions in Report ITU-R BT.961-2 [23].

Frequency assignment planning for TV in Bands I to V

Reception of analogue TV emissions is highly susceptible to interference from other emissions, whether from broadcasting transmitters or from the transmitters of sharing services. The assignment of frequencies to stations must be planned carefully if interference is to be avoided and spectrum use is to be efficient. In Band I, where the sky wave can, on occasion, propagate significant interference over distances of thousands of kilometres, international planning has been found desirable world-wide. However, in Bands III, IV and V propagation by the tropospheric wave limits significant interference to some hundreds of kilometres. Intensive use of broadcasting in Europe has made planning at an international level necessary, and the process has been extended to cover Africa. However, frequency assignment planning at UHF in Regions 2 and 3 is primarily a national process, with bilateral coordination between neighbouring countries to safeguard stations in border zones.

International frequency assignment plans for Region 1

The conference held in Stockholm in 1961 produced an agreement [24] which included a frequency assignment plan for TV (and also sound radio) broadcasting in Bands I to V in the European Broadcasting Area. This area is defined formally in RR S5.14; it comprises the area within Region 1, north of latitude 30°N and west of longitude 40°E, together with the parts of Syria, Iraq, Turkey and Ukraine which are outside those limits. This agreement superseded a plan produced by a conference held in 1952, also in Stockholm.

The Final Acts [25] of another conference, held in Geneva in 1963, contain an agreement and a frequency assignment plan for the African Broadcasting Area. This area is defined formally in RR S5.10–S5.13; it can be defined approximately as the part of Region 1 that is south of latitude 30°N. The TV assignments in the European plan are still in effect, amended from time to time as has been necessary, but the African plan for TV was revised at RARC-89 [26]. The technical parameters used in these planning processes are basically as set out below.

In an idealised form, that is for an indefinitely large geographical area, uniformly populated, the planning process would start with a hexagonal lattice of coverage areas, each served by a transmitting station, standard in its characteristics; see Section 4.1.3 above. Recommendation ITU-R BT.1123 [27] develops the lattice concept in some detail in the particular context of assignment planning for TV. An ideal distribution of assignments for a single programme can be drawn up in this way, using a repeating pattern of perhaps 12 channels; see Figure 4.6.

In a real planning situation, coverage areas do not form a regular lattice and transmitting stations are not uniform. Many stations are located where they will deliver a strong field strength to a centre of population. Where possible the stations are usually located on high ground, so serving a larger area without a more costly antenna mast and increasing the distance at which interference may be caused. The power of transmitters and the directional characteristics of their antennas are chosen to optimise the distribution of field strength over the accessible service area. When the scale of an ideal lattice has been adjusted to reflect the size of typical coverage areas, when the form of the lattice has been modified to represent the locations of the real transmitting stations and note has been taken of the power and directivity of these stations, the lattice will no longer be an ideal one. Nevertheless, a distribution of frequency assignments that would have been optimum for the ideal lattice is a good point of departure for the real planning process.

In a typical planning exercise, having made a plan of assignments to a real network of transmitting stations and coverage areas, based on an ideal lattice distribution with a repeating pattern of 12 channels broadcasting a single programme, it is necessary to ascertain whether the plan would allow the required protection ratios to be achieved. This stage requires the following.

a) The determination of the geographical separation between each transmitting station and the nearest point of the edge of the coverage area of any other station with which it might interfere. The edge of the coverage area will usually be where the median field strength from the wanted transmitter falls below the value quoted in Recommendation ITU-R BT.417-4 [28]. It may be sufficient to calculate the radial distance from the wanted transmitter at which this point is reached, using the available information on the transmitting antenna height and e.i.r.p. in the relevant direction; see Recommendation ITU-R BS.1195 [17]. It will often be preferable, however, to define the edge of the coverage areas by means of measured field strength contours where these are available.

b)   Propagation data to enable interfering field strengths to be calculated; see Recommendation ITU-R P.370-7 [29] and Appendix Section A2.2.2 below. The wanted-carrier-to-interference ratio that would arise at the input to a receiver at the edge of a coverage area can then be calculated, given assumptions about the receiving antenna characteristics; see, for example, Recommendation ITU-R BT.419-3 [30].

c)   Criteria to determine whether the calculated C/I ratios are acceptable. Recommendation ITU-R BT.655-4 [31] sets out recommended protection ratios for interference from other TV signals and from various other types of unwanted signal; see also Recommendation ITU-R BT.804 [32]. The requirements are complex. It is necessary to consider interference entering typical domestic receivers from transmitters operating in the same channel as the wanted transmitter (that is, co-channel interference), in a channel that overlaps the wanted channel, in the upper and lower adjacent channels and in the image channel.

When considering intermittent co-channel tropospheric interference from other TV signals, not necessarily identical in their emission characteristics, a basic protection ratio of 45 dB for a single interference entry is considered appropriate; a higher protection ratio is necessary against interference that does not fade. However, less stringent protection ratios may be used for TV carriers with small, precisely determined frequency differences. For co-channel interference from a TV emission with the same line frequency (625×25 Hz or 525×30 Hz), with carrier frequencies differing by a multiple of one-twelfth of the line frequency and with a carrier frequency tolerance of ± 500 Hz, the protection ratio may be reduced to about 28 dB; this operating mode is called 'non-precision offset'. By much more stringent control of carrier frequencies, a further reduction of the required protection ratio to around 24 dB is feasible; this mode is called 'precision offset'.

Depending on the results of the computations of interference levels, it may be concluded that the choice of a repeating pattern of 12 channels was basically sound, although minor adjustments to the scheme will often be necessary to eliminate localised peaks of interference. For example, it may be necessary to use carrier frequency offsetting or polarisation discrimination in places where this had not been assumed to be necessary in the first iteration. Where there are gaps in the coverage due, for example, to screening by hills, it may be necessary to add a cell to the lattice, assigned a 13th frequency. However, consistently low interference levels may be taken to imply that the use of fewer than 12 channels per programme might have made more efficient use of spectrum possible. Conversely, consistently high interference levels might imply that more than 12 channels are essential. After any necessary reiterations of the planning process, consideration could be given to planning assignments for additional programmes.

Digital TV broadcasting systems

Much effort has gone into the development of analogue TV broadcasting systems that would provide higher definition than current standard systems, the wide screen (aspect ratio = 16:9) and better reproduction of colour. The results have been systems that provide these improvements but only by increasing the necessary bandwidth per channel to a degree that is unattractive for satellite broadcasting and unacceptable for terrestrial broadcasting.

However, recent progress in the design of digital codecs for video signals which eliminate a large part of the redundancy involved in analogue transmission has made possible flexible digital systems with a reduced bandwidth per channel requirement. High picture definition is available with these systems but it is also feasible to trade-off bandwidth requirement for picture quality. Several TV channels can be multiplexed onto a common emission, enabling the trade-off to take the form of a choice between picture quality and the number of TV channels per multiplexed aggregate. Even more interesting is the use of statistical multiplexing, allowing the aggregate information transmission capacity of the emission to be switched flexibly to one programme in the multiplex or another, according to the instantaneous needs of each programme. A range of other facilities can be made available through the same multiplex system. And an important further advantage arising from digital modulation is reduced susceptibility to radio noise and interference.

Report ITU-R BT.2005 [33] is a selected bibliography of publications relevant to this revolution in TV broadcasting. Recommendation ITU-R BT.798-1 [34] lays down basic guidelines for the design of digital TV systems that are intended to be taken into use in Bands I to V; one of these guidelines is that systems should fit into the channels, 6, 7 or 8 MHz wide, that are used already for analogue systems in those bands. Recommendation ITU-R BT.1299 [35] summarises the basic elements recommended for a world-wide common family of digital TV systems, stressing in particular its statistical multiplexing capability and the use of the MPEG-2 transport stream. Recommendation ITU-R BT.1300 [36] identifies alternative transport systems, A and B, which the family includes, both covered by the MPEG-2 system and identifies the ATSC and ETSI standards which they incorporate. Recommendations ITU-R BT.1209-1 [37], BS.1196 [38] and BT.1301 [39] review the principles of other key sub-systems of this family of systems. Various other recommendations in the ITU-R BT Series discuss these sub-systems in greater detail.

Many countries are bringing digital TV systems into service, typically using vacant channels in a frequency band that is already in use for analogue broadcasting. Such channels may have been vacant because the interference level was too high to be used for analogue VSB TV, or because the use of an analogue transmission would have caused excessive interference elsewhere; for a low power, interference-resistant digital service the channel may be perfectly acceptable.

In time, when most of the interested public have equipped themselves to receive digital signals and due warning has been given, the Administration may withdraw the frequency assignments for analogue transmission. This will provide opportunities for expanding terrestrial TV services where required, frequency assignments having been replanned, nationally or internationally, and optimised according to the requirements of digital TV. It may, indeed, be feasible to re-allocate for other services spectrum no longer needed for TV.

Multipoint video distribution systems

Wide-band radio links are used in some countries to broadcast multiplexed analogue television programmes to the public; they are called multipoint video distribution systems (MVDS). Such systems compete with cable and satellite broadcasting systems distributing substantial numbers of programmes and having an economic advantage in some situations. Frequencies are assigned to these links in various microwave bands, typically in bands allocated internationally for the FS. The band 40.5–42.5 GHz has been recommended for MVDS in Europe but the shortness of the range that could be covered at such a high frequency, a few kilometres only in a temperate climate, seems likely to limit applications to relatively populous areas. See Report ITU-R BT.961-2 [23].

7.5.3 Sound broadcasting above 30 MHz

Analogue FM sound broadcasting at VHF

Broadcasting in Band II is almost invariably for analogue sound radio using FM. Various sets of emission parameters are in use; see Recommendation ITU-R BS.450-2 [40]. In a few countries the maximum frequency deviation is ± 50 kHz but elsewhere ± 75 kHz is used. Most systems provide stereophonic sound.

As with TV, Region 2 and most countries in Region 3 have not found a need for formal international frequency assignment planning in Band II. However, the Stockholm conference in 1961 [24] and the Geneva conference in 1963 [25] planned sound radio in Band II for the European and African Broadcasting Areas respectively. Those agreements were superseded by another [41], drawn up in Geneva in 1984, covering all of Region 1 and certain adjacent Region 3 countries.

National assignment planning for FM sound radio, whether it is carried out within the framework of a formal, Region-wide plan as in Region 1 or is national, supported as necessary by informal coordination between neighbouring Administrations, is similar in principle to planning for TV as described in Section 7.5.2. Basic planning data are to be found in Recommendation ITU-R BS.412-8 [42]. Reference may be made to Recommendation ITU-R BS.1195 [17] for guidance on transmitting antenna characteristics where specific data are not available. See Recommendation ITU-R BS.599 [43] for advice as to the directivity of receiving antennas.

Interference from broadcasting to aeronautical services at 108–137 MHz

Adjacent to broadcasting Band II there is an allocation for the ARNS at 108–117.975 MHz, used mainly for localiser signals of the instrument landing system (ILS) and VHF omnidirectional ranging (VOR) beacons. The frequency band 117.975–137 MHz is allocated for the AMS(R). Broadcasting emissions, especially those from transmitters located close to airports or the aircraft approach path to them, have been found to interfere with both ARNS and AMS services and are seen to be a danger to air navigation. Interference occurs by various mechanisms, including

- spurious emissions in the aeronautical bands from a broadcasting transmitter,
- intermodulation products generated between the carriers of several broadcasting transmitters, either within the transmitting systems or within the aircraft receivers, and
- desensitisation of aircraft receivers by powerful broadcasting carriers.

It is acknowledged that the problem arises, in part, from inadequate performance of some aeronautical receivers.

The problem has been under study since 1979; see RR Recommendation 704 [44] of WARC-79. Reports of the studies done during the 1980s are to be found in a CCIR publication [45]. RR Recommendation 714 [46] of WARC HFBC-87 calls for further study in ITU and ICAO, notes that ICAO had adopted new standards for aircraft receivers to improve their immunity to such interference and recommends that Administrations take action to protect aeronautical systems from interference of these kinds. Recommendation ITU-R IS.1009-1 [47] recommends technical bases for Administrations to use in this process, accepting that further refinement of the method is still required. Recommendation ITU-R IS.1140 [48] recommends test procedures that may be used to determine the characteristics of typical aircraft ARNS receivers for use at 108–118 MHz, a necessary part of that further refinement.

Digital sound broadcasting

FM sound broadcasting at VHF provides reception of much better audio quality for domestic receiving antennas at fixed locations than AM broadcasting at lower frequencies. However, the effective coverage area for stereophonic reception at fixed locations is considerably smaller than the area served with acceptable monophonic performance, and reception in moving vehicles is seriously degraded by multipath distortion.

RR Resolution 527, adopted at WARC-92, noted that digital systems that are now feasible for terrestrial sound broadcasting would give better audio quality with less powerful transmitters and more efficient spectrum use. The CCIR was called upon to study terrestrial digital audio broadcasting (T-DAB), having particular regard for its introduction in BS allocations in the VHF frequency range.

However, WARC-92 made new allocations for the BSS at 1452–1492 MHz. In some countries additional or alternative BSS allocations were made at 2310–2360 MHz or 2535–2655 MHz; see Section 7.5.1. The use of these bands is limited to satellite digital audio broadcasting (S-DAB). WARC-92 also made new allocations for the BS in these frequency bands, limited to T-DAB; see RR S5.345, S5.347, S5.393 and S5.418 and RR Resolution 528. These BS allocations were intended for complementing the service to be provided by the BSS. Work in ITU-R for S-DAB is focused on the use of these new allocations (see Section 10.4) but the T-DAB work covers both VHF and UHF bands. Report ITU-R BS.2002 [49] is a broad study of the issues arising in achieving a complementary introduction of digital audio broadcasting through the two media in the frequency bands between 1 and 3 GHz.

Recommendation ITU-R BS.774-2 [50] defines objectives for T-DAB systems. Particularly significant are:

- the frequency range 30–3000 MHz is covered, thus addressing the VHF bands in present use and also the new allocations intended for S-DAB;
- the importance given to reception by mobile receivers;
- a requirement for stereophonic reception, subjectively indistinguishable from reproduction by compact disc;
- better power and spectrum efficiency than analogue FM systems provide;
- flexibility and comprehensiveness, allowing trade-off, for example, between channel quality and number of channels and delivery of a wide range of broadcast services in addition to sound programmes;
- the necessity for the production of low-cost receivers to be feasible;

- the desirability of harmonising the specifications for terrestrial and satellite DAB.

Recommendation ITU-R BS.1114-1 [51] describes one T-DAB system that has been designed to meet these objectives. This is 'Digital System A', a product of the Eureka 147 project; see also ETSI Standard ETS 300 401. This system is also described as System A in Recommendation ITU-R BO.1130-1 [52], which relates to S-DAB; this recommendation also contains an incomplete description of another system identified as System B.

System A uses multiplex emissions in which multiple carriers are modulated by four-phase differential PSK in orthogonal frequency division multiplex (OFDM). Since the number of carriers per emission is very large, the symbol duration is relatively long, of the order of 1 ms, despite a high aggregate information rate. Consequently, the system is very insensitive to multipath distortion. Used for T-DAB, System A is also very insensitive to interference from synchronised emissions carrying the same programme signals but transmitted from distant stations. The possibility arises of a single frequency network, delivering the same programmes to an extensive area from many transmitting stations using the same frequency assignment. Typically each emission will provide an aggregate of six high quality stereophonic channels or the equivalent information rate used for other forms of signal.

System A is recommended for S-DAB in Recommendation ITU-R BO.1130-1 [52] for early implementation in the BSS (see Section 10.4 below) but it is recognised that other systems may also be recommended later. System A is recommended for T-DAB in Recommendation ITU-R BS.1114-1 [51]. Digital System A is being introduced into service in a number of countries in Europe and elsewhere. Various frequency bands will be used, depending upon the availability of spectrum in the various countries; some will use Band III, others the band 1452–1492 MHz and Bands I and II may also be used in some countries. In time some at least of the frequency assignments for FM sound broadcasting in Band II are likely to be withdrawn, allowing harmonisation of band usage to be considered.

7.6 References

1 'Service requirements for digital sound broadcasting to vehicular, portable and fixed receivers using terrestrial transmitters in the LF, MF and HF bands'. Recommendation ITU-R BS.1348, ITU-R Recommendations 1997 BS Series Supplement 1 (ITU, Geneva, 1997)
2 'Implementation of digital sound broadcasting to vehicular, portable and fixed receivers using terrestrial transmitters in the LF, MF and HF bands'. Recommendation ITU-R BS.1349, *ibid.*
3 'Digital broadcasting systems intended for AM bands'. Report ITU-R BS.2004. Published separately (ITU, Geneva, 1995)
4 'Factors influencing the limits of amplitude-modulation sound-broadcasting coverage in Band 6 (MF)'. Recommendation ITU-R BS.598-1, ITU-R Recommendations 1997 BS Series (ITU, Geneva, 1997)
5 'Final Acts of the Regional Administrative LF/MF Broadcasting Conference (Regions 1 and 3)', Geneva 1975 (ITU, Geneva, 1975)
6 'Final Acts of the Regional Administrative Broadcasting Conference (Region 2)', Rio de Janeiro 1981 (ITU, Geneva, 1981)

7 'Final Acts of the Regional Administrative Radio Conference to establish a plan for the broadcasting service in the band 1605–1705 kHz in Region 2', Rio de Janeiro 1988 (ITU, Geneva, 1988)

8 'Necessary bandwidth of emissions in LF, MF and HF broadcasting'. Recommendation ITU-R BS.639, ITU-R Recommendations 1997 BS Series (ITU, Geneva, 1997)

9 'Terms and definitions used in frequency planning for sound broadcasting'. Recommendation ITU-R BS.638, *ibid.*

10 'Radio-frequency protection ratios in LF, MF and HF broadcasting'. Recommendation ITU-R BS.560-4, *ibid.*

11 'Choice of frequency for sound broadcasting in the Tropical Zone'. Recommendation ITU-R BS.48-2, *ibid.*

12 'Maximum transmitter powers for broadcasting in the Tropical Zone'. Recommendation ITU-R BS.215-2, *ibid.*

13 'Transmitting antennas for sound broadcasting in the Tropical Zone'. Recommendation ITU-R BS.139-3, *ibid.*

14 'Single sideband (SSB) system for HF broadcasting'. Recommendation ITU-R BS.640-3, *ibid.*

15 'Channel spacing for sound broadcasting in Band 7 (HF)'. Recommendation ITU-R BS.597-1, *ibid.*

16 'Transmitting antennas in HF broadcasting'. Recommendation ITU-R BS.80-3, *ibid.*

17 'HF transmitting and receiving antennas, characteristics and diagrams'; Recommendation ITU-R BS.705-1. Also 'Transmitting antenna characteristics at VHF and UHF'; Recommendation ITU-R BS.1195. These two recommendations are published as a separate volume (ITU, Geneva, 1997)

18 'Synchronization and multiple frequency use per programme in HF broadcasting'. Recommendation ITU-R BS.702-1, ITU-R Recommendations 1997 BS Series (ITU, Geneva, 1997)

19 'Selectivity of receivers'. Recommendation ITU-R SM.332-4, ITU-R Recommendations 1997 SM Series (ITU, Geneva, 1997)

20 'Introduction of transmitters and receivers capable of both double sideband (DSB) and single sideband (SSB) modes of operation'. Recommendation 515 of WARC HFBC of 1987. See the Final Acts of that conference or Volume 3 of the RR, edition of 1990, revised 1994

21 'Protection ratios for 625-line television against radionavigation transmitters operating in the shared bands between 582 and 606 MHz'. Recommendation ITU-R BT.565, ITU-R Recommendations 1997 BT Series (ITU, Geneva, 1997)

22 'Television systems'. Recommendation ITU-R BT.470-5, ITU-R Recommendations 1997 BT Series Supplement 1 (ITU, Geneva, 1997)

23 'Terrestrial television broadcasting in bands above 2 GHz'. Report ITU-R BT.961-2, published separately (with Report BT.2003) (ITU, Geneva, 1995)

24 'Regional agreement for the European Broadcasting Area, Stockholm, 1961 (ITU, Geneva, 1961)

25 'Final Acts of the African VHF/UHF Broadcasting Conference', Geneva 1963 (ITU, Geneva, 1963)

26 'Final Acts of the Regional Administrative Radio Conference for the planning of VHF/UHF television-broadcasting in the African Broadcasting Area and neighbouring countries'. Geneva, 1989 (ITU, Geneva, 1989)

27 'Planning methods for 625-line terrestrial television in VHF/UHF bands'. Recommendation ITU-R BT.1123, ITU-R Recommendations 1997 BT Series (ITU, Geneva, 1997)

28 'Minimum field strengths for which protection may be sought in planning a television service'. Recommendation ITU-R BT.417-4, *ibid.*

29 'VHF and UHF propagation curves for the frequency range from 30 MHz to 1000 MHz. Broadcasting services'. Recommendation ITU-R P.370-7, ITU-R Recommendations 1997 P Series, Part 2 (ITU, Geneva, 1997)

30 'Directivity and polarization discrimination of antennas in the reception of television broadcasting'. Recommendation ITU-R BT.419-3, ITU-R Recommendations 1997 BT Series (ITU, Geneva, 1997)

31 'Radio-frequency protection ratios for AM vestigial sideband terrestrial television systems'. Recommendation ITU-R BT.655-4, *ibid.*
32 'Characteristics of TV receivers essential for frequency planning with PAL/ SECAM/NTSC television systems'. Recommendation ITU-R BT.804, *ibid.*
33 'Bit rate reduction for digital TV signals'. Report ITU-R BT.2005. Published separately (ITU, Geneva, 1995)
34 'Digital television terrestrial broadcasting in the VHF/UHF bands'. Recommendation ITU-R BT.798-1, ITU-R Recommendations 1997 BT Series (ITU, Geneva, 1997)
35 'The basic elements of a worldwide common family of systems for digital terrestrial television broadcasting'. Recommendation ITU-R BT.1299, *ibid.*
36 'Service multiplex, transport and identification methods for digital terrestrial television broadcasting'. Recommendation ITU-R BT.1300, *ibid.*
37 'Service multiplex methods for digital terrestrial television broadcasting'. Recommendation ITU-R BT.1209-1, *ibid.*
38 'Audio coding for digital terrestrial television broadcasting'. Recommendation ITU-R BS.1196, ITU-R Recommendations 1997 BS Series (ITU, Geneva, 1997)
39 'Data services in digital terrestrial television broadcasting'. Recommendation ITU-R BT.1301, ITU-R Recommendations 1997 BT Series (ITU, Geneva, 1997)
40 'Transmission standards for FM sound broadcasting at VHF'. Recommendation ITU-R BS.450-2, ITU-R Recommendations 1997 BS Series (ITU, Geneva, 1997)
41 'Final Acts of the Regional Administrative Radio Conference for the planning of VHF sound broadcasting (Region 1 and part of Region 3)', Geneva, 1984 (ITU, Geneva, 1984)
42 'Planning standards for FM sound broadcasting at VHF'. Recommendation ITU-R BS.412-8, ITU-R Recommendations 1997 BS Series Supplement 1 (ITU, Geneva, 1997)
43 'Directivity of antennas for the reception of sound broadcasting in Band 8 (VHF)'. Recommendation ITU-R BS.599, ITU-R Recommendations 1997 BS Series (ITU, Geneva, 1997)
44 'Relating to the compatibility between the broadcasting service in the band 100–108 MHz and the aeronautical radionavigation service in the band 108–117.975 MHz'. Recommendation 704. The Radio Regulations, edition of 1990, revised 1994 (ITU, Geneva)
45 'Compatibility between the broadcasting service in the band of about 87–108 MHz and the aeronautical services in the band 108–137 MHz'. (ITU, Geneva, 1991)
46 'Compatibility between the aeronautical mobile (R) service in the band 117.975–137 MHz and sound broadcasting stations in the band 87.5–108 MHz'. Recommendation 714. The Radio Regulations, edition of 1990, revised 1994 (ITU, Geneva)
47 'Compatibility between the sound-broadcasting service in the band of about 87–108 MHz and the aeronautical services in the band 108–137 MHz'. Recommendation ITU-R IS.1009-1, ITU-R Recommendations 1997 IS Series (ITU, Geneva, 1997)
48 'Test procedures for measuring aeronautical receiver characteristics used for determining compatibility between the sound-broadcasting service in the band of about 87–108 and the aeronautical services in the band 108–118 MHz'. Recommendation ITU-R IS.1140, *ibid.*
49 'Introduction of satellite and complementary terrestrial digital sound broadcasting in the WARC-92 frequency allocations'. Report ITU-R BS.2002. Published with ITU-R Reports BS.1203-1 and BS.2001 (ITU, Geneva, 1995)
50 'Service requirements for digital sound broadcasting to vehicular, portable and fixed receivers using terrestrial transmitters in the VHF/UHF bands'. Recommendation ITU-R BS.774-2, ITU-R Recommendations 1997 BS Series (ITU, Geneva, 1997)
51 'System for terrestrial digital sound broadcasting for vehicular, portable and fixed receivers in the frequency range 30–3000 MHz'. Recommendation ITU-R BS.1114-1, *ibid.*
52 'Systems for digital sound broadcasting to vehicular, portable and fixed receivers for broadcasting-satellite service (sound) bands in the frequency range 1400–2700 MHz'. Recommendation ITU-R BO.1130-1, ITU-R Recommendations 1997 BO Series (ITU, Geneva, 1997)

Chapter 8

The mobile services

8.1 Introduction

The mobile service (MS) includes the maritime mobile service (MMS), the aeronautical mobile service (AMS) and the land mobile service (LMS). Some frequency bands allocated for the AMS are reserved for air traffic control on major civil air routes and related purposes; these allocations are designated AMS(R), 'R' standing for 'Route' ; see RR paragraph S43.1. AMS allocations used for other purposes ('Off Route') are designated AMS(OR). RR S43.4 specifically forbids the use of exclusive AMS allocations for public correspondence, that is, for communication facilities for the use of passengers.

The spectrum allocated for these services is used for communication between mobile stations but more usually between a mobile station and a land station, the latter being the general term for a station in the mobile service that is operated at a fixed location. There are specific defined terms for the mobile and land stations of the MMS, the AMS and the LMS; see Table 8.1. A station that is movable may be considered to be a mobile station even if it cannot be operated whilst in motion. This may be a distinction of little practical importance when the mobile station is on a ship or an aircraft, but it leads to the existance to two quite different kinds of land mobile station, namely those that can be operated whilst in motion and those, perhaps better described as transportable stations, that can only be operated when at rest.

Table 8.1   Terms for stations in the mobile services

Service	Stations that operate at permanent locations	Stations that operate in motion or when halted
MS	Land station	Mobile station
MMS	Coast station	Ship station
AMS	Aeronautical station	Aircraft station
LMS	Base station	Land mobile station

Communication usually takes the form of a two-way telephone call or a telex, data or facsimile connection, carried by radio links that had been set up for the call and will be taken down as soon as the call has been completed. The radio channel then becomes available for another call, typically connecting another pair of stations. At present in the MMS and the AMS these links are often set up between operators acting for the end-users of the service but development is leading towards automatic call set-up and this stage has already been reached for almost all LMS systems above 30 MHz.

However, the LMS includes a great variety of applications which do not involve communication in the usually understood sense of the word. Such devices as short-range remote control, burglar and theft alarms and cordless telephones need very little transmitter power and use insensitive receivers. These devices are used in very large numbers but, given access to an adequate frequency band and constrained in respect of power, they neither cause nor suffer significant interference. Administrations usually authorise the use of frequency bands a few hundreds of kilohertz wide for such systems, mostly in the VHF or UHF range, there being no licensing formalities for individual devices. The bands chosen for this purpose may be standardised internationally, for example to facilitate international trade in the articles incorporating the devices, but such choices are not harmonised globally through the ITU.

8.2 The maritime mobile service below 30 MHz

The International Radiotelegraph Union was set up in 1906 to regulate the operation of radio services between ships and shore stations. When the ITU was founded in 1932 it took over the functions of the earlier body. The principal international body for the regulation of maritime operations nowadays is the IMO but the ITU still carries out important functions in maritime spectrum management, in collaboration with the IMO, above all for making regulatory provision to ensure the availability of clear channels for distress and safety communication. These provisions are outlined in this section and Section 8.7.

The ITU also publishes lists of coast stations, ships, radionavigation stations etc., giving their call signs, the frequencies they work on, the facilities they provide and so on; see RR Article S20. These publications are of operational value to shipping. Additionally, the RR and recommendations of ITU-R contain extensive regulatory material related to the operation of maritime radio services; as such it is outside the scope of this book but a brief outline is included where it may clarify some of the regulatory provisions.

There is some provision in the RR, although much less extensive, of regulations affecting the operation of aeronautical radio services. As with maritime services, another international organisation (ICAO in this case) has prime responsibility in this field. The MMS and AMS are the only services in which the ITU has a significant operational involvement.

MMS operating methods and traffic facilities

Hand-speed Morse telegraphy, operator-to-operator, transmitted by on–off keying of the radio carrier, was used in the early days of the MMS. Telephony followed, using AM DSB and extended into the PSTN. These basic communication techniques are still in general use, although Morse is often transmitted by keying a sub-carrier in the upper sideband of an SSB emission with suppressed carrier. SSB AM with suppressed carrier is always used for telephony below 30 MHz. Recommendation ITU-R M.1173 [1] gives basic performance characteristics for SSB transmitters used in the MMS. This recommendation is incorporated into the Radio Regulations by reference and its application is made mandatory by RR S52.181.

The use of Morse is tending to give way to narrow-band direct-printing (NBDP) telegraphy with ARQ on duplex circuits and FEC for unidirectional circuits (see Recommendations ITU-R M.625-3 [2], M.476-5 [3] and M.692 [4]) transmitted at data rates up to 100 bauds on FSK emissions with a frequency shift of 170 Hz. Narrow-band phase-shift keying (NBPSK) systems are also used for data rates up to 200 bauds; Recommendation ITU-R M.627-1 [5] outlines the radio frequency characteristics required of these systems. Recommendations ITU-R M.625-3, M.476-5 and M.627-1 are incorporated into the Radio Regulations, and their application is made mandatory, by reference in RR S51.41. There is provision in some frequency allocations for higher bit-rate data transmissions, facsimile, etc.; see Recommendation ITU-R M.1081 [6].

The traditional mode of operation in the MMS is as follows. In each frequency band a spot frequency is designated as the 'calling frequency'. If more than one type of transmission is used in the band, there may be a calling frequency for each type of transmission. These calling frequencies may also be used (see Section 8.7), for distress calls and the coordination of rescue operations. Ships or coast stations needing to communicate send a brief message on the calling frequency, identifying the sender and the station being called. The called station responds, also on the calling frequency. The two stations arrange to retune to working frequencies, they pass the traffic, then close down the links. When the initial call is made by a coast station, a stage is sometimes cut out of this procedure, the coast station designating the working frequencies to be used for passing traffic.

Operating in this way, coast stations publish schedules of times of day at which they will broadcast lists of the ships that they wish to contact and the frequencies that they will use. Ships' operators listen out for possible calls. Similarly, coast stations listen out for calls from ships. This procedure is labour-intensive and it is being superseded by the use of selective calling systems. The Sequential Single Frequency Selective-Calling System can be used by coast stations to call specific ships by a five-digit code number; see Recommendation ITU-R M.257-3 [7]. This recommendation is incorporated into the Radio Regulations by reference, for example, in RR S19.92 but it is an interim device, for use pending the general availability of digital selective-calling (DSC) systems which are usable coast-station-to-ship, ship-to-coast-station and ship-to-ship using a seven-digit code number. The basic principles of DSC systems are set out in Recommendation ITU-R M.493-9 [8]. Other recommendations in the ITU-R M Series are concerned with the elaboration of these principles for specific applications.

Akin to selective calling is automatic direct dialling from a ship into the PSTN and vice versa. Recommendation ITU-R M.1082-1 [9] contemplates end-to-end automation through DSC equipment. Recommendation ITU-R M.586-1 [10] describes a system optimised for operation with DSC equipment in the VHF MMS band; see Section 8.5.2 below. See also Recommendation ITU-R 1083 [11]. Report ITU-R 2009 [12] describes systems which provide an intermediate level of automation for use where DSC is not available; in general these systems are limited to permitting automatic ship–shore dialling once the radio link has been set up manually.

Major changes in the operation of maritime radio at MF and HF can be expected soon from the introduction of frequency adaptive systems; see RR Resolution 729.

Electromagnetic compatibility

Radio reception on board a ship is particularly susceptible to interference from electrical machinery, not least the radio transmitters that may be operating on the same ship. Recommendation ITU-R M.218-2 [13] lists measures that can be employed to minimise such interference.

The MMS frequency allocations below 526.5 kHz

A large part of the spectrum below about 150 kHz is allocated for the MMS, sharing with primary allocations for the FS, one of the radionavigation services, or both. Below 90 kHz the use of these MMS allocations is limited by RR S5.57 to coast stations transmitting telegraphy. With high transmitter power and efficient antennas, very long distances can be spanned in this frequency range. However, the use of these bands for the MMS is small and probably declining.

The band 415–526.5 kHz (415–525 kHz in Region 2) is allocated for the MMS, parts of the band being shared with the ARNS. Use of this band by the MMS is limited to radiotelegraphy, using Morse code or NBDP telegraphy. The calling frequency is 500 kHz. A number of frequencies within the band have been designated as working frequencies for ships' transmitters. Other frequencies are assigned to coast stations as working frequencies, the frequencies to be assigned being chosen and coordinated between neighbouring countries where necessary to minimise interference. A plan for MMS assignments in Region 1 covering the bands 415–495 kHz and 505–526.5 kHz was drawn up at an RARC in 1985; details of the plan are to be found in the Final Acts of the conference [14]. The long-established distress frequency for Morse was 500 kHz but its use is being phased out as the Global Maritime Distress and Safety System (GMDSS) is phased in; see Section 8.7.

The frequencies 490 and 518 kHz are assigned to coast stations for broadcasting maritime safety information (typically meteorological forecasts and navigational warnings) and the NAVTEX service respectively, using NBDP; see RR Resolution 339 and Recommendation ITU-R M.540-2 [15]. Such broadcasts are also planned for each of the MMS exclusive HF bands (see RR Appendix S15) and via INMARSAT satellites at about 1544 MHz.

The MMS allocations between 1605 and 4000 kHz

Its radio propagation characteristics make the frequency range between 1.6 and 4 MHz very suitable for many kinds of medium distance narrow-band service and the use that is made of it varies from Region to Region and from country to country. Most of this band is allocated for one or other of the mobile services but these allocations are also shared with other services. Only 2170–2194 kHz is allocated world-wide and virtually exclusively for the MMS. The calling frequency for maritime radiotelephony in this frequency range is 2182 kHz; it is also a long-established distress frequency for ships and aircraft using radiotelephony; see Section 8.7.

However, despite having to share spectrum in some sea areas with other services on land, the MMS makes substantial use of this frequency range for distances of up to a few hundred kilometres. There is a detailed international frequency band plan for maritime use in Region 1 for links using SSB for telephony and both Morse and NBDP telegraphy. For a plan for MMS use of the bands 1606.5–1625 kHz, 1635–1800 kHz and 2045–2160 in Region 1, see Reference [14]; see also RR S52.202–214. There is also band planning, though less, at a Regional level for Regions 2 and 3; see RR S52.11, S52.47–50, S52.165 and S52.215. The power of coast station transmitters used for telephony in these bands is limited to 5 kW peak envelope power north of latitude 32°N and 10 kW south of 32°N, except for the band 2170–2194 kHz where the limit is 400 W. The power limit throughout the band for ship transmitters used for telephony is also 400 W (see RR S52.184, S52.188 and S52.202).

The exclusive MMS allocations between 4 and 30 MHz

Radio links in the HF frequency range can span distances up to a few hundred kilometres by ground wave or many thousands of kilometres if the frequency is suitable for ionospheric reflection. Propagation via the ionosphere is unreliable and, basically for this reason, the MMSS is currently competing strongly for the long distance maritime traffic. Nevertheless, these terrestrial facilities are still important.

Ten frequency bands between 4 and 30 MHz are allocated exclusively world-wide to the MMS. Two pairs of these bands are rather narrow but they are closely adjacent in frequency and each pair is treated operationally as a single band. Thus, for operational purposes there are eight bands. As augmented by transfers of bandwidth from the FS agreed at WARC-79 and implemented by 1989, the MMS frequency limits are shown in Table 8.2. Each of the eight operational bands is divided into two parts, for ship transmitters and coast station transmitters respectively, and subdivided to segregate the various kinds of traffic. Most of these sub-bands are further divided, creating radio channel plans with channel bandwidths appropriate for the kind of traffic that the sub-band is to carry. Table 8.3 shows as an example the scheme for the 12 MHz band; the schemes for the other bands are similar in principle. Details of the subdivision and channelling of all of these bands are to be found in RR Appendix S17.

Table 8.3 shows that the 41 channels designated for telephony in the 12 MHz scheme, sub-bands (1) and (14), are paired. In these paired bands, channel 'n' shore-to-ship is

Table 8.2   The exclusive HF MMS allocations

Band name (MHz)	Band limits (kHz)
4	4063–4438
6	6200–6525
8	8195–8815
12	12 230–13 200
16	16 360–17 410
18/19	18 780–18 900 and 19 600–19 800
22	22 000–22 855
25/26	25 070–25 210 and 26 100–26 175

Table 8.3   The subdivision and channelling of the 12 MHz MMS band

	Sub-band (kHz)	Transmit from	Use	Channelling
(1)	12 230–12 353	Ship	Duplex telephony, paired with (14)	41 × 3 kHz
(2)	12 353–12 368	Ship & coast	Simplex telephony	5 × 3 kHz
(3)	12 368–12 420	Ship	Wide-band telegraphy, facsimile etc.	13 × 4 kHz
(4)	12 420–12 421.75	Ship	Oceanographic data transmission	5 × 0.3 kHz
(5)	12 421.75–12 476.75	Ship	Working frequencies for Morse telegraphy	110 × 0.5 kHz
(6)	12 476.75–12 549.75	Ship	NBDP and NBPSK telegraphy paired with (11)	146 × 0.5 kHz
(7)	12 549.75–12 554.75	Ship	Calling frequencies for Morse telegraphy	
(8)	12 554.75–12 559.75	Ship	As for (6)	10 × 0.5 kHz
(9)	12 559.75–12 576.75	Ship	NBDP, NBPSK and Morse telegraphy, unpaired	34 × 0.5 kHz
(10)	12 576.75–12 578.75	Ship	DSC calling	4 × 0.5 kHz
(11)	12 578.75–12 656.75	Coast	NBDP and NBPSK telegraphy paired with (6) and (8)	156 × 0.5 kHz
(12)	12 656.75–12 658.5	Coast	DSC calling	3 × 0.5 kHz
(13)	12 658.5–13 077	Coast	All unpaired services	
(14)	13 077–13 200	Coast	Duplex telephony, paired with (1)	41 × 3 kHz

always used in conjunction with channel 'n' ship-to-shore to form a duplex circuit. The 156 channel-pairs for NBDP or NBPSK telegraphy in sub-bands (6), (8) and (11), are treated similarly. RR Resolution 300 calls for Administrations to assign NBDP or NBPSK channels for transmission by their coast stations to assist in the orderly use of the paired channels. The five channels for simplex telephony, sub-band (2), are used sequentially in the two directions. In each of these cases, the selective calling system would typically be used to set up a call, using a working channel or paired channels designated by the coast station. Channels are allotted for DSC calling in both directions.

An international frequency allotment agreement has been drawn up for the coast station channels used for telephony, that is, for example, sub-band (14) in Table 8.3. The allotment plan, to be found in Section II of RR Appendix S25, allots each channel to a number of countries having need for an allotment, the countries sharing a channel being chosen with the aim of achieving an acceptably low probability of interference. Each Administration assigns its allotted channels to its coast stations. Some of the duplex telephony channel-pairs are identified for use for inter-ship communication; see RR Appendix S17, Part B, Section I. The procedure for amending the plan is in Section I of the Appendix.

RR S52.219 directs that the power of coast station transmitters used for telephony in this frequency range should be no more than is sufficient to cover the stations's service area, with a maximum of 10 kW peak envelope power. The corresponding limit for ships' transmitters is 1.5 kW; see RR S52.220. When used for telephony in the simplex mode or for inter-ship communication, the power limit is 1 kW; see RR S52.227.

For all of the other kinds of traffic that the scheme provides for, channels are prescribed for ships but not for coast stations. Coast stations are assigned frequencies, chosen with any necessary informal coordination between Administrations to avoid interference, within the band identified for unpaired services, namely sub-band (13). A coast station, setting up radio links for one of these facilities, designates the working channel on which the ship should transmit and identifies the working frequency within sub-band (13) on which it will itself transmit. For coast station transmitters used for telegraphy the power should not exceed 5 kW at 4 and 6 MHz, nor 10 kW at 8 MHz, nor 15 kW above 8 MHz; see RR S52.56.

The shared MMS allocations between 4 and 30 MHz

In addition to the exclusive MMS allocations reviewed above, 19 other bands are also allocated for the MMS, but they are all shared with other services, mostly the FS; see Table 8.4. For a few of these bands the RR provide some structure for their use by shipping.

a) Radio channel plans suitable for SSB telephony have been drawn up for the bands 4000–4063 kHz and 8000–8195 kHz, both of which are shared with the FS. These bands are used by the MMS as directed in RR Appendix S17, Part B, Section IC.

b) MMS use of 23 350–24 000 kHz is limited to inter-ship radiotelegraphy; see RR S5.157.

No doubt some considerable maritime use is made of some at least of the other shared HF bands, for example by naval ships and their coast stations. However, in

Table 8.4 *The shared HF MMS allocations (kHz)*

4000–4063	18 168–18 780
4438–4650	20 010–21 000
4750–4850 (Region 2 only)	23 000–23 200
5250–5450	23 350–24 000
5730–5900 (not Region 1)	25 010–25 070
8100–8195	25 210–25 550
10 150–11 175	26 175–27 500
13 410–13 570	27 500–28 000
13 870–14 000	29 700–30 005
14 350–14 990	

many of these bands the mobile allocation is secondary, sharing with a primary FS allocation; it may be supposed that the interference problem is considerable at some times of day.

8.3 The aeronautical mobile service below 30 MHz

The AMS(R) MF/HF frequency allotment plan

There are 12 frequency bands allocated exclusively for the AMS(R) between 2.8 and 22 MHz and all but one of these allocations are world-wide; they are listed in the first column of Table 8.5. A frequency allotment plan has been drawn up for these bands;

Table 8.5 *The exclusive HF AMS allocations (kHz)*

AMS (R)	AMS (OR)
2850–3025	3025–3155
3400–3500	3900–3950[1]
4650–4700	4700–4750
5450–5480 (Region 2 only)	5680–5730
5480–5680	6685–6765
6525–6685	8965–9040
8815–8965	11 175–11 275
10 005–10 100	13 200–13 260
11 275–11 400	15 010–15 100
13 260–13 360	17 970–18 030
17 900–17 970	
21 924–22 000	

[1] Note that 3900–3950 kHz is allocated for AMS (OR) in Region 1, for AMS without discrimination in Region 3 and for neither in Region 2.

see RR Appendix S27. The plan is based on the assumption that telephony and SSB modulation with suppressed carrier will be used, although telegraphic systems using no more bandwidth may also be used. The plan divides these allocations into channels 3 kHz wide and identifies each channel for one of four functions.

a)   Communication between any aeronautical station and an aircraft flying on any of a geographically defined group of long distance air routes, essentially international in character, extending through more than one country and requiring long distance communication facilities. These groups of routes are called Major World Air Route Areas.

b)   Communication between any aeronautical station and an aircraft flying on any of a geographically defined group of routes which are not classified as Major World Air Routes. Such a group of routes is designated a Regional and Domestic Air Route Area. These areas are divided into sub-areas.

c)   VOLMET allotments are made to aeronautical stations within a VOLMET allotment area for use in broadcasts to aircraft within the corresponding VOLMET reception area.

d)   WORLD-WIDE allotments provide for communication between an aeronautical station and an aircraft operating anywhere.

The VOLMET function is unidirectional. The other allotments are used bi-directionally in the simplex mode. The management of the WORLD-WIDE allotments rests with governments. The use of the other allotments is managed co-operatively by governments, mainly through ICAO. There is heavy pressure of traffic on these allotments in some parts of the world. Various programmes of action are urged upon governments to make more efficient their use of these HF assignments and to transfer as much as possible of the traffic which they now carry to the AMS(R) VHF allocations and to satellite systems using AMS(R)S allocations; see RR Resolutions 405 and 406 and RR Recommendations 401, 402 and 405.

The AMS(OR) MF/HF frequency allotment plan

There are 10 frequency bands allocated exclusively for the AMS(OR) between 3.0 and 18.0 MHz, and all but one of the allocations are world-wide; they are listed in the second column of Table 8.5. The prohibition of using exclusive AMS allocations for public correspondence (RR S43.4) applies here also.

A frequency allotment plan has been drawn up for these 10 bands. It was last revised at WARC-92; see RR Appendix S26. The plan is based on the assumption that telephony and SSB modulation with suppressed carrier will be used, although telegraphic systems using no more bandwidth may also be used. The plan divides these allocations into channels 2.8 kHz wide, their centres being separated by 3 kHz. The channels are allotted to countries, shared in a way that is calculated to limit co-channel interference to an acceptable level. See also RR Resolutions 411 and 412.

Other AMS allocations below 30 MHz

For Regions 2 and 3 there are several bands between 200 and 527 kHz in which the AMS has secondary allocations, typically shared with primary allocations to the ARNS. The aeronautical requirements for these bands are few and specialised.

Above 2 MHz in the MF range and at the lower and upper ends of the HF range there are a number of primary allocations for the AMS, including some for the AMS(R) and some for the AMS(OR), typically shared with primary allocations for the FS, the LMS and/or the MMS. Long distance interference may well make operating in these allocations difficult, but they are available and may be found usable for traffic overflowing from crowded bands in the MF/HF allotment plans.

8.4 The land mobile service below 30 MHz

There are many frequency bands allocated for the MS or the LMS, with a substantial aggregate bandwidth, spread over the spectrum between 2 and 30 MHz but all are shared with the FS. All the LMS allocations between 5.4 and 21 MHz are secondary. Furthermore, although the mobile allocations at 2300–2495/8 kHz, 3200–3400 kHz and 4750–4995 kHz are primary, they are shared, not only with the FS but also with the BS, which uses them for tropical broadcasting; see Section 7.3 above. Recommendation ITU-R BS.48-2 [16] urges that these bands should not avoidably be used for mobile services. Given substantial transmitter power and efficient antennas, the physical factors required for mobile communication at any time of day by sky wave propagation over distances of thousands of kilometres are available. It can, however, be foreseen that difficulties would arise in finding frequencies free from interference to assign for such a purpose.

Assignments in the primary mobile allocations below 5.4 MHz and above 21 MHz would not support continuous long distance communication by the sky wave alone but they might support communication over a distance of several hundred kilometres by the ground wave. Furthermore, if a frequency below 5.4 MHz were used by day and one above 21 MHz were used by night, the likelihood of sky wave interference to or from distant stations would be small. A base station operating in this mode, using a single channel SSB radio system, could provide a public service to vehicle-mounted mobile stations in an extensive, sparsely populated area.

Citizens' Band

In many countries a band around 27 MHz, allocated for the MS, is offered for use by anyone for mobile radiotelephony. This is called Citizens' Band (CB). Constraints are applied to the equipment used in this band and a fee may be charged for a licence to use it, but no other limitations are applied to participants provided that their emissions remain within the frequency limits of the band. Administrations consider that they have no responsibility for controlling interference between users.

8.5 Mobile services above 30 MHz

8.5.1 Frequency allocations and sharing

Below 30 MHz the principal frequency allocations for the MMS and the AMS are primary, world-wide and exclusive, and they are the subject of internationally agreed channelling plans and, in some cases, frequency allotment plans. This is necessary, because ships and aircraft of all nationalities travel the world and must be able to communicate easily with local stations wherever they may be. It is otherwise with the

LMS allocations below 30 MHz, but then the need for detailed international management of these bands does not arise; their use is national, even local, and the volume of use is small and probably shrinking, due to the development of mobile-satellite facilities.

Above 30 MHz the situation is quite different. The mobile services have a large number of allocations of very large aggregate bandwidth between 30 MHz and 275 GHz. However, with a couple of exceptions noted in Sections 8.5.2 and 8.5.3, these allocations are made, not for the MMS or the AMS or the LMS, but for the MS or the MS excluding the AMS. Almost without exception these allocations are also shared with the FS and many bands are additionally shared with other services, terrestrial and satellite. Finally, there has, until very recently, been little international planning of the way these bands are used. In short, at these higher frequencies, interference and planning have been seen mostly as a local problem, to be solved domestically by each Administration, in collaboration, where necessary, with the Administrations of neighbouring countries. The longer range interference problems raised by sharing between terrestrial and satellite services are avoided through sharing constraints and similar arrangements.

Thus, most of the allocations for the FS that are reviewed in Section 6.2.1 above are shared with the MS, although in some bands the MS allocation is secondary, whereas with few exceptions the FS allocations are primary. And the constraints on the FS that are summarised in Section 6.2.1 which facilitate sharing with other services apply equally to the MS in the same band if differences in the categories of allocations are taken into account.

Special mention must be made, however, of the sharing situation at 2025–2110 MHz and 2200–2290 MHz. These bands are allocated for the FS and the MS, sharing also with the SO, the EES and the SR. These 'science' services use these bands for inter-satellite links, among other applications. A group of sharing constraints has been adopted to facilitate this complex pattern of sharing; see item 4 in Section 13.4.1 below and item 4 in Section 6.2.1 above. The constraints on the FS which protect the science services cannot be applied to the MS. Accordingly, RR S5.391 requires Administrations to take the terms of Recommendation ITU-R SA.1154 [17] into account when assigning frequencies in these bands to stations of the MS. Recommendation ITU-R SA.1154 finds that low density MS applications, such as electronic news gathering (ENG), are acceptable sharing partners for the science services in these bands but sharing with high density MS applications such as a cellular radiotelephone network like IMT-2000 would not be acceptable.

In any country the way in which the use of a band shared by the FS and the MS is divided between the two services is decided by the Administration, although with due regard for the decisions made for neighbouring countries. Since, above 30 MHz, there has been only a quite limited need to standardise internationally the use which the mobile services make of their allocations, no more than a very general account can be given here of how these bands are used. Apart from the applications reviewed in Sections 8.5.2 and 8.5.3, ships and aircraft have relatively few assignments in this range of the spectrum.

However, the LMS has been expanding rapidly in most countries, starting in about 1980, mainly due to the growth of systems serving a wide range of vehicle- and hand-portable terminals for telephony, data, paging etc. In many countries these systems have ousted the FS from much of the spectrum between 30 and 960 MHz that is not used for broadcasting. The process is continuing, and it may be supposed that, before long, the FS will retain the use of relatively little bandwidth between 30 MHz and 3 GHz in many countries. Radio propagation factors make frequencies above 3 GHz unsatisfactory for most kinds of mobile system, and especially so for wide coverage land mobile systems in built-up neighbourhoods.

8.5.2 The MMS above 30 MHz

The MMS at VHF

The calling frequency for ships using the MMS VHF band for telephony is 156.8 MHz; it is also a distress and safety frequency (see Section 8.7 below). A narrow band surrounding that frequency, namely 156.7625–156.8375 MHz, is allocated exclusively for the MMS world-wide. The bands immediately adjacent to this exclusive band are also allocated for the MMS world-wide but they are shared with various other services. A radio channel plan has been drawn up for the MMS covering 156–156.7625 MHz, 156.8375–157.45 MHz, 160.6–160.975 MHz and 161.475–162.05 MHz, based on the use of FM in channels 25 kHz wide; see RR Appendix S18.

Administrations are required to reserve for the MMS the channels of this plan that they assign to their own coast stations for transmitting and receiving, and to avoid assigning other channels to stations of other services in locations where interference might be caused to the maritime services of neighbouring countries; see RR S5.226. These channels provide reliable, low-cost communication in coastal waters within about 50 km of a coast station and in harbour, with access from ships into the PSTN. Recommendation ITU-R M.489-2 [18] outlines the technical characteristics of the radio equipment used in this band; this recommendation is incorporated into the Radio Regulations by reference in RR S52.231 and elsewhere.

WARC Mob-87 noted that the bands covered by this channel plan were becoming congested at some locations and foresaw more general congestion as a problem for the future. RR Recommendation 318 asked for solutions to this problem to be sought. Recommendation ITU-R M.1084-2 [19] identified one way forward, quick and easy to implement, namely the use of FM equipment designed for a channel width of 12.5 kHz, recognising that this change would have to be implemented in a way that did not disrupt the continued operation of existing equipment having a band-width requirement of 25 kHz; see also Report ITU-R M.2010 [20].

WRC-97 revised the plan in RR Appendix S18. As revised, in addition to the chan-nels earmarked for calling and emergency use, 12 channels are for inter-ship communi-cation, 28 channels for public correspondence, mainly telephony but not excluding data and facsimile transmissions, and there is provision for the port operations and ship movement services. The revised plan does not differ radically from the previous version, agreed in 1987. However, Administrations with a need for more channels are permitted to introduce the use of narrow-band (12.5 kHz) FM equipment, subject to the avoidance of compatibility problems with 25 kHz equipment, especially problems involving the distress channel at 156.8 MHz, and subject to prior agreement with other Administrations whose ships might be affected.

For the longer term, Recommendation ITU-R M.1084-2 [19] foresaw that more advanced technology might be needed to meet the needs of the MMS in this part of the spectrum. RR Resolution 342 called for further study of the requirements of the service and ways of meeting them, for consideration at WRC-00. An interim answer on possible technical solutions is provided by Recommendation ITU-R M.1312 [21], which also draws attention to advances recently made in the provision of corresponding facilities in the LMS. One possible solution offered as an example by this recommendation would retain the 25 kHz channel plan for use with digital transmission, each radio channel being used by a TDMA system providing between four and eight digital telephone channels.

On-board communication for ships

There are no frequency allocations specifically for the MMS above 170 MHz. However, RR S5.287 identifies several spot frequencies around 457.5 and 467.5 MHz, in a MS frequency allocation, for on-board communication on ships. RR S5.288 contains another list of spot frequencies, differing in detail from those in the S5.287 list, which are preferred for this purpose in the territorial waters of the United States and the Philippines. The basic characteristics of the equipment used for on-board communication, with an occupied bandwidth of 25 kHz per channel, should be in conformity with Recommendation ITU-R M.1174 [22]. This recommendation is incorporated into the Radio Regulations by reference in RR S5.287 and S5.288. However, WRC-97 approved the introduction of equipment with an occupied bandwidth of 12.5 kHz for this purpose. RR S5.287 as revised by WRC-97 includes an additional list of spot frequencies for use with narrow-band equipment and RR Resolution 341 asks for basic characteristics for such equipment to be added to Recommendation ITU-R M.1174.

Other terrestrial maritime facilities above 30 MHz

Above 30 MHz there are many other large blocks of frequency allocated for the MS or the MS excluding the AMS but none is allocated specifically for the MMS. Some Administrations, no doubt, assign other VHF or UHF frequencies to ships and coast stations within their jurisdiction, taking whatever measures are seen to be required to avoid interference. For further reference to frequencies used by ships in emergencies, see Section 8.7.

8.5.3 The AMS above 30 MHz

The AMS(R) VHF frequency assignment plan

The band 117.975–137 MHz is allocated for the AMS(R); this allocation is, in effect, exclusive to the AMS(R) below 132 MHz. ICAO has drawn up a frequency assignment plan for this band and has incorporated it into its basic instrument, the Convention on International Civil Aviation. The plan is based on the use of AM telephony with DSB modulation. The aeronautical service has had the use of the band 118–132 MHz since 1947. Originally the channel bandwidth was 100 kHz. As receivers of improved selectivity have been taken into use, the plan has been modified; in particular the channel bandwidth has been reduced by stages to allow more channels to be provided. The allocation has also been extended upwards to 137 MHz. Currently a channel bandwidth of 8.33 kHz is in the course of staged introduction. These channels are used for short distance links, typically for air traffic control of aircraft en route but within range of VHF aeronautical stations and for the approach phase, landing and takeoff at airports.

There is a risk of interference from broadcasting transmitters operating below 108 MHz to aircraft ARNS receivers operating between 108 and 118 MHz, and a lesser risk of interference to AMS receivers operating above 118 MHz. See Section 7.5.3 above.

Public correspondence services for aircraft

No frequencies allocated for the AMS(R) or the AMS(OR) may be used to allow aircraft passengers to make telephone calls into the PSTN whilst in flight; see RR S43.4. If this facility is to be provided by terrestrial communication, links between the aircraft and the ground will operate in an allocation for the MS. Two situations arise, depending on whether the aircraft is or is not within range of a UHF radio station on land.

If the aircraft is over land or near to land, there may be an aeronautical station within range, equipped to provide the service and having access to the PSTN. At WARC-92 the bands 1670–1675 MHz and 1800–1805 MHz, allocated to MS but shared with other services, were identified for this purpose; see RR S5.380. Recommendation ITU-R M.1040 [23] defines the key characteristics of the Terrestrial Flight Telecommunication System (TFTS), recommended for use world-wide in this pair of bands. For Canada, the United States and Mexico the bands 849–851 and 894–896 MHz were allocated, also at WARC-92, for AMS and for the same purpose; see RR S5.318.

If there is no land-based station providing the TFTS or an equivalent facility within range of an aircraft it may nevertheless be feasible to get access to the PSTN via a ship. RR S41.1 and S51.68 to S51.70 permit aircraft to communicate with ships, using frequencies allocated for the MMS, for the purpose of public correspondence, and a ship might extend such a connection through a second radio link into the PSTN. However, such a service is operationally awkward to carry out and the option is seldom used, if ever, nowadays.

Other terrestrial aeronautical facilities above 30 MHz

In the VHF range and at higher frequencies there is a great deal of spectrum that is technically suitable for ground–air communication. None of this spectrum is allocated explicitly for the AMS. Much of it has a primary allocation for the MS but use for the AMS is explicitly excluded in some of these bands, in order to eliminate a need for precautions to be taken to avoid long distance interference to or from aircraft stations. Furthermore, all of these allocations are shared with primary allocations for other services. Nevertheless, a substantial bandwidth at VHF and UHF remains available and suitable for links with aircraft. Administrations, acting alone or in cooperation with other Administrations, can and do make assignments for AMS, to be used for airline communication and by private and military aircraft.

8.5.4   *LMS radiotelephone networks*

Private mobile radio networks

The basic function of a private mobile radio (PMR) network is to allow any of a fleet of motor vehicles to communicate, usually by telephony, with a base station, in pursuit of the business of the owner of the fleet. The first PMR networks were operated by public utilities, such as the police authorities in the 1930s using frequencies in the HF range. There were similar military networks. See also Section 8.4 above. Nowadays the VHF and UHF ranges are used.

An Administration prepares a frequency band, perhaps 20–40 MHz wide, for use by PMR networks by dividing it into two halves, one half for links from base stations to mobiles and the other for the mobile to base station direction. Each half-band is divided into channels, typically 12.5 or 25 kHz wide depending on the occupied bandwidth of the emissions to be used. The channels are numbered in sequence, each channel in the base-station-to-mobile half-band being associated with the corresponding channel in the mobile-to-base-station half-band, to provide a duplex or two-frequency simplex telephone circuit. The user is assigned a pair of frequencies for his network and, typically, buys and installs the radio equipment that the network uses at his base station and on his vehicles.

PMR networks are used, for example, by the emergency services (police, ambulance, fire), taxi fleets, public utilities and equipment servicing companies with teams of quick-response breakdown repairers in the field. Variants of the basic configuration, such as simplex circuit operation, a common base station serving several networks, data handling facilities, facsimile etc., can be implemented to meet users' specific requirements.

An Administration will require the radio equipment used for a PMR network to conform to a type approval specification. Recommendation ITU-R M.478-5 [24] contains basic guidance on the parameters that are desirable for the transmitters and receivers. Most systems use FM, although digital systems are coming into use, for example the TETRA system. Base stations are often clustered at topographically favourable sites, but in consequence, unwanted distortion products due to intermodulation between unwanted carriers in the input stages of a receiver may interfere with the reception of wanted signals; Recommendation ITU-R M.1072 [25] advises on means for reducing this interference. A PMR network may have a service area with a radius of up to about 40 km centred on the base station in the absence of screening by hills.

An applicant for a licence to operate a PMR network would tell the Administration the location of the proposed base station and the height of its antennas above the surrounding terrain. The desired service area would be defined and a forecast would be given of the extent of use of the network. This information enables the Administration to determine the transmitter power that the applicant will need, assign a channel-pair and license operation. These frequencies may be exclusive to this user, although they must often be shared with other users unless interference between the users denies them adequate access to the facility. The Administration might refuse to accept a service area that was abnormally large because this would reduce disproportionately the scope for re-using the same channels for other, geographically separate, networks.

Spectrum utilisation efficiency in PMR

There is a strong demand for PMR licences and there are also many other claims on the VHF/UHF spectrum. Most Administrations will have to give attention to efficiency in the use made of PMR bands, especially in metropolitan areas. This may involve action at several levels.

a)   The same channel-pair can be assigned to different networks with negligible inter-
     ference provided that the base stations are spaced by a sufficient distance. This
     distance will depend mainly on the frequency, the height of the base stations rela-
     tive to the local terrain and the nature of the terrain; see the Appendix, Section
     A2.2.2 below.

b)   Ways must be found for ensuring that narrow bandwidth radio equipment is used.
     Systems using emissions occupying a bandwidth between 20 and 30 kHz are wide-
     spread and a bandwidth of 50 kHz is not unknown. However, systems requiring
     only 12.5 kHz are also in common use, modern equipment needs only 6.25 or
     5.0 kHz and further reductions of bandwidth would be feasible. An Administra-
     tion can require new applicants for PMR licences to use modern, bandwidth-
     efficient equipment. A problem that may have to be addressed is how to per-
     suade the owners of existing networks to replace inefficient, old equipment with
     modern equipment while the old equipment is still usable. Licence fee policy
     might be designed to solve this problem; see further below.

c)   Some PMR networks carry heavy traffic, being in use for much of the time at their
     busy time of day. Many other networks carry little traffic. The inefficiency of spec-
     trum use that arises from network under-use can be reduced to some degree by
     assigning the same channel-pair to two or more networks with overlapping
     service areas. Selective calling devices and active channel detectors can be used
     to prevent a call from interrupting another call in progress on a shared channel-
     pair. Where there is a shortage of unassigned channels, Administrations should
     monitor the level of use of channels from time to time to find where channel-
     sharing assignments could be made. It may be found that some licensees make
     negligible use of their assignments, in which case their licences and their
     assignments might be withdrawn.

d)   With shared channels there will be many occasions, even when the total use of the
     channel pair is quite small, when one network is denied immediate access to the
     channel-pair because another network with the same frequency assignments is
     already using it. Users find this irksome. To combine high channel-pair utilisation
     efficiency with a very high probability of instant access, it is necessary to use
     systems designed to enable any of a group of networks to seize whichever of a
     group of channel-pairs is free at the moment when the call is to start; this principle
     is called 'trunking'. Trunking is not available with a simple PMR system, but see
     below.

Trunked PMR systems

Systems have been developed which combine some of the features of PMR networks
with trunked access to channels. For one trunked configuration, called public access
mobile radio (PAMR), a service area has a single base station that is operated by a
service provider. The service provider is licensed to control the use of a number of
channel-pairs, perhaps 100, including a two-way control channel linking all the
users' stations with the base station. A user wishing to set up a call sends a coded
group via the control channel that identifies the calling and the wanted stations. The
base station calls the wanted station, also via the control channel, and assigns a free
traffic channel-pair to the user stations to enable the call to proceed.

Recommendation ITU-R M.1032 [26] describes trunked systems which do not require a central control station. Both the systems described are intended primarily for short-range (typically 10 km radius) mobile-to-mobile links operating around 900 MHz. Each station has access to 80 channel-pairs, including control channels. Receivers standing-by are tuned to a control channel, recording which traffic channels have been taken into use. When a call is originated, the calling station identifies a free traffic channel-pair and addresses a coded group to the wanted station via the control channel, setting up the call on that traffic channel.

Public mobile radiotelephone access to the PSTN

There were radiotelephone systems in operation from around 1950 enabling access from vehicles into the PSTN, using frequencies in the VHF range. A single base station served a large, densely populated area and coverage did not, in general, extend into less populated areas. The subscriber at the mobile station tuned his station to the calling frequency of the base station, asked an operator for the wanted subscriber and the operator set up the connection. Such systems were not convenient for the user, they were not commercially successful and their extravagant use of spectrum would have made large scale implementation impossible.

By about 1980 developments in the technology of integrated circuit logic had made feasible the implementation of fully automatic systems. These systems provided region-wide or nation-wide coverage by means of a cellular arrangement of base stations and service areas, providing substantial frequency re-use. Analogue transmission and frequency modulation were used. An automatic control network maintained a system-wide register of the service area within which every standing-by mobile was located and set up connections as required, designating for use a free pair from a complement of working channels. If a mobile station moved from one service area to another while a call was in progress, the connection was handed off automatically from one base station to the next.

These first generation cellular systems were designed for operation in bands between 200 and 960 MHz, wherever enough spectrum, perhaps 20 or 30 MHz of bandwidth, that was not committed for broadcasting could be most readily cleared of the MS and FS systems that were already operating there. See CCIR Report 740 [27]. Little was done to achieve international standardisation of these systems, although Recommendation ITU-R M.624 [28] recommended standard practices for the registration of mobile station location. CCIR Report 742-2 [29] describes a number of these systems, all mutually incompatible in their basic characteristics; see also Report ITU-R M.742-4 [30]. Systems of this kind have been installed in many countries and the explosive growth of the traffic they have carried demonstrated the commercial importance of mobile radiotelephony. This growth of traffic was met, in part by assigning more spectrum to systems and in part by reducing the size of the coverage cells, so increasing the degree of frequency re-use.

Lack of standardisation limited competition between systems and between equipment manufacturers, and often made it impossible for a subscriber to use his mobile terminal when he was away from his home district or his home country. Recommendation ITU-R M.622 [31], prepared by the CCIR in 1986, reviewed the situation and recommended basic principles for future analogue systems intended for international use ('international roaming'). However, by the mid-1980s the course of development had turned to digital systems.

The GSM 900 digital system was developed under the aegis of CEPT to operate in the bands 890–915 MHz (mobile-to-base) and 935–960 MHz (base-to-mobile), mainly for vehicle-borne mobile stations. A parallel CEPT development with much in common with GSM is DCS 1800, operating in the bands 1710–1785 MHz or 1850–1910 MHz (mobile-to-base) and 1805–1880 MHz or 1930–1990 MHz (base-to-mobile) and optimised for small cells and hand-portable mobile stations. These systems have been installed widely in the European Union since 1991, where they are augmenting and will, no doubt, eventually replace the analogue systems.

There have been parallel developments of digital systems in other parts of the world. See Recommendation ITU-R M.1073-1 [32]. Some of these systems use, not the conventional frequency division multiple access to a block of spectrum, but spread spectrum modulation in the code division multiple access (CDMA) mode. A number of different speech coding techniques have been taken into use by these various systems; see Recommendation ITU-R M.1309 [33]. A parallel development of digital satellite-mobile systems for vehicle- and hand-portable earth stations is also in progress and some earth station equipment will be designed to be used with both terrestrial and satellite systems; see Recommendation ITU-R M.1221 [34] and Section 11.3 below.

The introduction of digital systems has probably brought with it some increase in opportunities for international roaming, particularly in countries using GSM, but the establishment of a single unified standard for all cellular systems is not seen at present as a practical objective. None the less, many countries have several competing systems, and it would be convenient for subscribers who travel extensively if at least one of the systems was compatible with a global standard specification. It would also be desirable for that standard system to be very versatile in the facilities it provides.

WARC-92 responded to the need for global, versatile, digital mobile radiocommunication, its RR Resolution 212 initiating preparations for the development and deployment of such systems. This project was initially called the Future Public Land Mobile Telecommunications System (FPLMTS); it has since been re-named International Mobile Telecommunications-2000 (IMT-2000). The name Universal Mobile Telecommunication System (UMTS) is often applied to such systems in Europe. The bands 1885–2025 and 2110–2200 MHz, already allocated world-wide for the MS, were chosen for this purpose; see RR S5.388. It was perceived that a satellite element would have to be integrated with the terrestrial system and a tentative choice of the frequency bands 1980–2010 and 2170–2200 MHz was made for it; MSS allocations will come into effect in these bands by 2000 or soon afterwards; see RR S5.389A. The resolution asked for work to be done in ITU-R and ITU-T to enable the characteristics of the systems to be defined.

The IMT-2000 frequency bands are also allocated for other services, in particular the FS and other mobile services, and they are in heavy use. Furthermore, it is foreseen that some countries may not wish to provide for IMT-2000. However, RR Resolution 212 resolves that the Administrations of countries that do wish to provide for the new system should prepare for it by assigning frequencies elsewhere for the systems now using frequencies in these bands that would not be compatible with the introduction of IMT-2000. Substantial work has already been done in ITU-R on the project and a number of recommendations have already been published; see Reference [35].

Sec in particular Recommendation ITU-R M.687-2 [36] which provides a detailed overview of the state of the development.

8.5.5 *Miscellaneous narrow-band mobile systems above 30 MHz*

There arc many kinds of mobile system which are valuable for niche users but which raise no special issues in spectrum management; most of these need not be specifically mentioned here. However, the following notes on selected facilities may be justified by a larger scale of use or broader application.

Radio-paging

In its simplest form, paging is a unidirectional system for transmitting a message to specified persons with hand-portable receivers, the persons being anywhere within a limited defined area, such as a hospital or a factory. Depending on the terrain, an induction loop system, a leaky feeder system (see Recommendation ITU-R M.1075 [37]) operating at VHF or UHF or a freely radiating low power VHF system will be suitable for such applications, providing a degree of privacy and minimal spectrum occupation.

Inevitably there has been an elaboration of the kinds of message that can be transmitted by paging systems, but in addition the basic paging facility has developed in three directions. Firstly, wide-area coverage is obtained with freely radiating radio systems, typically operating around 170 MHz or 470 MHz, typically using frequency assignments with a bandwidth of 25 kHz in frequency bands also used by PMR systems. Coverage may be city-wide from one transmitter, wider coverage being provided by transmitting the message via several or many transmitters. Recommendations ITU-R M.539-3 [38] and M.584-2 [39] recommend characteristics and codes for radio-paging systems intended for international coverage. The European Radio Messaging Service (ERMES) described in [38], is one such system.

Secondly, there has been an elaboration of the facilities which paging systems provide, especially systems which are limited to a single site and a single organisation. These additional facilities include two-way telephony and interconnection with wired communication systems. Thirdly, the unidirectional paging mode has been extended to become a limited-access broadcasting service delivering news messages and data to clients with hand-portable store-and-display receivers.

Cordless telephones and office systems

Low power radio links up to a few hundred metres long, operating in various frequency bands between 2 MHz and 2 GHz, may be used to connect telephones, work stations etc. to long distance communication facilities, replacing wired connections, for residential and office use. This principle may be extended, selection via the radio links taking the place of a private branch exchange or a local area network. There is a discussion of the principles that should desirably be followed in these systems, and a review of systems in use, in Recommendation ITU-R M.1033-1 [40].

Radio systems for news gathering, sporting and theatrical events

Theatrical and operatic performances involve substantial use of radio devices. The management of sporting events and the making of cinema films and TV programmes have intensive and complex radio needs, raising difficult spectrum management problems because such events are sporadic and to some extent unplannable. The use of mobile radio links, wide-band and narrow-band, in reporting in real time on news events is another significant spectrum requirement called electronic news gathering (ENG). The main applications for radio in activities of this kind are as follows.

- High performance radio microphones.
- Video and audio links carrying programme signals from the source to a local control room for editing and recording or onward transmission into the broadcasting network. This may include the use of ground-to-helicopter audio and video links for ENG and transportable earth stations (satellite news gathering (SNG), see Section 11.4).
- Talk-back audio systems for the control of operations.
- Data communication for the remote control of cameras, lights etc. and for the synchronisation of video cameras.

Where these needs are regular or predictable they can often be met by the assignment of frequencies in MS allocations in the usual way, with regular or short-term licences. It may also be convenient to assign frequencies for low-power mobile applications of this kind in bands allocated to the BS with a secondary MS allocation, for use in locations where the broadcasting channel is not active. However, to deal satisfactorily with the many requirements that are temporary, often arising at short notice and posing difficult problems of interference between the many radio devices in use on the same site, might demand expertise that an Administration is not likely to possess in-house. The Radiocommunications Agency, which is the Administration in the United Kingdom, has authorised a company with close involvement in broadcasting to manage a portfolio of assignments for use for these purposes on its behalf.

8.5.6 Wide-band transportable systems

Most of the many frequency bands above 3 GHz that are allocated for FS are also allocated for MS, or MS except for AMS. Some low power, short range, wide-band mobile devices may be assigned frequencies in these bands. However, the major MS use for these allocations is for assignments to transportable stations that are used to provide temporary wide-band links up to a few tens of kilometres long. These links are used, for example, to connect video, commentary and control signals from sporting events to a point where they can be transferred to programme centres via a channel in the PSTN or a permanent network used for distributing signals for broadcasting.

An Administration will reserve channels in FS radio channelling plans to be used by the MS for applications of this kind. The equipment of these transportable stations is physically similar to that of FS radio relay stations and the emissions are subject to the same mandatory constraints as FS emissions.

8.6 Licensing mobile radio stations

A principle common to all mobile service systems is that frequency assignments to the mobile stations are not notified to the ITU for registration in the MIFR. However, assignments, transmitting and receiving, to land stations and the areas within which their mobile stations move can be registered in the MIFR, and this provides indirect protection from foreign interference to the associated mobile stations. Both land stations and mobile stations must be licensed to use their assignments, although a single licence may cover a whole fleet of mobiles.

There is a wide variety of LMS devices; some are used in great numbers and, even so, their impact on spectrum may be quite small. If some administrative action is required to ensure that interference will not arise, it is appropriate that a fee should be charged to cover the cost of that action, but the cost per device will usually be quite small and the fee may be expected to be correspondingly small. However, there are also several kinds of land mobile system which have a large and rapidly growing requirement for spectrum, tending at present to deny access to spectrum for other desirable radio facilities. It is for consideration whether licensing policy, and in particular fee-charging policy, could and should be designed to optimise the use of this spectrum. PMR networks are one such case. Cellular networks and large scale PAMR networks are another, and it may be desirable to deal similarly with wide area paging and related systems.

Licence fees for PMR networks

PMR uses spectrum less efficiently than cellular and PAMR networks, but there is scope for substantial improvement in PMR efficiency through the general use of narrow bandwidth equipment. This is a matter of importance, since the demand for spectrum for cellular systems is large and increasing.

For some users the PMR network configuration is very advantageous. For example, the design of the terminal equipment may be optimised to suit the specific needs of some users. For others the option that all mobile stations may listen to all messages from the base station is operationally desirable. Other users prefer PMR because, for them, it costs less than cellular systems. Thus an Administration may find it expedient to set fees for PMR licences to encourage current PMR users to use narrow bandwidth equipment or to abandon the use of PMR in favour of cellular networks.

Fees for PMR licences are being raised in some metropolitan areas to discourage inessential use or to divert users to cellular networks. It may be supposed that such practices will spread, especially where the spectrum is being used intensively, but licence fees might have to be increased several-fold if much impact on the demand for PMR is to be achieved. However, since the purposes of spectrum management include enabling the public to obtain the greatest value from radio, it may be necessary to guide radio users into directions which will increase the number of users that can gain benefit from the medium.

Licence terms and spectrum pricing might be targeted for PMR as follows.

a) The Administration would, for an annual fee, issue licences having a period of five years, subject to the licensee observing the terms of the licence. An undertaking would be given that the frequencies assigned would remain unchanged during this period if this were feasible; if it were necessary to change the assigned frequencies within the period of the licence, they would be replaced by others that were within the tuning range of the licensee's equipment.

b) The 'basic' licence fee would be determined at a reference date, linked to the band-width of equipment in general use at that time, applicable in an area where the level of demand was substantial but not extremely high. Let this equipment be called the 'reference' equipment. Let this level of demand be called the 'standard' level. For areas of the highest demand the fee would be doubled. For areas of very light demand the fee would be halved. Another multiplier would take into account the number of mobile stations in the licensee's fleet.

c) As soon as another generation of equipment became available that was substan-tially more bandwidth-efficient, needing only $B\%$ of the bandwidth of the refer-ence equipment, the Administration would make available spectrum divided into radio channels appropriate for the new equipment. The new equipment would be licensed, for an annual fee which was equal to $B\%$ of the basic fee, give or take the various multipliers mentioned in stage (a) above. Starting at that same time, the annual fee for the reference equipment used in standard- or high-demand areas would be increased by perhaps 20% per year. There may be no justified case for raising fees in light-demand areas.

d) The licence fee for this second generation equipment would remain at the same level until yet newer, even more bandwidth-efficient, equipment became available. Then a new low licence fee would be fixed for the latest equipment, as in stage (c). Fees for the second generation equipment would be subject to an annual increase in standard- and high-demand areas and the annual licence fee penalties on the first generation equipment would be increased. And so on.

e) It might be feasible to renew licences at the end of their five-year periods, if the licensee wishes, without seriously impeding progress in spectrum efficiency. However, the basic licence fee set in stage (b) above should be high enough to ensure that, after a succession of annual increases, it will have become a consider-able sum. This should lead to a migration of PMR users from frequency bands channelled for out-of-date equipment to the newly channelled bands or to a cellu-lar mobile network. It would then become desirable for assignments for out-of-date equipment to be concentrated into some part of the frequency bands they have been using, to allow the vacated spectrum to be replanned for narrow-band equipment. Eventually the Administrative might find it desirable to cease renewing licences for seriously out-of-date equipment.

Licensing public cellular and large scale PAMR networks

Public access networks, with many channels and large, perhaps national, service areas are major telecommunication facilities, of great utility to their subscribers and with great potential for profit for their owners. Each network has a substantial spectrum requirement, and in aggregate they are among the major users of spectrum below 2 GHz. The Administration will be concerned to ensure that there is a progressive improvement in the efficiency with which these networks use spectrum. The Regulator will want to ensure that licensees make vigorous use of the opportunity that a licence gives for service to the public and that subscribers exert competitive pressure on the licensees, thereby regulating the price of facilities.

Under a liberal regime the Administration and the Regulator, conscious of a need to license service providers to offer more and/or new facilities to the public, may arrange preliminary discussions to identify the available technical options. Potential service providers, equipment manufacturers and the government agencies might participate in these discussions. A decision would be reached on the system design option to be implemented for the next generation of systems. Then the Administration and the Regulator would announce that blocks of spectrum could be allotted for new networks to a chosen common design specification, that proposals were invited for providing such systems to serve stated areas, and that it is expected that five-year or ten-year licences will be offered to the two, or possibly three service providers whose proposals are considered best. Proposals would be required to include:

- statement of the relevant technical, commercial and financial resources of the proposer,
- details of plans for 'rolling out' the network over the area to be served,
- facilities to be provided and preliminary intentions on tariffs,
- indications as to how the proposer would respond to an unexpectedly low, or unexpectedly high, level of demand,
- assurances that there will be no barriers to competition between the networks.

An annual licence fee might have been announced. Alternatively, proposers might be asked to state the annual sum of money that they would pay for a licence.

It may be supposed that several proposals would be received, some at least being satisfactory in general terms. If a fixed fee had been announced, it would be necessary to choose the successful proposals on the basis of the quality and credibility of the proposals. Such judgements are difficult and embarrassing for government agencies; they may be appealed against and appeals are a source of delay. An auction between a short list of proposers, each having made a proposal that was satisfactory in other respects, is less likely to be disputed and may yield more revenue to the government.

Successful proposers would then be licensed and their systems would be implemented. The networks of base stations would be planned and built. Frequencies, transmitting and receiving, for assignment to the base stations would be chosen and proposed to the Administration for formal assignment. In the absence of reasons for rejecting the proposals, the Administration would confirm the assignments and if appropriate, in due course, notify them to ITU for inclusion in the MIFR.

After some years, demand for the service might have outstripped the capacity of these networks. New equipment, more spectrum-efficient or offering desirable new features for the user, might meanwhile have become available. If so, new networks might be set up in the same way as before, using this later technology in different frequency bands, the new networks competing with one another and, for a while at least, with the networks of the earlier generation.

8.7 Distress and safety communication for mobile stations

The International Convention for the Safety of Life at Sea (SOLAS), signed under the aegis of the IMO in 1974 and amended in 1981 and 1983 [41] is the basis of the provisions for ensuring maritime safety that were in place up to January 1999. The regulatory provisions for this system, in as far as they needed to be promulgated in the RR, are mainly to be found in Part A of RR Appendix S13. The system used for initiating search and rescue action depends to a large extent on a listening watch, maintained by coast stations and also obligatory for all but small ships at sea, on 500 kHz for distress calls sent by Morse and on 2182 kHz for oral distress calls. These mandatory provisions are effective within range of coast stations and in busy sea lanes but they are precarious for ships in mid-ocean. The risks have been reduced in recent years by innovations, such as selective-calling systems, a distress watch at coast stations on HF calling frequencies and various satellite facilities. However, now that many ships no longer use Morse for their normal communications services, the maintenance in particular of the distress watch at 500 kHz may be seen as an expensive anachronism.

However, in 1988 there was a conference of the contracting governments to the SOLAS agreement of 1974 at which it was agreed that further and quite radical amendments should be made to that Agreement; see Reference [42]. These amendments have created the Global Maritime Distress and Safety System (GMDSS), the implementation of which began in 1992 and which were to have been completed by 1 February 1999. The major changes brought about by the introduction of the GMDSS are firstly the end of the use of Morse and of the distress watch at 500 kHz, secondly that the radio equipment which a ship is required to carry is to depend in future on the sea areas in which it sails, and thirdly there is to be major increase in the use of satellite systems. However, there will be a transitional period during which the new system will be fully activated but some of the provisions of the former system, not required in the GMDSS, will continue in operation; see, for example, RR Resolution 331.

As well as maritime emergencies, there are dangers by air and by land which may be relieved by radio communication. ICAO has made provision for emergencies involving aircraft in its management of the HF and VHF AMS(R) bands. Also RR S30.12 authorises land mobile stations in remote locations to use GMDSS frequencies for distress and safety communications; see also RR Resolution 209. For these reasons frequency bands in the immediate vicinity of 500 kHz and 2182 kHz are allocated exclusively for the MS as a whole. Similarly 406.0–406.1 MHz, 1544–1545 MHz and 1645.5–1646.5 MHz are allocated exclusively for the MSS, to be used for emergency services via satellite for any type of mobile earth station. Also the RR allow specific channels in MMS and AMS(R) bands to be used by both aircraft and ships cooperating in a search for a ship or an aircraft in danger.

The GMDSS came into effect in February 1999 for ships and aircraft with the necessary equipment. The systems which the GMDSS has displaced will remain in being for a while to serve ships which are not fully re-equipped. The radio regulations concerned with the GMDSS, in addition to various footnotes to the international table of frequency allocations (RR Article S5), are to be found in RR Articles S30–S34 and RR Appendix S15. It may, however, be noted that RR S4.9 makes it clear that these regulations are not intended to be exclusive; lacking the means of following the regulations, ships and aircraft in distress may use any available means of making their condition and location known.

The radio facilities involved in the GMDSS and in the corresponding provisions for aircraft in distress are as follows.

1   Ships transmit distress, urgency or safety messages using DSC equipment at 2187.5, 4207.5, 6312, 8414.5, 12 577 or 16 804.5 kHz or at 156.525 MHz. A DSC emergency message identifies the sending ship, its stated position, the time of the last up-

dating of that position and the nature of the emergency; see RR S34.2 and Recommendation ITU-R M.493-9 [8]. Frequencies have been identified for communication between ships in distress and those involved in search and rescue operations in the MMS exclusive bands, using speech and using NBDP, at 2, 4, 6, 8, 12, 16 MHz and in the 156 MHz band; see RR Appendix S15. Other frequencies, in the MF band and in all eight of the operational HF bands, are used for the distribution by NBDP of maritime safety information, including meteorological and navigational warnings. Ships equipped to use communication satellites operating in the bands 1645.5–1646.5 MHz and 1544–1545 MHz use this medium also for communication connected with search and rescue operations.

2   The frequencies at 3023 and 5680 kHz and also 121.5, 123.1 and 156.3 MHz are identified for communication between mobile stations involved in search and rescue operations involving aircraft.

3   Carrier frequencies of 121.5 and 243 MHz are used by emergency position-indicating radio beacons (EPIRBs), which are carried by aircraft but may be transferred to survival craft or arranged to float free from a sinking aircraft; see Recommendation ITU-R M.690-1 [43], incorporated into the Radio Regulations by reference. For maritime use float-free EPIRBs operate at 156.525 MHz, transmitting a DSC distress message; that is, the frequency and the basic message is the same as would be used by the mother ship in distress; see paragraph 1 above, RR S34.1 and Recommendation ITU-R M.693 [44]. RR Recommendation 604 expresses the view of ITU conferences that the use of EPIRBs had become complex and that ITU, IMO and ICAO should jointly review the role played by EPIRBs in search and rescue operations at sea; see also the next paragraph.

4   The bands 406.0–406.1 MHz and 1645.5–1646.5 MHz are allocated for MSS uplinks and used by satellite EPIRBs which are carried by ships but may be transferred to survival craft or arranged to float free from a sinking ship. There are two kinds of satellite EPIRB used by ships; see RR S34.1. One kind, described in Recommendation ITU-R M.633-1 [45], operates in the 406 MHz band and its signals are received by COSPAS-SARSAT satellites which are in low, highly inclined, orbits. RR Resolution 205 calls for action to ensure that the band 406.0–406.1 MHz is used for no other purpose. The other kind of EPIRB, described in Recommendation ITU-R M.632-3 [46], operates in the 1646 MHz band and its signals are received by geostationary satellites, typically those of INMARSAT; see also Recommendation ITU-R M.830 [47]. The reference to RR Recommendation 604 in item 3 above relates to this item also.

5   Ships carry transponders, called search and rescue radar transponders (SARTs) operating in the band 9200–9500 MHz which respond to radar interrogation signals and assist in the location of a ship in distress. SARTs may also be carried by survival craft. See Recommendations ITU-R M.628-3 [48] and M.629 [49].

8.8 References

1 'Technical characteristics of single-sideband transmitters used in the maritime mobile service for radiotelephony in the bands between 1606.5 kHz (1605 kHz Region 2) and 4000 kHz and between 4000 kHz and 27500 kHz'. Recommendation ITU-R M.1173, ITU-R Recommendations 1997 M Series Part 3 (ITU, Geneva, 1997)

2 'Direct-printing telegraph equipment employing automatic identification in the maritime mobile service'. Recommendation ITU-R M.625-3, *ibid.*

3 'Direct-printing telegraph equipment in the maritime mobile service'. Recommendation ITU-R M.476-5, *ibid.*

4 'Narrow-band direct-printing telegraph equipment using a single-frequency channel'. Recommendation ITU-R M.692, *ibid.*

5 'Technical characteristics of HF maritime radio equipment using narrow-band phase-shift keying'. Recommendation ITU-R M.627-1, *ibid.*

6 'Automatic HF facsimile and data systems for maritime mobile users'. Recommendation ITU-R M.1081, *ibid.*

7 'Sequential single-frequency selective-calling system for use in the maritime mobile service'. Recommendation ITU-R M.257-3, *ibid.*

8 'Digital selective calling system for use in the maritime mobile service'. Recommendation ITU-R M.493-9, *ibid.*

9 'International maritime MF/HF radiotelephone system with automatic facilities based on DSC signalling format'. Recommendation ITU-R M.1082-1, *ibid.*

10 'Automated VHF/UHF maritime mobile telephone system'. Recommendation ITU-R M.586-1, *ibid.*

11 'Interworking of maritime radiotelephone systems'. Recommendation ITU-R M.1083, *ibid.*

12 'Direct-dial telephone systems for the maritime mobile service'. Report ITU-R M.2009, ITU-R Reports M Series Part 3 (ITU, Geneva, 1995)

13 'Prevention of interference to radio reception on board ships'. Recommendation ITU-R M.218-2, *ibid.*

14 'Final Acts of the Regional Administrative Radio Conference for the planning of the MF maritime mobile and aeronautical radionavigation services (Region 1)', Geneva, 1985

15 'Operational and technical characteristics for an automated direct-printing telegraph system for the promulgation of navigational and meteorological warnings and urgent information to ships'. Recommendation ITU-R M.540-2, ITU-R Recommendations 1997 M Series Part 3 (ITU, Geneva, 1997)

16 'Choice of frequency for sound broadcasting in the tropical zone'. Recommendation ITU-R BS.48-2, ITU-R Recommendations 1997 BS Series (ITU, Geneva, 1997)

17 'Provisions to protect the space research (SR), space operations (SO) and Earth-exploration satellite services (EES) and to facilitate sharing with the mobile service in the 2025–2110 MHz and 2200–2290 MHz bands'. Recommendation ITU-R SA.1154, ITU-R Recommendations 1997 SA Series (ITU, Geneva, 1997)

18 'Technical characteristics of VHF radiotelephone equipment operating in the maritime mobile service in channels spaced by 25 kHz'. Recommendation ITU-R M.489-2, ITU-R Recommendations 1997 M Series Part 3 (ITU, Geneva, 1997)

19 'Interim solutions for improved efficiency in the use of the band 156–174 MHz by stations in the maritime mobile service'. Recommendation ITU-R M.1084-2, *ibid.*

20 'Improved efficiency in the use of the band 156–174 MHz by stations in the maritime mobile service'. Report ITU-R M.2010, ITU-R Reports M Series Part 3 (ITU, Geneva, 1995)

21 'A long-term solution for improved efficiency in the use of the band 156–174 MHz by stations in the maritime mobile service'. Recommendation ITU-R M.1312, ITU-R Recommendations 1997 M Series Part 3 (ITU, Geneva, 1997)

22 'Characteristics of equipment used for on-board communications in the bands between 450 and 470 MHz'. Recommendation ITU-R M.1174, *ibid.*

23 'Public mobile telecommunication service with aircraft using the bands 1670–1675 MHz and 1800–1805 MHz'. Recommendation ITU-R M.1040, *ibid.*

24 'Technical characteristics of equipment and principles governing the allocation of frequency channels between 25 and 3000 MHz for the FM land mobile service'. Recommendation ITU-R M.478-5, ITU-R Recommendations 1997 M Series Part 1 (ITU, Geneva, 1997)

25 'Interference due to intermodulation products in the land mobile service between 25 and 3000 MHz'. Recommendation ITU-R M.1072, *ibid.*

26 'Technical and operational characteristics of land mobile systems using multi-channel access techniques without a central controller'. Recommendation ITU-R M.1032, *ibid.*

27 'General aspects of cellular systems. CCIR Report 740'. Recommendations and Reports of the XVIIth Plenary Assembly of the CCIR, Annex to Volume VIII (ITU, Geneva, 1990)

28 'Public land mobile communication systems location registration'. Recommendation ITU-R M.624, ITU-R Recommendations 1997 M Series Part 1 (ITU, Geneva, 1997)

29 'Public land mobile telephone systems'. CCIR Report 742-2, Recommendations and Reports of the XVIIth Plenary Assembly of the CCIR, Annex to Volume VIII (ITU, Geneva, 1990).

30 'Public land mobile telephone systems'. Report ITU-R M.742-4, ITU-R Reports M Series Part 1 (ITU, Geneva, 1995)

31 'Technical and operational characteristics of analogue cellular systems for public land mobile telephone use'. Recommendation ITU-R M.622, ITU-R Recommendations 1997 M Series Part 1 (ITU, Geneva, 1997)

32 'Digital cellular land mobile telecommunication systems'. Recommendation ITU-R M.1073-1, *ibid.*

33 'Digitally coded speech in the land mobile service'. Recommendation ITU-R M.1309, *ibid.*

34 'Technical and operational requirements for cellular multimode mobile radio stations'. Recommendation ITU-R M.1221, *ibid.*

35 The ITU-R recommendations to date on IMT-2000 are to be found in ITU-R Recommendations 1997 M Series Part 2 (ITU, Geneva, 1997)

36 'Future Public Land Mobile Telecommunications Systems (FPLMTS)'. Recommendation ITU-R M.687-2, *ibid.*

37 'Leaky feeder systems in the land mobile service'. Recommendation ITU-R M.1075, ITU-R Recommendations 1997 M Series Part 1 (ITU, Geneva, 1997)

38 'Technical and operational characteristics of international radio-paging systems'. Recommendation ITU-R M.539-3, *ibid.*

39 'Standard codes and formats for international radio-paging'. Recommendation ITU-R M.584-2, ITU-R Recommendations 1997 M Series Part 1 Supplement 1 (ITU, Geneva, 1997)

40 'Technical and operational characteristics of cordless telephones and cordless telecommunication systems'. Recommendation ITU-R M.1033-1, ITU-R Recommendations 1997 M Series Part 1 (ITU, Geneva, 1997)

41 'International convention for the safety of life at sea' (The International Maritime Organization, London, 1986)

42 'Amendments to the 1974 SOLAS Convention concerning radiocommunications for the Global Maritime Distress and Safety System' (The International Maritime Organization, London, 1989)

43 'Technical characteristics for search and rescue radar transponders'. Recommendation ITU-R M.690-1, ITU-R Recommendations 1997 M Series Part 4 (ITU, Geneva, 1997)

44 'Technical characteristics of VHF emergency position-indicating radio beacons using digital selective calling (DSC VHF EPIRB)'. Recommendation ITU-R M.693, *ibid.*

45 'Transmission characteristics of a satellite emergency position-indicating radio beacon (satellite EPIRB) system operating through a low polar-orbiting satellite system in the 406 MHz band'. Recommendation ITU-R M.633-1, ITU-R Recommendations 1997 M Series Part 5 (ITU, Geneva, 1997)

46 'Transmission characteristics of a satellite emergency position-indicating beacon (satellite EPIRB) system operating through geostationary satellites in the 1.6 GHz band'. Recommendation ITU-R M.632-3, *ibid.*

47 'Operational procedures for mobile satellite networks or systems in the bands 1530–1544 MHz and 1626.5–1645.5 MHz which are used for distress and safety purposes as specified for GMDSS'. Recommendation ITU-R M.830, *ibid.*

48 'Technical characteristics for search and rescue radar transponders'. Recommendation ITU-R M.628-3, ITU-R Recommendations 1994 M Series Part 4 (ITU, Geneva, 1994)

49 'Use of the radionavigation service of the frequency bands 2900–3100 MHz, 5470–5650 MHz, 9200–9300 MHz, 9300–9500 MHz and 9500–9800 MHz'. Recommendation ITU-R M.629, *ibid.*

Chapter 9

The fixed-satellite service

9.1 Frequency allocations and sharing

9.1.1 Introduction

There is a code of letter symbols for identifying in broad terms the frequency ranges in which FSS and MSS systems operate, as follows:

Code letter	Frequency range (GHz)
L	1–2
S	2–3
C	4–6
Ku	10–15
Ka	17–31

This code is too imprecise for regulatory purposes but it is widely used in the literature and it is used where convenient in this book.

The basic application of the FSS is for the uplinks and downlinks of systems which serve earth stations at fixed locations. At present almost all of these systems use geostationary satellites, although one system uses a few satellites in inclined elliptical orbits and new systems in prospect plan to use large constellations of satellites in low circular orbits. Some systems are used for government purposes, including military use but most are commercial. In addition, RR S1.21, in defining the FSS, permits the inclusion within the service of feeder links, such as those that supply programme signals from an earth station at a fixed location to a broadcasting satellite.

For various reasons it is not satisfactory for all of these various applications to be mixed together in the same frequency bands. A large amount of spectrum, including about 55% of all the spectrum between 2.5 and 31 GHz, is allocated for the FSS and to a considerable extent it has been possible to segregate the applications into different frequency bands. In some cases the reservation of a band for a specific application has been made mandatory by the RR, forming a separate service within the FSS, but some linkages between bands and applications have arisen out of usage. However, it has not been possible to segregate completely the more recently introduced applications, despite the dual use of many frequency bands, for the uplinks of one application and the downlinks of another, and various means have been devised to make mixing applications within a frequency band workable.

With one exception, mentioned in Section 9.1.2, item 5 below, the FSS allocations are all primary. Some of the FSS allocations in the international table of frequency allocations are shared with primary allocations for other satellite services. In general, but with exceptions and special provisions also noted in Section 9.1.2, FSS networks are coordinated with one another and with the systems of sharing satellite services by the procedures reviewed in Section 5.2 above. Section 9.2 below also deals with the sharing of spectrum between GSO FSS networks.

Almost all FSS allocations are shared with primary allocations for terrestrial services, almost always the FS, often the MS and sometimes other terrestrial services as well. Here again, with a few special provisions mentioned in Section 9.1.2, the constraints and coordination procedures reviewed in Section 5.2 are sufficient to control interference between systems of these satellite and terrestrial services.

9.1.2 The FSS allocations, applications and sharing

The FSS allocations in Ku band and Ka band are numerous and their usage for the various applications is complex. Figures 9.1 and 9.2 show these allocations diagrammatically, identifying the principal applications using the numbers that are also used below in reviewing the applications. The nine applications are as follows.

1   Commercial FSS C band GSO networks.
2   Commercial FSS Ku band GSO networks.
3   The FSS allotment plan.
4   Government FSS networks.
5   Commercial Ka band GSO networks.
6   Commercial non-GSO FSS networks.
7   BSS feeder links.
8   MSS feeder links.
9   Miscellaneous other FSS allocations.

1 Commercial FSS C band GSO network allocations

A world-wide allocation at 3400–4200 MHz (downlinks), a Region 1 allocation at 5725–5850 MHz (uplinks) and world-wide allocations at 5850–6725 MHz and 7025–7075 MHz (both uplinks) are used by large numbers of FSS GSO networks, most occupying a bandwidth of 500 MHz for uplinks and 500 MHz for downlinks. RR S22.2 applies in these bands and coordination is in accordance with RR Article S9.

The band 7025–7075 MHz is allocated for feeder downlinks from non-GSO MSS satellites; see RR S5.458B and item 8(b) below. Interference between earth stations by Mode 4 (Figure 5.2) may arise unless sites for the earth stations used for these purposes are chosen with care and RR S5.458C requires consultation to avoid this interference when new earth stations and new assignments for GSO uplinks in this band are contemplated. Interference from MSS satellite transmitters to the FSS satellite receivers may occur via Mode 5B (Figure 5.3); to limit this interference RR S22.5A applies a constraint to the aggregate spectral PFD with which any MSS constellation may illuminate the GSO and its vicinity; see also Recommendation ITU-R S.1256 [1].

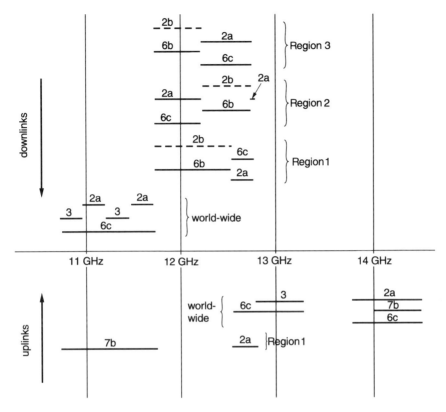

Figure 9.1 The allocations for the FSS, 10.7–14.5 GHz, and their applications. For the code to the applications, see the text in Section 9.1.2

2 Commercial FSS Ku band GSO network allocations

a) World-wide allocations for FSS downlinks at 10.95–11.2 GHz and 11.45–11.7 GHz and for uplinks at 13.75–14.5 GHz, plus several Regional allocations, are used by FSS Ku band networks. The Regional bands are as follows.

- Region 1: 12.5–12.75 GHz (uplinks and downlinks).
- Region 2: 11.7–12.2 GHz (downlinks) and 12.7–12.75 GHz (uplinks).
- Region 3: 12.2–12.75 GHz (downlinks, also see RR S5.491).

RR S22.2 applies in these bands. Many geostationary networks are operating in these bands, most occupying a bandwidth of 500 MHz for uplinks and 500 MHz for downlinks. In a number of countries the Regional downlink bands are not used for terrestrial services and there is no allocation for the FS or the MS in the uplink band 14–14.3 GHz in the framed part of the international table of frequency allocations. These allocations, virtually exclusive for the FSS where these conditions apply, are preferable for FSS networks serving earth stations in urban locations.

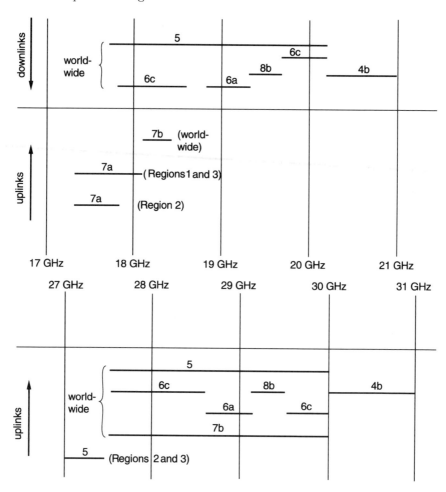

Figure 9.2   The allocations for the FSS, 17.3–21.2 GHz and 27–31 GHz, and their applications. For the code to the applications, see the text in Section 9.1.2

WRC-97 agreed a provisional procedure whereby non-GSO systems of the FSS might also operate in any of these bands, using the same directions of transmission as geostationary networks; see item 6(c) below.

The bands 10.95–11.2 GHz and 11.45–11.7 GHz are also allocated in Region 1 for FSS uplinks, limited by RR S5.484 to use for BSS feeder links; see item 7(b) below.

RR Article S21 does not provide a constraint on the PFD set up at the Earth's surface by FSS satellite transmitters in Region 2 operating in the band 11.7–12.2 GHz, to protect FS receivers. RR S5.488 requires uses of the band 11.7–12.2 GHz for FSS downlinks to be agreed with other Administrations that would be affected and Recommendation ITU-R SF.674-1 [2] recommends a criterion for the need to coordinate in the form of a value of downlink PFD. RR

S5.485 permits transponders of these satellites to be used for satellite broadcasting subject to various conditions.

A frequency band allocated and planned for the BSS around 12 GHz in one Region may be allocated for the FSS in another Region. This situation arises at 12.5–12.7 GHz (Region 1), 11.7–12.2 GHz (Region 2) and 12.2–12.7 GHz (Region 3). RR Appendix S30, Article 7, prescribes a coordination procedure, additional to the procedure of RR Article S9, which may have to be carried out by an Administration proposing to set up an FSS network in such a band. See also RR S5.487 and Recommendation ITU-R S.1067 [3], and also Section 10.3.3 below.

In Region 3, sharing the FSS allocation at 12.5–12.75 GHz, there is an allocation for the BSS which is not included in the BSS frequency assignment plan. Possible problems of compatibility between the two services are mitigated by a sharing constraint on broadcasting satellites which limits the power of BSS emissions; see RR S5.493.

Between 13.75 GHz and 14.5 GHz the FSS allocation is shared with primary allocations for the RLS and the RN. There are also some secondary sharing allocations. Sharing constraints imposed by RR S5.502 on the FSS and the RLS facilitate coordination between these services; see also RR S5.503A and Recommendation ITU-R S.1068 [4]. A sharing constraint limits the power of RN transmitters to a value that should not cause unacceptable interference to the FSS; see RR S5.504 and Recommendation ITU-R M.496-3 [5]. Having regard to sharing between the FSS and the SR, see RR S5.503, S5.503A and Recommendation ITU-R S.1069 [6]. There is a reference to the secondary LMSS allocation in item 1(b) of Section 11.1.2 below.

b) The bands 11.7–12.5 GHz (Region 1), 12.2–12.7 GHz (Region 2) and 11.7–12.2 GHz (Region 3) are allocated for the BSS. Frequency assignment plans have been agreed for these bands; see RR Appendix S30 and Section 10.3 below. However, RR S5.492 permits BSS channels planned for a country to be assigned by that country for FSS downlinks if they are not required for broadcasting, provided that they cause no more interference and require no more protection from interference than the planned broadcasting emissions. This may provide additional spectrum for commercial Ku band FSS systems, although RR S5.492 foresees that these bands will be used principally for satellite broadcasting.

WRC-97 agreed a provisional procedure whereby non-GSO systems of the FSS might also operate downlinks in these bands; see item 6(b) below.

3 Allocations for the FSS allotment plan

World wide allocations 4500–4800 MHz, 10.7–10.95 GHz and 11.2–11.45 GHz (all for downlinks) and 6725–7025 MHz and 12.75–13.25 GHz (both for uplinks) are used in the FSS spectrum and orbit allotment agreement; see RR S5.441 and Section 9.3 below. RR S22.2 applies in these allocations.

The band 6725–7025 MHz is allocated for feeder downlinks from non-GSO MSS satellites; see RR S5.458B and item 8(b) below. The same interference hazards arise and the same constraints apply here as at 7025–7075 MHz; see item 1 above.

The bands 10.7–10.95 GHz and 11.2–11.45 GHz are also allocated for uplinks for BSS feeder links; see item 7(b) below. The possibility of Mode 4 interference (Figure 5.2) arises and coordination may be necessary to ensure that earth station sites are chosen to avoid this. Provision has also been made for the introduction, on a provisional basis, of downlinks of non-GSO networks of the FSS in these bands; see item 6(c) below.

4 *The government FSS network allocations*

a) World-wide allocations 7250–7750 MHz (downlinks) and 7900–8400 MHz (uplinks) for the FSS are used mainly for government, including military, FSS networks. These bands are shared with terrestrial services and the usual sharing constraints apply. The bands 7250–7375 MHz and 7900–8025 MHz are also allocated for the MSS, using the same directions of transmission as the FSS, but subject to coordination with Administrations concerned; see RR S5.461 and Section 5.3 above.

There are sharing allocations for the MetS at 7450–7550 MHz and 8175–8215 MHz using the same directions of transmission as the FSS. At present some of the MetS satellites operating in the 7450–7550 MHz band are non-GSO but this usage is being phased out; see RR S5.461A and item 5(c) in Section 13.4.1 below.

There is an allocation for the EES at 8025–8400 MHz. The direction of transmission is the reverse of that used in this band by the FSS and therefore the risk arises of interference between earth stations and between satellites (Modes 3 and 4, Figure 5.2). RR S22.5 places a constraint on the PFD with which an EES satellite transmitter may illuminate the GSO. RR S5.462A applies a severe constraint on EES satellite PFD at the ground in Regions 1 and 3 but see item 5(b) in Section 13.4.1 below.

b) World-wide allocations 20.2–21.2 GHz (downlinks) and 30.0–31.0 GHz (uplinks) are also used for government FSS networks. These bands are also allocated for the MSS, operating in the same directions of transmission as the FSS.

In a number of countries there are also sharing allocations for the FS and the MS in these bands, primary at 20 GHz and secondary at 30 GHz. However, RR S5.524 withholds any downlink PFD constraint from the satellite services at 20 GHz whereas RR S5.542 applies the usual RR Article S21 constraint on terrestrial transmitting stations at 30 GHz.

5 *Commercial Ka band GSO network allocations*

World-wide FSS allocations at 17.7–20.2 GHz (downlinks) and 27.5–30.0 GHz (uplinks), plus uplink allocations in Regions 2 and 3 at 24.75–25.25 GHz and 27.0–27.5 GHz, are available for FSS Ka band GSO networks but sharing with other applications of the FSS are likely to place limits on access. In these bands coordination is in general by the procedures of RR Article S9 and RR S22.2 applies; the exceptions are indicated below.

An allocation for the BSS which includes 17.7–17.8 GHz comes into effect in Region 2 in 2007. From that time, the allocation of FSS downlinks will become, in practice, secondary with respect to the BSS; see RR S5.517.

The band 17.3–18.4 GHz world-wide is allocated for FSS uplinks, its use being limited by RR S5.516 and S5.520 to feeder links to broadcasting satellites. See Recommendation ITU-R S.1063 [7] and also items 7(a) and 7(b) below. In the parts of this band which are used for feeder links, it will be necessary to coordinate the downlinks of FSS networks in accordance with RR Appendix S30A Article 7 as well as RR Article S9.

For the bands 17.8–18.6 GHz, 27.5–28.6 GHz and 29.5–30.0 GHz, WRC-97 agreed a provisional procedure whereby non-GSO systems of the FSS might share with GSO FSS systems, using the bands in the same directions of transmission as the GSO networks, coordinating by the procedures of RR Article S9 and with RR S22.2 applying; see item 6(c) below.

The band 18.1–18.3 GHz is allocated for downlinks from geostationary MetS satellites by RR S5.519. Assignments to these satellites will have to be coordinated in the usual way with downlinks from geostationary FSS satellites and feeder links to BSS satellites. See also item 6 in Section 13.4.1.

In Region 2 there are primary allocations at 18.6–18.8 GHz for passive sensors of the EES and the SR. In Regions 1 and 3 the corresponding allocations are secondary. RR S5.523 asks Administrations to limit as much as possible the PFD of FSS downlinks in this band, in order that the quality of measurements made by spaceborne passive sensors should not be avoidably degraded. See also item 3 in Section 13.2.3, below.

The bands 18.8–19.3 GHz (downlinks) and 28.6–29.1 GHz (uplinks) are also allocated for non-GSO FSS networks; see RR S5.523A and item 6(a) below. New FSS networks will be coordinated in accordance with RR Resolution 46 and RR S22.2 will not be applicable.

The bands 19.3–19.7 GHz and 29.1–29.5 GHz are also allocated for feeder links for non-GSO MSS networks, shared with FSS GSO networks in the same directions of transmission. The rules for the frequency coordination procedure and the application of RR S22.2 are complex; see RR S5.523B, S5.523C, S5.523D, S5.523E and S5.535A. See also item 8(b) below. Responding to RR Resolution 121, Recommendation ITU-R S.1324 [8] provides a method for calculating levels of interference to and from links used for these applications; see also Recommendations ITU-R S.1325 [9] and S.1328 [10]. RR S5.541A requires the use of adaptive uplink power control or another method of fade compensation. See also Recommendation ITU-R S.1255 [11].

There is concern that high levels of interference will occasionally enter networks with uplinks close to 30 GHz, due to rain attenuation at the uplink earth station. See for example Recommendation ITU-R S.1255 [11]. To minimise this interference, three world-wide space-to-Earth FSS allocations have been made solely for the implementation of adaptive uplink power control, namely at 27.500–27.501 GHz, 27.501–29.999 GHz and 29.999–30.000 GHz. See RR S5.538 and S5.540. The second of these allocations has secondary status.

The bands 19.7–20.2 GHz (downlinks) and 29.5–30.0 GHz (uplinks) are shared with allocations for GSO MSS networks, secondary in Regions 1 and 3 at 19.7–20.1 GHz and 29.5–29.9 GHz but otherwise primary. It is foreseen that there could be economic advantages for satellite networks operating in these bands if they could be used for both FSS and MSS networks but there is concern that this might considerably reduce the total number of FSS networks that could operate in these bands; see RR Recommendations 715 and 719. Recommendation ITU-R S.1329 [12] is an interim report on the work called for in RR Recommendation 719. It contains analytical material on interference involving spread spectrum emissions and data on FSS and MSS networks that operate in this part of the spectrum. For the present it recommends that these data be taken into account in the design of dual-service systems and asks for more data to be collected. Various *ad hoc* measures intended to mitigate the problem are set out in RR S5.525 to S5.529.

There is no world-wide or Region-wide allocation for terrestrial services in the bands 19.7–20.2 GHz and 29.5–30.0 GHz, but RR S5.524 and S5.542 make primary allocations for the FS and the MS in a number of countries at 19.7–20.2 GHz and corresponding secondary allocations at 29.5–30.0 GHz. However, whereas the usual sharing constraints on terrestrial transmitter power are applied in the latter band, no spectral PFD limit is applied to satellite service downlinks in the former band.

The band 24.75–25.25 GHz (Regions 2 and 3 only) is available for FSS uplinks but RR S5.535 gives priority to feeder links to broadcasting satellites.

The band 27.0–27.5 GHz (Regions 2 and 3 only) is shared with a primary allocation for the ISS. RR S5.537 withholds the application RR S22.2 where ISLs serving non-GSO satellites are coordinated with GSO FSS networks; see also item 3 in Chapter 14.

The band 27.5–30.0 GHz may also be used for feeder links for BSS satellites; see RR S5.539, Recommendation ITU-R S.1063 [7] and item 7(b) below.

6 Commercial non-GSO FSS network allocations

a)   There are world-wide allocations at 18.8–19.3 GHz (downlinks) and 28.6–29.1 GHz (uplinks) for non-GSO FSS networks, sharing with GSO FSS networks and operating in the same directions of transmission; see RR S5.523A, RR Resolution 132, Recommendations ITU-R S.1328 [10] and S.1324 [8], and item 5 above. In future, the coordination of both GSO and non-GSO networks will be carried out under the procedures of RR Resolution 46 and RR S22.2 will not apply.

b)   New downlink allocations were made by WRC-97 for non-GSO FSS networks, sharing with BSS allocations at 11.7–12.5 GHz in Region 1, at 12.2–12.7 GHz in Region 2 and at 11.7–12.2 GHz in Region 3; see RR S5.487A. Pending the outcome of studies being undertaken by the ITU-R, protection of services of the BSS will be in accordance with the terms of RR Resolution 538, namely RR S22.2 is applicable, the equivalent PFD at the Earth's surface due to all the satellites in the FSS constellation is to be within the constraint given in the resolution and frequency coordination is to be based on Article S9.

c)   In addition, RR Resolution 130 expresses the wish of WRC-97 that more spectrum be allocated for non-GSO FSS networks, although with all proper care to avoid unacceptable interference to satellite networks using geostationary satellites. The conference decided to set up a provisional regime for managing sharing between GSO and non-GSO networks in a number of bands, subject to further consideration and possible revision at WRC-00. The main elements of this regime involve the continued applicability of RR S22.2 and frequency coordination in accordance with RR Article S9, together with the application of constraints on the power etc. of non-GSO networks. Established FSS allocations in which this provisional regime is to be implemented are as follows.

  • Uplinks: 12.5–13.25 GHz, 13.75–14.5 GHz, 27.5–28.6 GHz and 29.5–30.0 GHz (all world-wide).
  • Downlinks: 10.7–11.7 GHz, 17.8–18.6 GHz and 19.7–20.2 GHz (all world-wide), plus 12.5–12.75 GHz in Region 1, 11.7–12.2 GHz in Region 2 and 12.2–12.75 GHz in Region 3.

Note that RR S22.5B applies a sharing constraint to the aggregate PFD of a constellation of non-GSO FSS satellites in the direction of the GSO in the band 17.8–18.1 GHz.

7   BSS feeder link allocations

a)   The world-wide FSS uplink allocations at 14.5–14.8 GHz, the Region 1 and Region 3 allocations at 17.3–18.1 GHz and the Region 2 allocation at 17.3–17.8 GHz are limited to use for feeder links for BSS satellites; see RR S5.510 and S5.516. Countries in Europe are excluded from the use of the 14.5–14.8 GHz band. Frequency assignment plans for these bands for feeder links to broadcasting satellites transmitting around 12 GHz are to be found in RR Appendix S30A. See also Section 10.3 below.

b)   Footnotes to the frequency allocation table identify other FSS uplink allocations which may be used for feeder links for broadcasting satellites, either with or to the exclusion of uplinks used for other purposes, but with no explicit requirement that the feeder links be used in accordance with the planned assignments of RR Appendix S30A. The frequency bands involved are 10.7–11.7 GHz (Region 1) (see RR S.484), 14.0–14.5 GHz (except for countries in Europe; see RR S5.506), 24.75–25.25 GHz (Regions 2 and 3; see RR S5.535) and 27.5–30.0 GHz (world-wide; see RR S5.539). See also item 6(c) in Section 10.1.

8   MSS feeder link allocations

a)   No frequency allocations have been made specifically for assignments for feeder links for GSO networks of the MSS; the frequency bands that are used for GSO FSS networks are available and are used for this purpose

b)   However, new FSS assignments in several world-wide allocations are limited to feeder links serving non-GSO MSS networks. The allocations, which are considered in more detail in item 7 in Section 11.1.2, are as follows:

   • 5091–5250 MHz (uplinks); see RR S5.444A and S5.447A,
   • 5150–5216 MHz (downlinks); see RR S5.447B,
   • 6700–7075 MHz (downlinks); see RR S5.458B and items 1 and 3 above,
   • 15.43–15.63 GHz (uplinks and downlinks) (see RR S5.511A),
   • 19.3–19.7 GHz (downlinks) and 29.1–29.5 GHz (uplinks) (see RR S5.523C, S5.523D, S5.523E and item 5 above).

   Frequency coordination in the bands at 5, 6 and 15 GHz is in accordance with RR Resolution 46 and RR S22.2 does not apply. The rules applicable at 19 GHz are complex, but they are set out in the footnotes referenced above.

9   Miscellaneous other FSS allocations
In addition to the bands listed in items 1 to 8 above, there are also the following FSS allocations. Clearly marked applications have not yet emerged for these bands, although progress has been made in preparing a regulatory framework for the use of some of them. Except where the contrary is indicated, these allocations are world-wide.

a)   There are FSS allocations at 2500–2655 MHz (downlinks) and 2655–2690 MHz (uplinks and downlinks) in Region 2 and 2500–2535 MHz (downlinks) and 2655–2690 MHz (uplinks) in Region 3. It should be noted that these FSS allocations are part of a complex pattern of band sharing involving the MSS, the BSS, the FS, the MS etc. and RR S5.415 applies constraints to the FSS to facilitate sharing.

b) There is a world-wide allocation for the FSS at 37.5–40.5 GHz plus one limited to Regions 2 and 3 at 40.5–42.5 GHz, both bands being for downlinks and shared with other services. One of the sharing services at 37.5–40.5 GHz is the FS and consideration is being given to using this allocation for so-called high-density applications; see Section 6.2.4 above. There is some concern that the constraint on the PFD at the Earth's surface allowed for the FSS in this band by RR Article S21 Section V will be too light for compatibility with this FS application and RR Resolution 133 calls for study of how the band might best be shared by the services for which it is allocated. The 40.5–42.5 GHz band, which was allocated to the FSS at WRC-97, is provisionally to enter into effect on 1 January 2001, subject to the outcome of studies of its compatibility with the services for which the band was previously allocated; see RR S5.551E and RR Resolution 134. The objectives of these studies are set out in RR Resolutions 128 and 129; see also RR S5.551B.

c) There is an FSS allocation at 42.5–43.5 GHz (uplinks) and another, also for uplinks, at 47.2–50.2 GHz. RR Resolution 122 calls for studies to be made of sharing the latter band between the various services for which it is allocated. Note RR S5.552, which urges Administrations to reserve the band 47.2–49.2 GHz for feeder links for broadcasting satellites transmitting in the band 40.5–42.5 GHz.

d) There is another FSS allocation for uplinks at 50.4–51.4 GHz. RR Resolution 643 from WRC-95 had asked for studies to be made to identify frequency bands between 50 and 70 GHz that were suitable for allocation to the ISS. Recommendation ITU-R S.1327 [13] offers various possibilities, one of which is 50.4–51.4 GHz, sharing with FSS uplinks. Recommendation ITU-R S.1326 [14] comments in more detail on this latter sharing possibility.

e) A dozen more bands have been allocated for the FSS between 52 and 250 GHz.

9.2 Geostationary FSS networks

9.2.1 Technical factors affecting spectrum utilisation efficiency

Permission for an earth station to make use of the facilities of a satellite is often conditional on the station meeting various technical standards, especially those involving the gain and key polarisation characteristics of the antenna and the receiving figure-of-merit. These requirements would be imposed by the satellite service provider in order that the satellite may be used efficiently.

Section 4.2.4 identifies several other requirements that all geostationary satellites must observe, these requirements having been incorporated in the RR in order that the GSO might be available to as many satellite networks as is feasible. Of these, the most important is the requirement that satellite transmitters can be switched off by tele-command. There are, in addition, requirements for satellite east–west station-keeping and antenna beam pointing accuracy. These requirements were first developed for application in the FSS and they remain of special importance for that service. However, for the FSS, the satellite service under the greatest demand, there is also a need for high standards to be achieved in other respects.

The main factors which determine how much interference enters one network from another, and which indirectly determine how many FSS GSO networks can operate in the same pair of frequency bands, are

- the off-beam gain of the earth station antennas,
- the gain of the satellite antennas towards land areas on Earth outside the intended service area,
- the distribution of power in the spectrum of earth station and satellite emissions, and
- the way in which susceptibility to interference is distributed across the spectrum of the emissions received by earth stations.

Earth station antenna radiation characteristics

The antenna of a typical earth station in a GSO network has a narrow main lobe along the axis of symmetry of the antenna, the maximum value of gain G_0 ranging from about 40 dBi to about 65 dBi, depending on the size of the antenna and the operating frequency. Surrounding this main lobe and concentric with it there will be a series of narrow sidelobes, which in principle fill the spherical space of which the antenna is the centre. The peak gain of the sidelobes decreases as ϕ, the angle relative to the axis of the antenna (angle ϕ in Figure 5.1) rises. For a large, well-designed antenna the gain of the sidelobe peaks may have fallen to 80 dB below the maximum gain of the main lobe at an angle where ϕ equals 60°; for smaller antennas the range of gains is likely to be less. If the GSO is to be used efficiently, it is important that the gain should decline rapidly to a low minimum value.

The principal purpose of Recommendation ITU-R S.465-5 [15] is to provide agreed values of earth station off-beam antenna gain, for use in the calculation of prospective interference in frequency coordination in the absence of measured data. It provides a reference radiation pattern for the peak values of sidelobe gain for antennas operating between 2 and 30 GHz. It is recommended that it be assumed that G_ϕ, the peak gain of sidelobes at an angle $\phi°$ from the axis, is given by

$$G_\phi = 32 - 25 \log \phi \text{ dBi} \tag{9.1}$$

where ϕ lies between $\phi_{min}^°$ and 48°
  $\phi_{min} = 1°$ or $100\lambda/D$ degrees, whichever is the greater, and
  λ is the wavelength.

Where ϕ is greater than 48° it should be assumed that

$$G_\phi = -10 \text{ dBi} \tag{9.2}$$

See Figure 9.3, curve A. It is recommended that a somewhat higher value of G_ϕ should be assumed for smaller antennas ($D < 100$) in networks coordinated prior to 1993 (see Note 4 to the recommendation).

If the antenna to be coordinated is known to have sidelobes of higher gain, those measured values are used, instead of the recommended values, in calculating the interference which the earth station will cause to another station but the recommended values are used when calculating the level of interference that is likely to be received at the station; see, for example Recommendation ITU-R S.466-6 [16].

A design objective for earth station antennas has been established in Recommendation ITU-R S.580-5 [17], in order to bring about an improvement in the sidelobe suppression of new antennas in the directions of greatest concern in the long term. It is recommended for newly procured antennas that at least 90% of the sidelobes

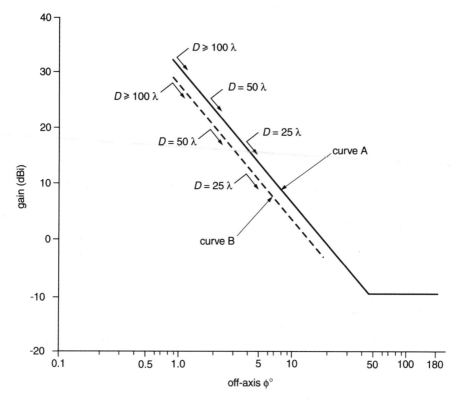

*Figure 9.3   A reference radiation pattern (curve A) for the sidelobe gain of present-day earth
station antennas, based on Recommendation ITU-R S.465-5 [15]. Curve B is a
design objective for new antennas for sidelobes directed towards the area of the sky of
greatest concern (see Figure 9.4 and Recommendation ITU-R S.580-5 [17]). λ is
the wavelength*

which illuminate an area of the sky, between ϕ_{min} and 20° to east and west of the
nominal longitude of the satellite and 3° to north and south of the GSO, as seen
from the earth station (see Figure 9.4) should not exceed the following limits of gain
for big antennas, such that D/λ is greater than 150,

$$G_\phi = 29 - 25 \log \phi° \, \text{dBi} \qquad (9.3)$$

and $\phi_{min} = 1°$.

See curve B in Figure 9.3. Outside the specified area of the sky, it is recommended
that equations (9.1) and (9.2) should be taken as the objective. For smaller antennas,
such that D/λ is between 50 and 150, the design objective is that the gain towards any
part of the defined segment of the sky should not exceed

$$G_\phi = 32 - 25 \log \phi° \, \text{dBi} \qquad (9.4)$$

and $\phi_{min} = 100 \, \lambda/D$.

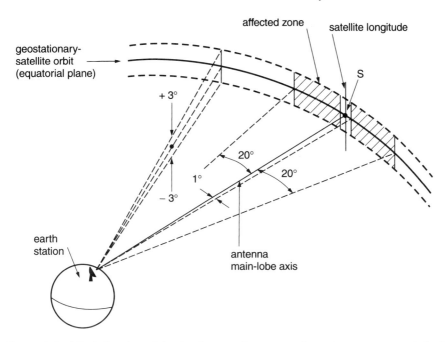

Figure 9.4   The 'affected zone' in the sky, seen from an earth station antenna, to which the sidelobe gain objective (curve B in Figure 9.3) relates (after Figure 1 of Recommendation ITU-R S.580-5 [17])

It is important that earth station antennas should have low sidelobe gain and in particular low gain in the sidelobes that are directed towards the GSO but this is not the only factor of importance. Interference levels are determined primarily by the e.i.r.p. delivered through earth station antenna sidelobes to other satellite networks. Raising the gain of the main lobe of the earth station antennas, thereby reducing the required earth station and satellite transmitter power, also reduces the power fed through sidelobes which causes interference. By requiring system designers to ensure that their earth stations achieve a single performance criterion, which combines the effects of antenna sidelobe gain and main lobe gain, economic benefits can be linked with interference reduction. A multidimensioned criterion of this kind is called a generalised parameter.

Recommendation ITU-R S.524-5 [18] sets maximum permissible levels of off-axis e.i.r.p. spectral density radiated more than 2.5° from the axis of FSS earth station antennas (other than VSATs and planned TV feeder links, limits for which are considered in Sections 9.2.2 and 10.3 respectively) operating at 6 and 14 GHz. The recommendation is rather complicated because it has been found necessary to recommend different constraints for 14 and 6 GHz and for different types of emission, some with the off-axis power integrated over any 4 kHz sampling band, others integrated over 40 kHz and yet others for which the constraint is linked with the power of the carrier of the emission. The constraints are expressed in the recommendation in the form of equations that relate ϕ to an appropriate measure of sidelobe radiation level; the

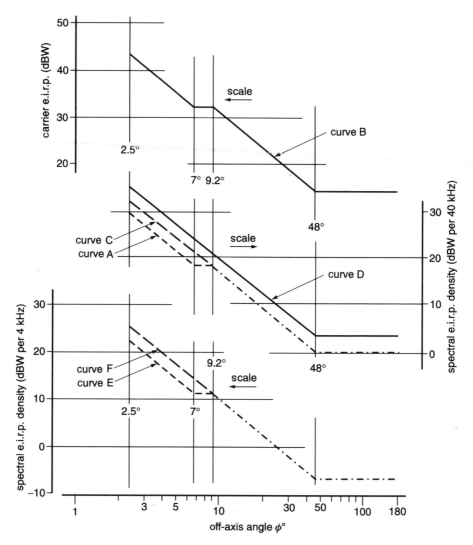

Figure 9.5   Maximum levels of off-axis e.i.r.p. from earth station antennas, based on Recommendation ITU-R S.524-5 [18]. The curves are defined in the text

constraints are presented in graphical form in Figure 9.5. The figure shows the constraints for the following six situations.

- Curve A: emissions operating at 14 GHz, excluding FM analogue video signals, sampled over any 40 kHz band. FM-TV emissions with carrier energy dispersal may exceed curve A by 3 dB, provided that the total power of the emission does not exceed curve B.

- Curve B: the power of the carrier of FM-TV emissions at 14 GHz, operating without carrier energy dispersal but with programme material or an appropriate test pattern always present.
- Curve C: voice-activated telephony single-channel-per-carrier (SCPC) FM emissions operating at 6 GHz.
- Curve D: voice-activated telephony SCPC/PSK emissions operating at 6 GHz.
- Curve E: emissions operating at 6 GHz, other than those of curves C and D (but see Curve F).
- Curve F: earth stations, with emissions as for Curve E, using antennas brought into use before 1988.

Satellite antenna gain characteristics

Orbit and spectrum would be used efficiently if, outside the required footprint and allowing a small margin for imperfect satellite station-keeping and incorrect beam pointing, the gain fell away quickly, so minimising the potential for interference with other networks serving other areas of the Earth.

Satellite antennas are designed to provide not less than a chosen value of gain in every direction that reaches the Earth within a desired footprint, the coverage area. The cross-section of the beam of a simple antenna is circular or elliptical. However, the radiation pattern of such an antenna, extended outwards from the edge of the footprint to include all rays that are intercepted by the Earth, requires a rather complicated mathematical expression to describe it. To provide optimally for more complicated coverage areas, antennas are used with distorted feeds, distorted reflectors or multiple feeds to produce a beam with an intricately shaped cross-section and these have radiation patterns that are even more complex to describe.

It is not feasible at present to define realistic satellite antenna radiation patterns that could be used in estimating interference levels; it is necessary to use measured radiation patterns. However, Recommendation ITU-R S.672-4 [19] provides a six-segment radiation pattern for use as a design objective for antennas providing circular beams, tentatively extendible to antennas with elliptical beams but subject to experimental verification. This may lead in time to a reduction of satellite radiation outside the wanted service area. It is not at present feasible to offer a design objective for more complex antennas.

Polarisation isolation

Many satellite networks provide frequency re-use by orthogonal polarisations. In such cases the system designer or the satellite service provider would impose appropriate standards of on-axis polarisation purity for earth station antennas. For networks that operate in a single polarisation mode it may also be desirable for the cross-polar response of antennas, transmitting and receiving, to be low, possibly providing some discrimination against interference.

At the centre of the main lobe of most earth station antennas the ratio of the gain in the wanted polarisation mode to the gain in the orthogonal mode is high but this high ratio seldom extends into the outer parts of the main lobe and it is usually quite low in the sidelobes. For satellite antennas designed for dual polarisation, high polarisation discrimination is required in all parts of the main lobe that serve the footprint to be covered, but the discrimination is likely to be degraded elsewhere in the radiation pattern. Furthermore, the polarisation state of radio waves in the troposphere may be degraded, especially when scatter-mode propagation is involved.

For these reasons the mandatory procedure for determining the need for frequency coordination between earth stations and terrestrial stations (RR Appendix S7) makes no allowance for polarisation discrimination and no general standards of polarisation isolation have been made mandatory by the ITU. However, the procedure for determining whether coordination between a pair of satellite networks is necessary provides for a limited allowance to be made in some circumstances (see Section 2.2.3 of RR Appendix S8) if the Administrations involved have signified their willingness for this to be done.

In other cases the Administrations may conclude that an allowance for polarisation discrimination can safely be made in calculating levels of interference when detailed coordination is in progress. If so, it will be desirable, perhaps necessary, to base interference calculations on measurements of the polarisation characteristics of the satellite antennas. Recommendation ITU-R S.731 [20] provides a reference cross-polarised radiation pattern for the sidelobes of earth station antennas; the pattern is illustrated by Figure 9.6. Recommendation ITU-R S.736-3 [21] provides a method for making such calculations.

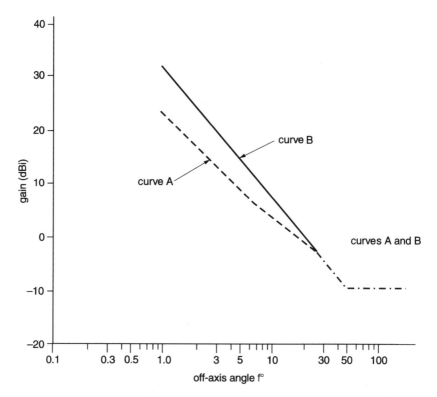

Figure 9.6   A reference radiation pattern (curve A) for the gain of cross-polar sidelobes of an earth station antenna having a main reflector diameter not less than 100 times the wavelength, based on Recommendation ITU-R S.731 [20]. Curve B is the corresponding pattern for the required polarisation, repeated from Figure 9.3

Non-uniformity of emission spectra

GSO networks of the FSS would use spectrum and orbit more efficiently if the spectral energy of the emissions they transmit were distributed uniformly across the frequency band. This ideal might be approached if all emissions were digitally modulated and all the earth stations had antennas with comparable figures of merit (G/T ratio). The trend in telecommunications is towards digital systems but there are many analogue emissions still, including most video links. These analogue links use FM emissions which have a strong concentration of spectral energy at the carrier frequency and analogue telephone links are also particularly susceptible to interference from emissions which have strong concentrations of spectral energy. Finally, the FSS, unlike most other radio services, serves a wide variety of users, for some of which the optimum earth station antenna is big with a high figure of merit whereas a small antenna with a low figure of merit may be essential for the economic viability of others.

The loss of efficiency of orbit and spectrum use that arises from the non-uniformity of the spectrum of individual emissions is mitigated to a considerable extent by carrier energy dispersal; see below. Some further improvement in efficiency can be obtained by choosing frequencies for assignment to emissions in such a way that strong concentrations of interference power in one network do not coincide in frequency with emissions that are highly susceptible to interference in other networks. This harmonisation of frequency assignment plans could be implemented in the course of the frequency coordination process or alternatively by band segmentation. Concepts such as microsegmentation, macrosegmentation and the use of generalised network parameters to facilitate efficient orbit utilisation are discussed in Recommendations ITU-R S.1002 [22] and S.742-1 [23] and some of these concepts have been included in the management of the frequency and orbit allotment agreement, reviewed in Section 9.3.

Carrier energy dispersal

RR S21.16 and S21.17 apply limits to the downlink spectral PFD with which a satellite may illuminate any part of the Earth where the frequency band is also allocated with equal rights for FS or MS. This constraint is applied to prevent unacceptable interference to terrestrial systems via interference mode 8 (see Figure 5.5). Since almost all FSS allocations are shared with equal rights with the FS, this constraint applies to almost all FSS emissions. See also Section 5.2.3 and Figure 5.6.

The downlink carrier power within a sampling bandwidth of 4 kHz that this constraint permits in a frequency allocation below 15 GHz is sufficient for an FM emission with a narrow baseband which is received at an earth station having a high figure-of-merit. Above 15 GHz a 1 MHz sampling bandwidth allows the use of wider basebands or service to smaller earth station antennas. For most FM emissions, however, this constraint requires carrier energy dispersal techniques to be used. Digital emissions do not have such dense concentrations of spectral energy at the carrier frequency but relatively strong spectral lines tend to arise when a repetitive digital sequence is being transmitted, as may happen, for example, when a TDM telephony aggregate is carrying little traffic. Thus energy dispersal may often be necessary for PSK emissions too.

Fortunately, quite simple measures provide a dispersal of energy peaks that is satisfactory for the protection of terrestrial services and with almost no detriment to the FSS network. Recommendation ITU-R S.446-4 [24] recommends that it be used. For this purpose it is usual to add to the baseband of an FDM/FM emission a triangular waveform of sufficient amplitude, having a period of a few milliseconds. For FM video emissions the period is made equal to the duration of one field or one frame; that is between 16 and 40 ms. Where it is necessary to apply energy dispersal to a digital emission, a pseudo-random digital sequence is usually generated at the transmitting earth station and added to the modulating signal. Both the triangular waveform and the pseudo-random sequence can easily be removed from the signal at the receiving earth station. For networks serving earth stations which have a very low figure-of-merit, the degree of energy dispersal that is necessary may be greater than these relatively simple methods can provide, in which case spread spectrum modulation may have to be used.

These techniques of energy dispersal are primarily designed to reduce interference to terrestrial services but they may also facilitate frequency coordination between satellite networks. Recommendation ITU-R S.446-4 [24], in fact, recommends that means for energy dispersal should be available for both purposes. However, it will be seen from Recommendation ITU-R S.671-3 [25] that carrier energy dispersal of an FM video emission by means of a triangular waveform at the frame rate provides little reduction of the interference suffered by a narrow-band emission such as SCPC; in fact its principal effect may be to increase the number of emissions suffering interference.

9.2.2   Very small aperture terminals

Large numbers of earth stations operating in C band or Ku band with antennas having a diameter between 3 m and 1 m have come into use since the mid-1980s. Typically they are located at the end-user's premises, often on the roof of a down-town office building. The traffic loading per station is light in most cases and consequently the pressure to minimise equipment cost is high. Seldom are there any staff with radio technical skills on site. These stations are called very small aperture terminals (VSATs).

This combination of circumstances, which does not arise elsewhere in the FSS at present, has certain consequences.

a)   The antennas of FS stations are often to be found at or above roof level in towns. If the VSAT and FS stations use the same frequency band, frequency coordination may be particularly difficult. It is therefore desirable for VSAT links to be assigned frequencies in frequency bands that are not used in the same area for terrestrial services. The band 14.0–14.5 MHz for uplinks and either 12.5–12.75 GHz (Regions 1 and 3) or 11.7–12.2 GHz (Region 2) for downlinks are more likely to fulfil these requirements.

b)   Having antennas of relatively low gain, VSAT transmitter power tends to be high compared with terrestrial systems of comparable bandwidth and the off-axis e.i.r.p. density also tends to be large. Recommendation ITU-R S.728-1 [26] sets limits on the off-axis e.i.r.p. density which may be radiated by a VSAT. Recommendation ITU-R S.727 [27] sets standards for the axial ratio of antenna polarisation. Recommendation ITU-R S.726-1 [28] sets limits on the spurious emission levels that a VSAT may radiate in various frequency ranges.

c)   It is recognised that malfunction of a VSAT's equipment or storm damage to its antenna may cause the station to disrupt the operation of its own network, or interfere strongly with unrelated networks. Recommendation ITU-R S.729 [29] lists minimal protective facilities that should be provided to reduce the likelihood of such events. See also Recommendation ITU-R S.725 [30].

9.2.3 *International spectrum management for the FSS in the GSO*

Except for the frequency bands considered in Sections 9.3 and 9.4, the use of the GSO for the FSS, as for other services, is managed by the application of the frequency co-ordination and assignment registration procedures set out in RR Articles S9 and S11; see also Section 5.2. The criterion that is used to determine whether it is necessary to coordinate a proposed new network with existing networks, contained in RR Appendix S8, was developed originally for the FSS. It is recognised that the criterion is not fully satisfactory (see Section 5.2.2 above) and studies continue in ITU-R Study Group 4, seeking ways by which the criterion might be made more effective or easier to apply. The basic method set out in Recommendation ITU-R S.738 [31] does not differ in major substance from that of RR Appendix S8. However, ways of facilitating the processes of coordination, appropriate to the circumstances of the FSS, are discussed in Recommendations ITU-R S.739 [32] and S.740 [33].

In the course of frequency coordination it may be found that a satisfactory solution could be reached if the position of the satellite of another network, in operation or soon to be launched, could be moved to another longitude. This may be impracticable, in particular because the antennas of the satellite could not cover the whole of the required service area from the other orbital location. However, small changes of satellite longitude may sometimes yield a significant reduction of interference. Recommendation ITU-R S.670-1 [34] recommends, as a design objective, that a satellite should be designed so that it could operate satisfactorily anywhere within an orbital arc $\pm 2°$ relative to the nominal location, if the configuration of the service area allows this degree of flexibility. Satellite owners are encouraged to increase this flexibility to $\pm 5°$ where possible.

Section 5.2.7 above reviews a examination of the state of work on the frequency coordination of geostationary networks, mostly of the FSS, that had been carried out by the ITU Radiocommunication Bureau in mid-1997. The examination revealed that large numbers of coordinations had been initiated but delays were heavy and relatively few coordinations were being successfully concluded. It is difficult to estimate to what extent this situation arises because the GSO is already carrying as many, or almost as many, satellite networks as is feasible. Saturation may be a substantial factor in the C and Ku bands in certain arcs of the geostationary orbit and certain coverage areas. What is certain, however, is that much delay and many unsuccessful attempts to coordinate arise from causes that are avoidable, albeit at a cost. Some of these causes had already been documented in outline; see for example RR Recommendations 715 and 719 and Recommendations ITU-R S.744 [35] and S.1254 [36].

It seems likely that much of the difficulty being experienced in coordination at the present time arises from the following practices.

a)   An FSS satellite may be equipped to operate in more than one pair of frequency bands, most typically in the C and Ku bands. If, as in this typical case, both band pairs may already be heavily loaded with operating networks and planned new networks that have already been coordinated, finding a single orbit location where the proposed new satellite can be successfully coordinated for all of the frequency bands it is equipped to use will be protracted and may prove to be impossible.

b)   A satellite may be equipped for two applications of the FSS, one using frequency bands which are managed by coordination and the other using bands that are subject to a plan agreement. It is quite likely to be impossible to coordinate the first application for the orbital position previously planned for the second application. The reverse strategy, involving obtaining agreement to a modification of the plan to allow the use of an orbital position that can be coordinated, is more likely to be feasible but such negotiations may be protracted and complex. Similar problems may arise, for example, if a Ka band GSO FSS network is proposed, occupying a band that is subject to more than one management regime in the different parts of the band which it is to occupy.

c)   The coverage area of the antennas of a satellite may extend far beyond the territory which it will, in fact, serve, making coordination more difficult.

d)   An Administration may announce a plan to set up satellite networks using several orbital locations and enter into coordination processes for all of them when in fact its intention is to set up only one network in the firmly foreseeable future.

Of these practices, the first three are attractive to satellite service providers because they can be expected to save space segment cost or extend the potential space segment market and the fourth may lead to an early conclusion of at least one of the coordination operations. Only the last is contrary to the intentions of the RR. But all four waste the efforts of the Administrations that are drawn into the coordination process and obstruct the coordination procedure. Furthermore, by constraining the choice of orbital location or by needlessly extending coverage areas, they all lead to less efficient use of the GSO. WRC-97 made changes to the RR with the intention of curbing some of these practices, and WRC-00 is to give further consideration to the matter.

9.3 The spectrum and orbit allotment agreement

Section 5.4 refers to the concern, mainly of developing countries, for equitable access to the GSO for their FSS networks. To meet this concern, a spectrum and orbit allotment agreement was concluded at WARC Orb-88 to enable all countries to set up a geostationary satellite system of their own if and when they found a need for one.

The agreement is contained in RR Appendix S30B. It allots to each ITU member-country a transmission bandwidth of 800 MHz, with dual polarisation if required, covering the whole of its territory, and a location in the GSO from which it can be served. More than one orbital location is allotted if the territory is so widespread in longitude that it cannot be served adequately from a single location. There is provision for countries to convert their national allotments into a joint allotment, forming what is called a 'sub-Regional system'.

Five world-wide frequency allocations are used for this purpose, in C band at 4500–4800 MHz (downlinks) and 6725–7025 MHz (uplinks), and in Ku band at 10.7–10.95 GHz and 11.2–11.45 GHz (downlinks) and 12.75–13.25 GHz (uplinks). It will be noted that all of these frequency bands are shared with primary allocations for the FS and the MS. It will usually be necessary for the FSS assignments to conform with sharing constraints, and for earth station assignments to be coordinated with terrestrial station assignments, as set out in Section 5.2.3 above.

The basis of the plan

For use in constructing the plan, it was necessary to make broad assumptions about the characteristics of the networks that would use it. The characteristics assumed are stated in Annex 1, Section A of RR Appendix S30B; in brief they are as follows.

a) Earth station antennas will be 7 m and 3 m in diameter at C and Ku bands, respectively, the aperture efficiency will be 70% and the peak sidelobe gain envelopes will conform with equations (9.1) and (9.2). The receiver noise temperatures will be 140 K and 200 K at 5 and 11 GHz, respectively.

b) Satellite antennas will have beams of elliptical or circular cross-section. The half-power beamwidth will be just sufficient to cover the territory of the country or the sub-Regional system receiving the allotment, allowing for beam pointing errors not exceeding 0.1°, although for a small country the beamwidths will not be assumed to be less than 1.6° and 0.8° at C and Ku bands, respectively. The noise temperatures of satellite receivers will be 1000 K and 1500 K at 7 and 13 GHz, respectively.

c) Dual polarisation may be used and no polarisation discrimination was assumed in estimating interference levels between networks.

d) The nominal satellite location will be within an orbital arc, any part of which can be seen from all of the territory in question with a minimum angle of elevation of between 10° and 40°, depending on the climate in the service area. Special rules apply for countries at high latitudes and for very large countries.

e) No assumptions were made about the choice of carrier frequencies or modulation methods. The carrier powers, averaged over the occupied bandwidths of emissions, are assumed to be sufficient to provide a total C/N ratio not worse than 16 dB for 99.95% of the year at 4 GHz and 99.9% of the year at 11 GHz.

With these assumptions, a nominal orbital location was computed for one satellite for every country (and more than one if essential for complete national coverage) and for the proposed sub-Regional systems. The planning objective was to ensure that the ratio of carrier power to total interference from other satellite networks would be not worse than 26 dB, even if all the planned networks were implemented. In the event WARC Orb-88 did not find ways of ensuring that every allotment would achieve the carrier-to-total interference objective and RR Resolution 105 provides for further effort to be given to the improvement of the plan to enable that objective to be attained.

Flexibility

Among the key features of the FSS are the variety of the facilities that it provides, the range of network characteristics that are used and the rate of innovation. The allotment plan was devised to accommodate this diversity. Furthermore, since the agreement was intended to remain in force for at least 20 years, means were provided for amending it from time to time to allow allotments to be added to it for new states as they come into being and for newly agreed sub-Regional systems. The arrangements made to meet these needs affect both emission parameters and orbital positions.

There is no requirement that networks set up to use these allotments should adopt the characteristics assumed at the planning stage. Nevertheless, interference must be kept within acceptable limits even when all the allotments have been taken into use. Thus, when an Administration proposes to implement its allotment the parameters that it proposes to use are tested for compatibility with the reference set of parameters on which the plan is based by means of a system of four generalised parameters. Furthermore, the acceptance of the proposals is facilitated if the concept of macrosegmentation has been applied to the frequency assignment plan.

The generalised parameters are defined in Section B of Annex 1 of RR Appendix S30B (see also Recommendation ITU R S.1002 [22]). Briefly their functions are as follows.

- Parameter A is a measure of the potential interference from an earth station transmitter to the receiver of a foreign satellite, analogous to the off-axis spectral e.i.r.p. density limit of Recommendation ITU-R S.24-5 [18] which is discussed in Section 9.2.1.
- Parameter B is a measure of the susceptibility of the satellite receiver to interference from foreign earth stations.
- Parameter C is a measure of the potential interference from a satellite transmitter to a foreign earth station.
- Parameter D is a measure of the susceptibility of an earth station receiver to interference from a foreign satellite.

Macrosegmentation, as the term is used in this plan agreement, involves ensuring that emissions with high peaks of spectral power are unlikely to interfere with emissions in other networks which are particularly susceptible to narrow bands of intense interference. For the purposes of this agreement, macrosegmentation is considered to have been applied if the upper 60% of each frequency band is used for high density carriers and the lower 40% is used for low density carriers; see RR Appendix S30B, Annex 3B.

The original orbital position allotted to a country in the plan is not necessarily the longitude at which the satellite will be stationed when the allotment is taken into service. Initially, before designing the satellite has begun, the orbital arc $\pm 10°$ relative to the original planned position is called the predetermined arc (PDA) and it may be assumed that the satellite will eventually take up a position within that PDA. If designing has not begun 20 years after the entry of the agreement into force, the width of the PDA will be increased to $\pm 20°$. The width of the PDA is reduced in stages as indicated below once implementation begins, becoming zero when the network enters service.

Implementation

When an Administration proposes to take its allotment into use, it advises the Radiocommunication Bureau, notifying the service area (which may be smaller than the area assumed when the plan was drawn up), the characteristics of the earth station antennas and satellite antennas and general information on the power spectra of the emissions.

If the coverage area as indicated by the characteristics of the satellite antennas does not extend beyond the area assumed when the plan was drawn up, if the generalised parameters revealed by the notified characteristics are also in accordance with the planning values, and if the concept of macrosegmentation has been implemented, the network is considered compliant with the plan. If the value of generalised parameters B and/or D reveal excessive susceptibility to interference, the proposal is accepted but the Administration is warned of the risk of sub-standard performance. If the values of generalised parameters A and/or C reveal excessive potential interference to other networks, or if macrosegmentation has not been implemented, the coordination process set out in RR Appendix S30B Article 6 Section 1A and Annex 4 is applied to protect other allotments, unless the proposals are modified. If the coverage area of the proposed antennas extends beyond the planned coverage area, the proposals are declared non-compliant with the agreement.

When the proposals have been accepted by the Bureau, the nominal orbital position contained in the plan is reviewed in the light of additions to the plan, networks that have already been established and networks that are being prepared for operation. A new PDA is fixed, $\pm 5°$ relative to the planned orbital position, revised as has been found necessary, and the designing of the satellite can begin.

In due course the Administration notifies the frequency assignments for the satellite to the Bureau. The optimum nominal orbital position is considered again, to take the latest situation in orbit into account, and finalised. The PDA is reduced to zero. The Administration notifies the ITU of the frequency assignments to the earth stations and coordinates them with terrestrial stations in other countries in accordance with RR Articles S9 and S11. The network begins operation and the frequency assignments are entered in the MIFR.

The plan of the agreement was drawn up as if no FSS networks had been established in the frequency bands used for the plan. In fact, a number of networks had been notified to the ITU before the planning deadline. A procedure for dealing with the problem that was thus created for countries with allotments incompatible with these pre-established networks is set out in RR Appendix S30B Article 6 Section 1B.

9.4 Non-GSO FSS systems

The GSO has been used for almost all FSS networks so far, but satellites in inclined highly elliptical orbits have also been used in the FSS over many years for coverage areas which this type of orbit can serve more effectively than the geostationary orbit. Moreover, new FSS applications are currently emerging for which satellites in low Earth orbits would be best. It may be noted that the various technical requirements summarised in Section 9.2.1 above, apart from the need for remote control of transmitters, do not apply to non-GSO networks.

Inclined, highly elliptical, orbits for the FSS

An orbit with its orbital plane inclined at an angle of 63.4° to the Earth's equatorial plane, an argument of perigee of 270°, an eccentricity of 0.75 and a period of half the sidereal day is sketched in Figure 9.7. Such an orbit is relatively stable. The perigee is about 500 km above the Earth's surface but a satellite in this orbit spends several hours of each orbital period relatively close to a point in the sky, as seen from the Earth, at the same longitude of its apogee, at a latitude of about 60° N and about 40 000 km above the ground. A satellite in such an orbit has some characteristics like

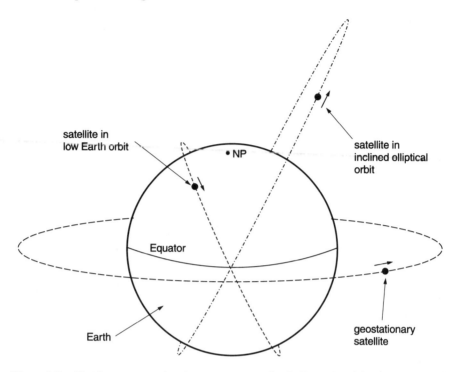

Figure 9.7   Sketches, not to scale, of geostationary, inclined elliptical and low orbits round the Earth

those of a geostationary satellite but it provides better coverage of a large service area at a high latitude which is extensive in longitude, such as Russia. Orbits such as these have been used for Molniya satellites of the Intersputnik system since 1965.

Satellites in orbits like these have been operated in frequency bands mainly used by geostationary satellites. They are coordinated in frequency in accordance with the applicable elements of the RR Article S9 procedure. RR S22.2 (quoted in Section 5.2.1 above) applies but this requirement has not raised problems, since the Molniya satellites are used for their main purposes only when they are far away from the equatorial plane.

Low Earth orbits for the FSS

The low Earth orbits (LEO) to be used for the FSS are circular, with radii in the range 7450 ± 300 km; that is, their heights above the Earth's surface will range between about 780 and about 1400 km. The inclination of the orbital plane relative to the equatorial plane typically ranges from 90° (that is, a polar orbit) to about 50°, the latter being usable if coverage is not required at very high latitudes. Such an orbit is sketched in Figure 9.7. At these heights the size of the instantaneous coverage area of a satellite is much smaller than that provided from high orbits, but continuous coverage is available using constellations containing a large number of satellites in phased orbits.

Seen from an earth station such satellites move rapidly across the sky but the transmission loss between an earth station and a satellite is about 30 dB less than that which arises with satellites in geostationary or elliptical, 12-hour orbits, greatly facilitating the design of hand-portable earth stations.

The FSS allocations 18.8–19.3 GHz (downlinks) and 28.6–29.1 GHz (uplinks) are available for both GSO and non-GSO networks, both being coordinated by the procedures of RR Resolution 46. RR S22.2 does not apply in these bands; see RR S5.523A, RR Resolution 132 and Sections 5.2.1, 5.2.5 and item 6(a) in Section 9.1.2 above. In addition to these allocations, tentative provision has been made for the access of non-GSO networks to other frequency bands subject to various management regimes; see items 6(b) and 6(c) in Section 9.1.2.

9.5 National regulation and spectrum management in the FSS

Service regulation

FSS networks are used commercially for the following main purposes:

- international links between national PSTN networks;
- telecommunications links, ranging in capacity from multi-channel trunk connections to single telephone or data channels, forming part of the PSTN or a special purpose data network;
- occasional audio or video channels, mainly linking programme material from sporting, cultural and news events into broadcasting systems;
- private business networks of many kinds; and
- regular audio or video channels, distributing broadcasting programme material to terrestrial radio broadcasting stations and the head-ends of cable broadcasting networks. By a widely tolerated abuse of the intention of the RR, such TV facilities are also widely used to provide 'direct-to-home' (DTH) satellite broadcasting; see Section 10.1 below.

It may happen that a single company owns or controls satellites and some or all of the earth stations that use them. However, in the general case these two roles are separate; there are satellite service providers and earth station owners. In many cases a single satellite is used simultaneously for several, possibly all, of the purposes listed above. Some earth station owners use the facilities that the satellite network provides for the purposes of their own businesses and the others are telecommunications service providers, making satellite facilities available to corporate users or to the public, typically through the PSTN.

Unless governments rule otherwise, satellite service providers compete internationally to serve earth station owners, subject to whatever geographical limitations may arise from the coverage areas of the satellite antennas and non-uniformities of national or Regional frequency allocation. Likewise, telecommunication service providers owning earth stations may be required to compete to serve end-users. Finally, the FSS may compete with terrestrial radio and cable to provide a transmission medium for all the purposes that these various media can serve. Thus, the FSS is a powerful instrument for the liberalisation of the telecommunication service industry.

It will be necessary for the Regulator and the Administration to collaborate in ensuring that permission to implement the national allotment of spectrum and orbit, and permission to seek to set up satellites for coordination by the other procedures, is granted to appropriate organisations for approved purposes. It will also be necessary for these two agencies to do whatever lies within their proper responsibilities to ensure that access to satellite services is available to telecommunications service providers and direct satellite users. In pursuing both of these roles, regard will be paid to national frequency allocations and the other radio services that operate within their own country in the frequency bands which a foreign satellite network uses.

Spectrum management

In principle an Administration has the same kinds of responsibilities and functions with regard to frequency assignments for the FSS as for other services. The Administration assigns frequencies for transmitting and receiving at any satellite in its charge and notifies the assignments to the ITU. It therefore plays a leading role in the coordination of those assignments with those to other satellites using the same frequency bands, although the technical staff of the satellite service provider usually participate extensively in the coordination of the satellite.

Similarly, the Administration assigns and notifies frequencies for transmitting and receiving at earth stations within its jurisdiction. Consequently it must harmonise its assignments with those it has made to terrestrial stations within its jurisdiction; in frequency bands which are used by satellite systems for both uplinks and downlinks there may also be a need to harmonise assignments made to earth stations with those made to other earth stations. The Administration may also have to coordinate assignments to earth stations with the Administrations of neighbouring countries that may have stations within interference range. Finally, the Administration must ensure that the characteristics and use of satellites for which it is responsible and earth stations within its jurisdiction are in keeping with the provisions of the RR.

However, four factors which arise in satellite services, above all in the FSS, complicate the realisation of these frequency assignment functions. The electrical characteristics of the satellite transponders, antennas and emissions must be taken into account when choosing frequencies for assignment and in coordinating. Secondly, many satellite emissions are uplinked within the jurisdiction of one Administration and downlinked within the jurisdiction of other Administrations. Thirdly, in virtually all present-day satellite networks there is a rigid relationship between the frequency of an uplink carrier and the frequency of the downlink carrier that relays its signal; consequently the choice of an uplink frequency cannot be dissociated from a downlink choice. Finally, the tuning ranges of earth station receivers and transmitters may not be as wide as those of the satellite transmitters and receivers.

A consequence of these factors is that whilst the choice of frequencies for assignment to a link remains ultimately a responsibility of Administrations, the satellite service provider is also involved. Other bodies may be involved as well, such as the owners of the earth stations and indirectly the Administrations of countries that may suffer interference from the earth station.

9.6 References

1 'Methodology for determining the maximum aggregate power flux-density at the geostationary-satellite orbit in the band 6700–7075 MHz from feeder links of non-geostationary satellite systems in the mobile-satellite service in the space-to-Earth

direction'. Recommendation ITU-R S.1256, ITU-R Recommendations 1997 S Series (ITU, Geneva, 1997)

2 'Power flux-density values to facilitate the application of Article 14 of the Radio Regulations for FSS in relation to the fixed-satellite service in the 11.7–12.2 GHz band in Region 2'. Recommendation ITU-R SF.674-1, ITU-R Recommendations 1997 SF Series (ITU, Geneva, 1997)

3 'Ways of reducing the interference from the broadcasting-satellite service into the fixed-satellite service in adjacent bands around 12 GHz'. Recommendation ITU-R S.1067, ITU-R Recommendations 1997 S Series (ITU, Geneva, 1997)

4 'Fixed-satellite and radiolocation/radionavigation services sharing in the band 13.75–14 GHz'. Recommendation ITU-R S.1068, *ibid.*

5 'Limits of power flux-density of radionavigation transmitters to protect space station receivers in the fixed-satellite service in the 14 GHz band'. Recommendation ITU-R M.496-3, ITU-R Recommendations 1997 M Series Part 4 (ITU, Geneva, 1997)

6 'Compatibility between the fixed-satellite service and the space science services in the band 13.75–14 GHz'. Recommendation ITU-R S.1069, ITU-R Recommendations 1997 S Series (ITU, Geneva, 1997)

7 'Criteria for sharing between BSS feeder links and other Earth-to-space or space-to-Earth links of the FSS'. Recommendation ITU-R S.1063, *ibid.*

8 'Analytical method for estimating interference between non-geostationary mobile-satellite feeder links and geostationary fixed-satellite networks operating co-frequency and codirectionally'. Recommendation ITU-R S.1324, *ibid.*

9 'Simulation methodology for assessing short-term interference between co-frequency, codirectional non-geostationary-satellite orbit (GSO) fixed-satellite service (FSS) networks and other non-GSO FSS or GSO FSS networks'. Recommendation ITU-R S.1325, *ibid.*

10 'Satellite system characteristics to be considered in frequency sharing analyses between geostationary-satellite orbit (GSO) and non-GSO satellite systems in the fixed-satellite service (FSS) including feeder links for the mobile-satellite service (MSS)'. Recommendation ITU-R S.1328, *ibid.*

11 'Use of adaptive uplink power control to mitigate codirectional interference between geostationary satellite orbit/fixed-satellite service (GSO/FSS) networks and feeder links of non-geostationary satellite orbit/mobile-satellite service (non-GSO/MSS) networks and non-GSO/FSS networks'. Recommendation ITU-R S.1255, *ibid.*

12 'Frequency sharing of the bands 19.7–20.2 GHz and 29.5–30.0 GHz between systems in the mobile-satellite service and systems in the fixed-satellite service'. Recommendation ITU-R S.1329, *ibid.*

13 'Requirements and suitable bands for operation of the inter-satellite service within the range 50.2–71 GHz'. Recommendation ITU-R S.1327, *ibid.*

14 'Feasibility of sharing between the inter-satellite service and the fixed-satellite service in the frequency band 50.4–51.4 GHz'. Recommendation ITU-R S.1326, *ibid.*

15 'Reference earth-station radiation pattern for use in coordination and interference assessment in the frequency range from 2 to about 30 GHz'. Recommendation ITU-R S.465-5, *ibid.*

16 'Maximum permissible level of interference in a telephone channel of a geostationary-satellite network in the fixed-satellite service employing frequency modulation with frequency-division multiplex, caused by other networks of this service'. Recommendation ITU-R S.466-6, *ibid.*

17 'Radiation diagrams for use as design objectives for antennas of earth stations operating with geostationary satellites'. Recommendation ITU-R S.580-5, *ibid.*

18 'Maximum permissible levels of off-axis e.i.r.p. density from earth stations in the fixed-satellite service transmitting in the 6 and 14 GHz frequency bands'. Recommendation ITU-R S.524-5, *ibid.*

19 'Satellite antenna radiation pattern for use as a design objective in the fixed-satellite service employing geostationary satellites'. Recommendation ITU-R S.672-4, *ibid.*

20 'Reference earth-station cross-polarized radiation pattern for use in frequency coordination and interference assessment in the frequency range from 2 to about 30 GHz'. Recommendation ITU-R S.731, *ibid.*

21 'Estimation of polarization discrimination in calculations of interference between geostationary-satellite networks in the fixed-satellite service'. Recommendation ITU-R S.736-3, *ibid.*

22 'Orbit management techniques for the fixed-satellite service'. Recommendation ITU-R S.1002, *ibid.*

23 'Spectrum utilization methodologies'. Recommendation ITU-R S.742-1, *ibid.*

24 'Carrier energy dispersal for systems employing angle modulation by analogue signals or digital modulation in the fixed-satellite service'. Recommendation ITU-R S.446-4, *ibid.*

25 'Necessary protection ratios for narrow-band single-channel-per-carrier transmissions interfered with by analogue television carriers'. Recommendation ITU-R S.671-3, *ibid.*

26 'Maximum permissible level of off-axis c.i.r.p. density from very small aperture terminals (VSATs)'. Recommendation ITU-R S.728-1, *ibid.*

27 'Cross-polarization isolation from very small aperture terminals (VSATs)'. Recommendation ITU-R S.727, *ibid.*

28 'Maximum permissible level of spurious emissions from very small aperture terminals (VSAT)'. Recommendation ITU-R S.726-1, *ibid.*

29 'Control and monitoring function of very small aperture terminals (VSATs)'. Recommendation ITU-R S.729, *ibid.*

30 'Technical characteristics for very small aperture terminals (VSATs)'. Recommendation ITU-R S.725, *ibid.*

31 'Procedure for determining if coordination is required between geostationary-satellite networks sharing the same frequency bands'. Recommendation ITU-R S.738, *ibid.*

32 'Additional methods for determining if detailed coordination is necessary between geostationary-satellite networks in the fixed-satellite service sharing the same frequency bands'. Recommendation ITU-R S.739, *ibid.*

33 'Technical coordination methods for fixed-satellite networks'. Recommendation ITU-R S.740, *ibid.*

34 'Flexibility in the positioning of satellites as a design objective'. Recommendation ITU-R S.670-1, *ibid.*

35 'Orbit/spectrum improvement measures for satellite networks having more than one service in one or more frequency bands'. Recommendation ITU-R S.744, *ibid.*

36 'Best practices to facilitate the coordination process for fixed-satellite service satellite networks'. Recommendation ITU-R S.1254, *ibid.*

Chapter 10

The broadcasting-satellite service

10.1 Introduction

The terrestrial broadcasting service, by definition, consists of transmissions that are intended for reception by members of the general public; see RR S1.38. The definition of the BSS is broader, embracing individual reception, often called direct broadcasting by satellite (DBS), but also community reception. It is assumed that individual reception will involve simple receiving installations with small antennas, requiring a strong signal from the satellite and very effective protection from interference. Community reception, serving the public typically through cable distribution systems, is assumed to involve more complex receiving installations with larger antennas which do not need such powerful signals. See RR S1.39, S1.129 and S1.130.

The uplinks that are used to feed programme signals from earth stations at fixed locations to the broadcasting satellite are called feeder links. These feeder links are transmitted in uplink allocations for the FSS.

Frequency band planning

WARC-79 recognised that the GSO is likely to be used for the BSS and that there would be advantages in implementing satellite broadcasting in the framework of a plan of satellite locations and channel frequencies. Accordingly it was resolved that the agree ment of a plan should normally precede taking a new frequency allocation for the BSS into use; see Resolution 507. However, it was recognised that an Administration might wish to start using a band before a plan for it had entered into force, and RR Resolution 33 provides a procedure for coordinating and registering a frequency assignment in that event.

Programme material provided by the BSS

So far only one plan agreement has been drawn up for the BSS, using a group of alloca-tions around 12 GHz and feeder link bands at 14 and 17 GHz; see Section 10.3 below. There are no constraints on the kind of material that is broadcast in these allocations. The plan was based on the parameters of analogue, standard definition (625 or 525 lines per frame) colour TV but a planned assignment may be used for TV with other parameters or sound broadcasting systems instead, provided that it causes no more interference, nor needs more protection from interference, than the TV system assumed in planning. There are constraints on the use of some other BSS allocations, some being limited to systems for community reception, high definition television (HDTV) or digital sound programmes.

Limitations of this kind have been foreseen to be necessary in order to provide a technical basis for planning, above all for defining the channel bandwidth for single channel analogue emissions. Such constraints will probably be found to be less necessary in the future, when multiplexed digital systems may be used to provide a mixture of kinds of programme multiplexed into a high bit rate aggregate; see, for example, Recommendation ITU-R BO.1294 [1].

Frequency allocation sharing

Most BSS frequency allocations are shared with the BS. The intention of this provision has been to facilitate the use of terrestrial broadcasting to fill gaps in satellite coverage, within the service area of the satellite broadcast, for example owing to the obstruction of a propagation path by hills, large buildings and so on. Used in this way, interference between the two services should not arise.

However, all BSS allocations are shared with other services and in particular with the FS and the MS, the BSS allocations being primary but the sharing allocations being, in most cases, also primary. The same sharing situation arises in the FSS allocations that are available for feeder links to the broadcasting satellites. To prepare the way for planning the 12 GHz BSS bands and the associated feeder link bands, the allocations for sharing services were in effect down-graded. Footnotes to the international table of frequency allocations and the terms of RR Appendices S30 and S30A require the emissions of stations of services which are neither the BSS nor the BS to avoid causing interference to broadcasting signals. The footnotes also deny these stations protection from BSS interference; see Section 10.3.3. No doubt there will be a similar curtailment of the status of allocations for other services in the bands that are planned for the BSS in the future. Broadcasting satellites operating in unplanned bands, coordinated in accordance with RR Resolution 33 (see above), will not have these advantages.

Frequency allocations for the BSS and feeder links

The BSS allocations and the bands available for feeder links for broadcasting satellites are as follows.

1 Television broadcasting at 700 MHz
RR S5.311 permits broadcasting by satellite of analogue TV using FM in the band 620–790 MHz, subject to a constraint on the PFD at the Earth's surface quoted in that paragraph (see also RR Recommendation 705). Use of this allocation is subject to the agreement of Administrations operating services (mainly the BS, FS and MS) also having allocations in that band.

2 BSS allocations between 1452 and 2670 MHz

a) There is a world-wide allocation for the BSS at 2520–2670 MHz. The use of this allocation is limited to national and regional systems for community reception, subject to agreement with other affected Administrations by the mechanism of RR S9.21 (see RR S5.416 and Section 5.3 above). Frequency coordination is to be in accordance with RR Article S9. The band is shared with various other services, satellite and terrestrial, most having equal primary status. The constraint on the PFD of BSS emissions at the Earth's surface is as shown for this frequency range in Figure 5.6. See also the allocation made by RR S5.418 in the next paragraph.

b) There is an allocation for the BSS and also the BS at 1452–1492 MHz, world-wide excepting for the USA, where the corresponding allocations are at 2310–2360 MHz; this latter band is also allocated for the BSS and the BS in India and Mexico; see RR S5.344 and S5.393. For several countries listed in RR S5.418 there is another allocation for the BSS and also the BS at 2535–2655 MHz (see the previous paragraph). These allocations to the BSS and the BS are limited to S-DAB and T-DAB as appropriate, the BSS being planned in accordance with RR Resolution 528; see RR S5.345, S5.393 and S5.418 and Section 10.4 below.

3 BSS allocations around 12 GHz

a) The 12 GHz planned bands: BSS allocations between 11.7 and 12.7 GHz, differing in detail in the three Regions, and FSS bands for feeder links around 17 and 14 GHz, were used for frequency assignment plans drawn up between 1977 and 1988; see Section 10.3. RR S5.492 permits these planned assignments to be assigned for downlinks of the FSS provided that such transmissions do not cause more interference nor require more protection from interference than the planned BSS transmissions. However, it is expected that these bands will be used mainly for the BSS. WRC-97 allocated these bands also for the FSS, to be used for downlinks from non-GSO systems in accordance with the terms of RR Resolution 538; see item 6(b) in Section 9.1.2 above.

b) There is an allocation for the FSS in Region 2 at 11.7–12.2 GHz, but RR S5.485 permits FSS satellite transponders operating in this band to be assigned for satellite broadcasting, subject to constraints on the power of the downlink emissions and limits on interference requirements. It is expected that this band will be used principally for the FSS.

c) The BSS allocation in Region 3 at 12.5–12.75 GHz, shared with FSS downlinks, is limited by RR S5.493 to community reception and a constraint on downlink PFD. The procedures of RR Resolutions 33 and 34 apply until a frequency plan is agreed for the BSS.

4 BSS allocations, 17–22 GHz

a) In Region 2 an allocation for the BSS at 17.3–17.8 GHz comes into effect in the year 2007. From that date, downlinks for networks of the FSS, for which the band 17.7–17.8 GHz is also allocated, may not cause harmful interference to systems of the BSS and will have no protection from interference from the BSS; see RR S5.517. The band 17.3–17.8 GHz is also allocated in the uplink direction for planned feeder links to broadcasting satellites serving Region 2 in the 12.2–12.7 GHz band; there are criteria in Section 1 of Annex 4 of RR Appendix S30A for determining whether coordination is required between those feeder links and these BSS emissions; see also RR S5.515.

b)   The BSS allocations in Regions 1 and 3 at 21.4–22.0 GHz enter into effect in the year 2007. RR Resolution 526 records the decision of WARC-92 that these allocations should be used for HDTV using wide RF bandwidth. In 1992 the MUSE system of HDTV was in service in Japan; this system is described in Recommendation ITU-R BO.786 [2]. The HD-MAC system, had also been developed; see the signal specification in Recommendation ITU-R BO.787 [3]. See also Report ITU-R BO.1075-2 [4]. RR Resolution 525, also of WARC-92, recognised that the development of HDTV equipment for the mass market was not complete at that time but it provided an interim procedure whereby experimental systems might operate at 21.4–22.0 GHz in Regions 1 and 3 before 2007. The resolution foresaw that a plan would be agreed for the operation of the BSS in these bands at some time after 2007, and that this would give the BSS allocation a higher status than those of the FS and MS for which the bands are allocated already.

5 *BSS allocations above 22 GHz*

a)   There is a world-wide allocation for the BSS at 40.5–42.5 GHz. A sharing allocation was made for the FSS (downlinks) in Regions 2 and 3 in this band at WRC-97 but there was some unease about the feasibility of this sharing arrangement. RR Resolution 134 placed a delay on the entry into force of the FSS allocation, pending the outcome of studies called for by RR Resolutions 128 and 129; see RR S5.551B, S5.551D and S5.551E and also item 9(b) in Section 9.1.2 above.

b)   There is a world-wide allocation for the BSS at 84–86 GHz.

6 *Feeder links for the BSS*

a)   Specific bands, at 14 and 17 GHz, allocated for the FSS have been reserved and planned for the uplinks to satellites broadcasting in the planned BSS allocations around 12 GHz; see Section 10.3.

b)   RR S5.552 urges that the FSS uplink allocation at 47.2–49.2 GHz be reserved for feeder links to satellites broadcasting in the band 40.5–42.5 GHz.

c)   Footnotes to the frequency allocation table identify other FSS uplink allocations for use for feeder links to broadcasting satellites, either together with or to the exclusion of uplinks used for other purposes. There is no explicit requirement that the feeder links will be used to serve satellites broadcasting in a specified frequency band. Thus Region 1 allocation 10.7–11.7 GHz and world-wide allocation 18.1–18.4 GHz (see RR S5.484 and S5.520) may only be used in the Earth-to-space direction for BSS feeder links. However, a world-wide allocation 14.0–14.5 GHz (limited for feeder links to countries outside Europe), Regions 2 and 3 allocation 24.75–25.25 GHz and world-wide allocation 27.5–30.0 GHz (see RR S5.506, S5.535 and S5.539) may be used for BSS feeder links as well for as other uplinks. See also item 7 in Section 9.1.2 above.

Technical standards by regulation

The technical standards for geostationary satellites noted in Section 4.2.4 apply to broadcasting satellites except where more stringent requirements are set in the 12 GHz plan (see Section 10.3.1). There are no specific limits at the time of writing on the spurious emissions radiated by satellite transmitters but RR Appendix S3 Section II imposes a limit on new satellites launched after the year 2002 and on all satellites after 2012; see Section 4.2.3. Even so, it is likely that spurious emissions from broadcasting satellites will remain a matter of concern for radio astronomy, since the transmitter power required for satellite broadcasting is high and especially so for satellites that are fully compliant with the 12 GHz plan. For satellites broadcasting in this frequency range, the problem is aggravated for radio astronomy by the coincidence in frequency of second harmonic radiation from satellites broadcasting at about 11.9 GHz and the particularly important RAS allocation at 23.6–24.0 GHz.

'Direct to home' satellite broadcasting

Implementation of satellite broadcasting in the 12 GHz planned bands, fully compliant with the planned parameters, has been slow. Some planned frequency assignments have been implemented with less demanding 'interim' parameters; see RR Resolutions 42 and 519. Other planned assignments have been taken into use after changes in the planned service area have been negotiated, making implementation more economically viable. But most of the extensive implementation of satellite broadcasting that has taken place in recent years has used frequency allocations, not of the BSS, but of the FSS.

FSS networks have been used extensively and are still being used for the distribution of television programme signals to terrestrial broadcasting stations and to the head-ends of cable broadcasting networks, using C band or Ku band for downlinks. In the mid-1980s, large numbers of unlicensed receive-only earth stations, with antennas up to a few metres in diameter, were being installed at their homes by members of the public to intercept these programme signals for their domestic enjoyment. This practice became known as direct-to-home (DTH) satellite broadcasting.

Some broadcasters welcomed the transnational outlet that the FSS offered for their programmes. Other broadcasters disapproved of the loss of copyright for material that DTH involved, but this problem was resolved, where particular sensitivities arose, by the emergence of effective conditional access devices. Administrations disapprove of breaches of regulations, not least one involving the use by the public of frequency bands that are already used extensively for FS and FSS services, but they have been reluctant to take up a posture that would be very visible to the general public and likely to be seen as pointlessly legalistic. Thus many Administrations have tolerated the emergence of DTH although without undertaking to resolve problems of interference from legitimate users of the band. Finally, satellite service providers have welcomed a major new market. Recent satellite designs for the FSS, with DTH broadcasting in view, have transmitters with increased power which, taken together with improvements in domestic receiver design, have substantially reduced the required domestic antenna size. DTH broadcasting is probably the dominant activity in the FSS today.

10.2 Spectrum management and regulation of satellite broadcasting

Spectrum management for DBS in planned bands at 12 GHz

Several DBS frequency assignments at 12 GHz, typically five, all to be transmitted from the same orbital location, have been planned for each country. When the Administration has accepted a proposal from a service provider for bringing some or all of those assignments into use, it will be necessary to plan the feeder link earth station and design the satellite, or the relevant sub-systems of a shared satellite, so that firmly based emission and antenna parameters can be calculated. Then the Administration notifies the parameters of the proposed emissions to the ITU Radiocommunication Bureau in accordance with the administrative procedures summarised in Section 10.3.1 below.

When acceptable parameters have been registered in the MIFR, the Administration makes the assignments to the feeder link earth station, the satellite receivers and the satellite transmitters and licenses their use. The Bureau is notified when a registered assignment is brought into use, and the date is added to the entry in the Register. It will clearly be necessary for the Administration to ensure that no assignments are made within its jurisdiction for stations of other services which would interfere with the satellite broadcasting service. It is likely to be necessary for the Administration to coordinate with the Administrations of neighbouring countries to ensure that the BSS service is protected from interference; see Section 10.3.3 below.

The Administration will, no doubt, make a charge for a DBS licence. Having regard to the potential profitability of the service in favourable markets and the limited availability of planned assignments, the charge may be substantial. There may indeed be competition for the opportunity to provide the service.

Spectrum management in unplanned BSS allocations

Most frequency bands allocated for the BSS are, at present, unplanned. The RR do not encourage the use of these allocations in advance of the planning process. Perhaps this is because the presence of systems already operating in a band when the time comes to draw up a plan for it makes planning more complex, may make the plan less efficient and may give an inequitable share of the future use of the band to particular countries. However, RR Resolution 33 provides a procedure for coordinating frequencies for broadcasting in unplanned BSS bands.

There are two elements in the RR Resolution 33 procedure. The purpose of one element is the coordination of frequencies proposed for the downlink broadcasting emissions with the assignments already made and registered for uplinks and downlinks for other satellite systems operating in the same frequency band. The procedure is based on RR Articles S9 and S11, with minor variations.

The other element of RR Resolution 33 provides for the coordination of the broadcasting emissions with foreign stations of sharing terrestrial services. There is no close equivalent of this process in RR Article S9. The procedure is simple in principle. The Administration proposing the broadcasting emissions provides the necessary information to the ITU and this information is published in the 'Weekly Circular'. An Administration that foresees a prospect of interference to its services is required to inform the proposing Administration and the ITU of its concerns. It may be that, after detailed coordination, it will be found that the interference will not be unacceptable or that a solution to the problem can be found. For the proposal to go forward in the absence of a solution might lead to interference complaints and a costly change in the proposals for broadcasting; see Section 4.5.

The feeder link to the broadcasting satellite will operate in an FSS allocation and it will be coordinated by the procedure of RR Articles S9 and S11; see Sections 5.2.1 and 5.2.3 above.

Spectrum management for DTH broadcasting

In regulatory terms, DTH broadcasting is done by networks of the FSS. It is subject to the constraints that are applied to FSS networks by RR Articles S21 and S22, most significantly perhaps the constraint on the PFD at the Earth's surface applied by RR S21.16. Satellites, feeder link earth stations, representative receiving earth stations and the links between them are coordinated, and their assignments registered, in accordance with RR Articles S9 and S11; see Sections 5.2.1 and 5.2.3 above.

Licensing satellite broadcasters

A government will usually wish to regulate the use which broadcasting service providers make of the medium. For DBS there is a close analogy with terrestrial broadcasting. Access to radio frequencies, feeder link and downlink, is firmly under the control of the government of the country receiving most of the broadcasting coverage, although this may not be true of the outer fringes of the coverage area. For DTH however, circumstances may differ significantly from the terrestrial situation.

Thus any Administration on Earth may take responsibility for a satellite used for DTH broadcasting and for assigning and coordinating the frequencies which it receives and transmits. Feeder link earth stations may be located anywhere within sight of the satellite. The service area of the downlink beams may include the territory of several or many countries; the consent of their Administrations is not necessary. Finally, whereas there may be a shortage of assignable frequencies for DBS, there may be ample for DTH. Thus the control which a Regulator can exercise over the DTH broadcasting that reaches his country is quite weak. The approach of the United Kingdom to the regulation of satellite broadcasters is offered as an illustration what may be done.

Broadcasting regulation in the United Kingdom

The function of setting up a satellite and a feeder link earth station for the transmission of broadcasting programmes by DBS or DTH is closely analogous to that of the 'local delivery service' provider in the scheme of terrestrial broadcasting regulation which is contained in the Broadcasting Act of 1990; see Section 7.1.2 above. However, the Act, which also deals with service provision for satellite broadcasting, makes no provision for service delivery; the Administration has responsibilities in this area flowing from the Wireless Telegraphy Act of 1949.

The Act of 1990 recognises a need to license two kinds of satellite broadcasting service for television, namely a 'domestic satellite service' and a 'non-domestic satellite service'. In the terminology used in this book, a domestic satellite service is television by DBS, the feeder link earth station being in the United Kingdom. Such services are effectively under regulatory control, since the assignment of BSS frequencies from the plans of Appendices S30 and S30A is a function of the Administration. A non-domestic satellite service is television by DTH, receivable in the United Kingdom and/or elsewhere, for which the feeder link earth station is located in the United Kingdom or in another country that has been prescribed by the United Kingdom government. This category also includes a DTH service for which the feeder link earth station is elsewhere but the content of the programme material is, in whole or in part, controlled from the United Kingdom. The Act of 1990 takes into account the reality that the control of the United Kingdom Regulator over non-domestic satellite services is rather nebulous.

The Independent Television Commission would grant 15-year licences to broadcasting service providers for domestic satellite services by DBS, following procedures similar to those applied to terrestrial programme three; see Section 7.1.2 above. Fees would be based on auctions plus a sum related to income from advertising if the degree of competition justifies it. The terms on which the Commission may grant a television service provider's licence for DTH, including the fee (if any) are at the Commission's discretion.

The Broadcasting Act of 1996 authorises the licensing of service providers to assemble digital programmes supplied by broadcasting service providers into multiplexes for transmission.

The Act of 1990 gives powers to the Independent Television Commission to enforce the terms of a domestic satellite service licence, including the acceptability of programme contents. The Act of 1996 extends these powers to cover licensed non-domestic services.

The Act of 1990 authorises the Radio Authority to license sound radio programmes by satellite which are transmitted from feeder link earth stations in the United Kingdom or are controlled, as to programme content, from the United Kingdom. This role is similar to that of the Independent Television Commission with regard to satellite television broadcasting.

Finally, the Act of 1990 defines a third kind of satellite programme service, namely a 'foreign satellite service'. A foreign satellite service is one that is capable of being received in the United Kingdom but the feeder link earth station is outside the United Kingdom. The Commission has no role in the licensing of a foreign satellite service. However, if there should be sufficiently grave complaints about the programme material that is broadcast, there is a mechanism for prosecuting persons in the United Kingdom who are involved in various ways with the service. This procedure applies to both TV and sound broadcasting by satellite.

10.3 The 12 GHz frequency assignment agreement for TV

10.3.1 The agreement and the plans

The sequence of planning processes

WARC-71 allocated various frequency bands to the BSS and its Resolution Spa2-2 called for early action in preparing frequency plans for the service. This haste did not arise solely from an expectation that satellite broadcasting would come quickly into widespread use; indeed many Administrations thought that the technology required for low-cost satellite broadcasting was very immature. But there were two other reasons for early planning. For some Administrations the urgent reason for planning the bands that the BSS was to share with various terrestrial services was that an agreed plan for BSS would facilitate the use for the FS and the MS of spectrum not required in a given location for satellite broadcasting. Other Administrations were concerned that developing countries might not be able to get equitable access to the medium if planning were delayed until satellite broadcasting had begun.

WARC SAT-77 was convened to prepare plans for the BSS in the three Regions in allocations around 12 GHz. It was agreed that all satellites operating in these bands must be geostationary; this decision now forms RR Resolution 506. It was also agreed that the plans would take the form of defined orbital locations and frequency channels that a named Administration would be free to assign for broadcasting to specified service areas within its own jurisdiction at any time. However, a considerable number of Administrations, in particular in Region 2, wanted the planning process to be delayed pending progress in technical development that could be expected to enable more realistic parameters to be employed in planning.

In the event, plans were prepared and agreed at WARC SAT-77 for downlinks in the band 11.7–12.5 GHz and 11.7–12.2 GHz for Regions 1 and 3 respectively, and regulations that would be required to manage the use of the plan were drafted and agreed. These plans and regulations, together forming the agreement, became the first version of RR Appendix S30. Planning for Region 2 was postponed. WARC SAT-77 recognised that plans for feeder links, complementary to the downlink plans, were also needed, but there were no suitable FSS allocations at that time that could be used for the purpose without unacceptable harm to the development of the FSS and WARC SAT-77 had no authority to make new allocations.

The ITU Administrative Council decided that an RARC should meet in 1983 to prepare a plan for Region 2. WARC-79 decided that the Region 2 plan for downlinks should be within the frequency range 12.1–12.7 GHz and allocated 17.3–18.1 GHz world-wide for FSS uplinks, the use of this band being limited to feeder links for broadcasting satellites. However, there was concern that radio propagation degradations in the troposphere would make that band unsatisfactory for feeder links from high rainfall areas. Accordingly, another FSS uplink allocation was made at 14.5–14.8 GHz, to be used for the planned feeder link assignments of countries that wished to use it, although countries in Europe were excluded from that option; see RR S5.510.

RARC-83 prepared plans for Region 2 for downlinks in the band 12.2–12.7 GHz and feeder links in the band 17.3–17.8 GHz. However, there had been progress since 1977 in the design of domestic receivers, in particular low noise amplifiers. There had also been other significant advances in knowledge of the effect of rain on wave polarisation and in the design of computer software for plan optimisation. Consequently there are significant differences between the results of the 1977 and 1983 planning exercises. The downlink plan was incorporated by WARC Orb-85 into RR Appendix S30 and the Region 2 feeder link plan became the first version of RR Appendix S30A. Finally, one of the main tasks of WARC Orb-88 was to prepare feeder link plans for Regions 1 and 3 in the bands 17.3–18.1 GHz and 14.5–14.8 GHz and to edit the various plans and regulations into a harmonious whole in revised RR Appendices S30 and S30A.

The plans

All countries, including the smallest, stated a requirement for satellite broadcasting facilities or were assumed to have a requirement. A few groups of countries asked for at least one of their planned assignments to be given coverage of the whole group but the great majority of assignments were to be limited to national coverage. For large countries a separate group of assignments was planned for each time zone. In Region 1, and especially in the arc of the GSO that serves Africa and Europe, the density of requirements is highest because there are so many separate states to be served. It was agreed that the Region 1 plan should provide an equal number of channels for each country or time zone, the number to be as large as was feasible. Maximising the number of channels calls for system technical standards that are stringent, uniform and detailed. WARC SAT-77 found that it could provide five TV channels per country and per time zone. The other Regions accepted similar basic technical standards, each country specifying the number of channels that should be planned for it.

The technical assumptions and standards which form the basis for planning are set out in Annex 5 of RR Appendix S30 and Annex 3 of RR Appendix S30A. They are too extensive and detailed to be quoted here in full but the following outline is indicative of the whole.

1 Satellite station-keeping

Satellites must remain within $\pm 0.1°$ of their nominal longitude. In Regions 1 and 3 their north–south perturbation must also be limited to $\pm 0.1°$; in Region 2 this standard is recommended but not required.

2 Satellite antenna beams

The deviation of an antenna beam, transmitting (for broadcasting) or receiving (for the feeder link), from its nominal direction must not exceed 0.1°. The maximum angular rotation of a beam about its nominal orientation is to be $\pm 2°$ in Regions 1 and 3 but $\pm 1°$ in Region 2. Required service areas, which might be smaller than the national territory but were not to go beyond it except for agreed joint coverage beams, were stated by Administrations. It was assumed that the beam would be the smallest beam of circular or elliptical cross-section that could enclose the required service area at its half-power gain contour, allowance being made for the permitted angular deviations quoted above. For the smaller service areas it was assumed that the half-power beamwidth would be not less than 0.6° in Regions 1 and 3, nor 0.8° in Region 2. Stringent limits are placed on the satellite antenna gain outside the 3 dB contour.

3 Polarisation

Both senses of circular polarisation are used for feeder links and downlinks, providing frequency re-use.

4 Signal and radio channel standards

It was assumed that 625-line or 525-line analogue colour television systems would be used, such as are described in Recommendation ITU-R BT.470-4 [5] for terrestrial use. The baseband which frequency modulates the carrier would consist of the luminance waveform plus sub-carriers for chrominance and sound signals in the same frequency relationships relative to the luminance signal as they bear to the carrier in VSB emissions used terrestrially. A typical 625-line system used in this way has a necessary bandwidth of 27 MHz; for 525-line systems the necessary bandwidth is 24 MHz.

Recommendation ITU-R BO.650-2 [6] subsequently recommended the use of one of the multiplex analogue components (MAC) family of systems as an alternative to the conventional systems covered by Recommendation ITU-R BT.470-4 [5]. There are brief descriptions of the MAC systems in Recommendation ITU-R BT.650-2. However, the conventional systems are most commonly used; the MAC systems give better performance but they are more expensive to interface with conventional domestic receivers.

Recently various digital television systems, compatible with the plan's channel standards, have been developed and some have been brought into use. Recommendation ITU-R BO.1211 [7] describes a framing structure, channel coding and modulation methods for a flexible multiplexed system, closely akin to those of the system described for terrestrial use in Recommendation ITU-R BT.1209-1 [8]; see Section 7.5.2. Recommendation ITU-R BO.1294 [1] places this system in context with two other similar systems which are also in use and recommends that one of these three systems should be used for digital satellite television.

5 Frequency plan

For Regions 1 and 3, predominantly using 625-line TV, the frequency separation between planned carrier frequencies was 19.18 MHz. For Region 2, mainly using 525-line TV, the corresponding figure was 14.58 MHz. Thus allowance had to be made for substantial interference from adjacent channels.

6 Orbital plan

Satellites are to be located somewhat to the west of the longitude of the service area, so that they would remain in full sunshine until past midnight at all times of year, perhaps reducing the battery capacity required for use during the eclipse season. For Region 1, clusters of satellites using both senses of polarisation are to be located at nominal intervals of 6° of longitude and a similar, though somewhat less regular, arrangement is used for Region 3.

RARC-83 planned with more flexible software and was less confident than WARC SAT-77 about the degree of polarisation discrimination that would be obtainable. It was found that more channels could be obtained for Region 2 by using a larger number of satellite clusters, with less regular orbital spacing. However, each Region 2 cluster was divided into two sub-clusters, one 0.2° to the east of the nominal position of the cluster for satellites using left-hand circular polarisation and the other sub-cluster 0.2° to the west of the nominal position of the cluster and using right-hand circular polarisation.

7 Receiving earth stations

It was assumed in planning for Regions 1 and 3 that the figure of merit (G/T) of domestic receiving systems for individual reception would be 6 dB(K^{-1}) and that the half-power beamwidth of the antenna would be 2°. This combination would be achievable, for example, with an antenna reflector 80 cm in diameter and a noise temperature of 1400 K. The corresponding values for Region 2 are 10 dB(K^{-1}) and 1.7°. Given these G/T assumptions and the emission parameters implied above, minimum values of PFD in the service area of minus 103 dB(W/m^2) and minus 107 dB(W/m^2) would be required for 99% of the time to provide the performance objective in Regions 1 and 3 and Region 2 respectively, taking predicted radio propagation degradations into account. It may be noted that the satellite transmitter power required to serve a large country with a rainy climate, using the assumptions applied to Regions 1 and 3, may be very large.

8 *Feeder link earth stations*

For planning purposes it was assumed that the transmitting antennas at the feeder link earth stations would have a diameter of 5 m at 17 GHz. The corresponding size for 14 GHz feeder link earth stations was 6 m. However, provision was also made for smaller antennas. There are stringent limits on off-beam radiation. It was found to be important for the uplink signals reaching the various satellites forming an orbital group to be relatively uniform in power level despite attenuation due to heavy rain in the vicinity of a feeder link earth station. Various options, including adaptive control of feeder link transmitter power, were prescribed for ensuring this uniformity.

The administrative provisions for managing the plans

Article 4 of RR Appendix S30 and Article 4 of RR Appendix S30A provide procedures whereby an Administration may seek the agreement of all other Administrations that would be affected to proposals for a modification of the plans for the BSS downlinks and the feeder links respectively.

When an Administration intends to bring into use a frequency assignment for satellite broadcasting in a band covered by these plans, it notifies the ITU Radiocommunication Bureau in accordance with Article 5 of RR Appendix S30. The notification is to be made not more than three years nor less than three months before the assignment is to be brought into use. The procedure for notifying a proposed feeder link frequency assignment in a planned feeder link band is to be found in Article 5 of RR Appendix S30A. These notifications must include full details of the system to be used. The details are published in the ITU 'Weekly Circular'.

If, on examination, the Bureau finds that the notification is in full conformity with the plan or is in conformity except for certain specified characteristics (see paragraph 5.2.1(c) of Article 5) which will have no adverse effect on the interests of other Administrations, the assignment is entered in the MIFR. If departures from the planned characteristics go beyond these specified exceptions, the intended system may nonetheless be agreed to fall within the definition of an 'interim system' in RR Resolution 42 (see also RR Resolution 519), allowing the assignment to be given a qualified entry in the MIFR. Otherwise, the Bureau will not register the notification until the proposing Administration has successfully used the plan modification procedure to align the plan with the proposals.

Other articles of RR Appendices S30 and S30A deal with the regulation of sharing of the planned bands with other services; see Section 10.3.3 below.

10.3.2 The implementation and evolution of the agreement

The take-up of the frequency assignments planned for the BSS at 12 GHz has been limited and some of the assignments that have been implemented have had the non-standard characteristics of 'interim systems'. During the same period of time, DTH satellite broadcasting has grown extensively, despite the questionable security of the allocations that are used for it.

With hindsight, this outcome can be seen to have had several possible causes. A theme that dominated government thinking when satellite broadcasting was first given serious consideration was limitation of the coverage of the broadcasts, as completely as is technically feasible, to the territory of the licensing Administration, in the absence of the agreement of the countries that would be affected by wider coverage. In 1982, Resolution 37/82 of the United Nations General Assembly demanded that such limits should be applied and this policy is central to the plans produced by the ITU in 1977–1988.

Some, at least, of the reasons for this attitude are obvious. In political, cultural and religious matters, TV broadcasting is potentially very persuasive. In the politically polarised world of the 1970s and 1980s, governments did not want to lose ultimate control over the medium. Furthermore, TV is an effective and potentially profitable medium for commercial advertising, not to be surrendered freely to foreign broadcasters. Thirdly, great importance is attached by many governments to safeguarding equitable national access to satellite communication and satellite broadcasting (see Section 5.4 above), and how can national rights be defined in a plan except in national terms?

Given governmental choice of national service areas, DBS has not been attractive to broadcasters. Public service broadcasting, predominant until recently, mainly terrestrial and national in coverage, has now been augmented in many countries by commercial broadcasting, not necessarily having exclusively national connections and often preferring wide coverage. Secondly, the stringent technical requirements, mainly due to the pessimistic technical assumptions on which the plans were based and the constraints on coverage areas, raise the cost and diminish the market for commercial DBS. Thirdly, the wealthier countries, which might be expected to have made substantial use of DBS, were the first to make extensive use of terrestrial TV broadcasting and some have since acquired DTH and cable broadcasting facilities.

Revision of RR Appendices S30 and S30A

There have been many proposals for using the Article 4 procedure of the two Appendices for amending specific planned assignments but the procedure has been found difficult to implement. There is also a clear need for the basic technical provisions of the plans as a whole to be reconsidered, especially those for Regions 1 and 3. Much new technical data for use in planning has come to hand since WARC SAT-77; see, for example, Report ITU-R BO.812-4 [9] and Recommendations ITU-R BO.1213 [10], BO.1295 [11], BO.1296 [12] and BO.1297 [13]. Recommendations ITU-R BO.1213, BO.1295, BO.1296 and BO.1297 have all been incorporated into the Radio Regulations by reference in RR Appendix S30, S30A or both. However, the principle of national coverage, in the absence of inter-governmental agreements to the contrary, was restated in RR Resolution 536 of WRC-97; see also RR S23.13.

RR Resolution 524 of WARC-92 set in motion further studies to prepare the way for revising the plans for Regions 1 and 3. WRC-95 prepared an interim report, and proposed revised values for some of the basic planning parameters (see RR Resolution 531 and RR Recommendation 521 respectively). WRC-95 instructed the Radiocommunication Bureau to draft amendments to the two Appendices taking the new parameters into account and incorporating amendments that had been proposed by Administrations to the extent possible.

In general WRC-97 accepted the amendments flowing from the initiatives agreed at WRC-95 although recognizing that care had to be taken to ensure that the integrity of assignments planned at RARC-83 for Region 2 was not damaged thereby. RR Appendices S30 and S30A were revised accordingly. RR Resolutions 53 and 533 provided for some of the revised parameters to be implemented for newly notified systems forthwith. WRC-97 also recognised that a more far-reaching revision of the plans for Regions 1 and 3 might be timely, to serve various objectives, most notably to increase the number of channels provided to about 10 per service area and to prepare for the introduction of digital TV. Accordingly, various further technical studies were initiated. An inter-conference representative group was set up to consider the way forward and report to WRC-00; see RR Resolution 532.

10.3.3 Frequency allocation sharing under the agreement

Sharing in BSS frequency bands

The frequency bands planned for the BSS around 12 GHz are shared with terrestrial services, mainly the FS, the BS and the MS (except the AMS) mostly with primary allocations, but RR S5.487 and S5.490 forbid stations of the sharing services to cause harmful interference by mode 6 (see Figure 5.4) to satellite broadcasting which is in accordance with the BSS plans. The coordination procedure set out in Article 6 of RR Appendix S30 is carried out, in addition to any action that may be appropriate with regard to RR Articles S9 and S11, before an assignment to a terrestrial station is registered in the MIFR, to ensure that harmful interference will not be caused to a present or future satellite broadcasting system which is in accordance with the plan. A qualified registration in the MIFR will be made for a terrestrial assignment that would interfere with a satellite broadcast that has not begun service if an undertaking is given to respect the broadcast in accordance with RR S4.4 when the latter is taken into service.

There are allocations for downlinks from GSO FSS networks in frequency bands which are planned for the BSS in other Regions. The bands 12.5–12.7 GHz (Region 1), 11.7–12.2 GHz (Region 2) and 12.2–12.7 (Region 3) are involved. The FSS satellite transmitters might interfere with BSS domestic receivers in other Regions by mode 2 (Figure 5.1). A procedure for identifying and evaluating such interference is given in Article 7 of RR Appendix S30; this procedure is additional to coordination by the procedure of RR Article S9. RR Resolution 73 makes special provision regarding the band 12.2–12.5 GHz to make good a flaw in the RR Appendix S30 procedure. See also Recommendations ITU-R S.1063 [14], S.1065 [15] and S.1066 [16] and also item 2(a) in Section 9.1.2 above.

WRC-97 made a new primary allocation for downlinks of non-GSO FSS networks in the frequency bands around 12 GHz that are allocated and planned for the BSS; see RR S5.487A and item 6(b) in Section 9.1.2 above. These FSS downlinks are another potential source of interference to satellite broadcast reception via mode 2 (Figure 5.1). RR Resolution 538 and RR S22.5C apply a provisional mandatory constraint to the equivalent PFD at the Earth's surface in a sampling bandwidth of 4 kHz from all the satellites of a non-GSO constellation, to protect the BSS and the other services that share these frequency bands. The Resolution also provides an interim procedure for any necessary coordination arising from these new allocations and calls for this constraint and the procedure to be re-examined in the ITU-R for possible revision at WRC-00.

Finally, WRC-97 made a provisional arrangement whereby non-GSO FSS systems might operate uplinks world-wide in a band which includes 12.5–12.7 GHz, planned for the BSS in Region 2, in accordance with the terms of RR Resolution 130; see also item 6(c) in Section 9.1.2 above.

Sharing in feeder link bands

The planned feeder link bands at 14 and 17 GHz are shared with various terrestrial services which also have primary allocations but these do not cause harmful interference to the feeder links, being constrained by power limits designed to control interference mode 9 (see Figure 5.4); see Section 5.2.3 above. However, an Administration considering the assignment of a frequency in one of these bands to a terrestrial receiving station will wish to take into account possible interference from a feeder link earth station in an adjacent country by interference mode 7 (see Figure 5.4).

This presents no particular problem once the location and characteristics of the earth station are known. Article 6 of RR Appendix S30A provides a procedure for requiring an Administration to declare the intended position of, and the basic data for, a feeder link earth station that has not, at that time, been brought into service, to enable effective terrestrial use of the frequency band to proceed. If these data are not supplied, the receiving assignment to the terrestrial station may be registered in the MIFR and the feeder link earth station, when it is built, must be located where it will not cause harmful interference to the terrestrial station.

In Region 2 the feeder links for the 12 GHz plan in the band 17.3–17.8 GHz will share, after 2007, a band allocated for the downlinks of a different group of BSS systems. If and when a plan is drawn up for BSS assignments in this band, some rules will be required to limit interference between these two broadcasting applications. Between 2007 and the date of entry into effect of that plan the BSS will have the same status in that band as the FSS; see the next paragraph.

Primary allocations for downlinks of FSS GSO networks also share all or parts of the bands at 17 GHz that have been planned for feeder links to broadcasting satellites in all three Regions. This creates the possibility of two modes of interference. In the first, an FSS satellite transmitter might cause interference at the broadcasting satellite receiver (mode 3, Figure 5.2). The coordination procedures set out in Article 7 of RR Appendix S30A and RR Article S9 will identify and so avert a risk of unacceptable interference. The second potential interference mode is from the feeder link earth station transmitter to an FSS earth station receiver (mode 4, Figure 5.2). This is the complement of interference from a feeder link earth station transmitter to a terrestrial service receiver (see above) and is treated in the same way. If the site for the feeder link earth station has not been chosen by that time, Article 7 of RR Appendix S30A enables the Administration coordinating the FSS earth station to require the Administration responsible for the feeder link earth station to declare the location and characteristics that may be assumed for it.

Finally, there are primary allocations for the FSS, limited to feeder links for broadcasting satellites at 14.5–14.8 GHz and assignments are planned in this band for some countries in Regions 1 and 3. These bands are shared with primary allocations for the FS and the MS (except the AMS) and potential interference problems are dealt with in the same ways as for terrestrial allocations in the 17 GHz band.

10.4 Satellite sound broadcasting

A planned assignment in the BSS bands around 12 GHz may be used for broadcasting a group of sound programmes instead of a TV programme provided that this emission causes no more interference and can tolerate as much interference as the planned emission; see paragraph 3.1.3 of Annex 5 of RR Appendix S30. Some limited use is being made of this option and Recommendation ITU-R BO.712-1 [17] contains information on three multichannel sound systems that have been developed for this application. The domestic reception of sound broadcasts conforming to the plan for TV requires a similar receiving installation, in particular an antenna of substantial gain.

However, a major use of sound broadcasting is for reception using hand-portable and vehicle-borne receivers. The quality of MF AM broadcasting is no longer generally acceptable and VHF FM broadcasting is not satisfactory for mobile reception because of the effects of multipath propagation. Section 7.5.3 above refers to the various objectives that have led to interest in T-DAB; the same objectives are sought for S-DAB; see Recommendation ITU-R BO.789-2 [18]. However, satellite broadcasting to mobile receivers at 12 GHz using a receiving antenna of substantial gain is not at present feasible; practicable satellite broadcasting using low gain receiving antennas requires the use of a much lower frequency.

WARC-92 allocated the band 1452–1492 MHz for BSS, the allocation being worldwide except for the USA; the corresponding allocation for the USA is at 2310–2360 MHz and this allocation also applies in India and Mexico; see RR S5.345 and S5.393 and item 2 in Section 10.1 above. There is an additional BSS allocation at 2535–2655 MHz in the countries listed in RR S5.418. All of these bands are also allocated for the BS, the FS and other services. The use of these BSS and BS allocations is limited to DAB; see RR S5.345. The S-DAB assignments are to be planned internationally as required by RR Resolution 528. It is intended that terrestrial broadcasting in the same frequency bands would be used to complement the coverage of the satellite systems.

Recommendation ITU-R F.1242 [19] provides a revised radio frequency channel plan for digital systems of the FS, designed to harmonise with the use of the new BSS allocation at 1452–1492 MHz; see item 3(a) in Section 6.2.1 above. Pending planning there is provision in RR Resolution 528 for taking the upper 25 MHz of any of the allocations into use for S-DAB in accordance with the terms of RR Resolution 33. The resolution permits T-DAB to be introduced in the BS allocations, subject to coordination with other Administrations whose services may be affected. Report ITU-R BO.2006 [20] addresses the problems of using the BS to complement the BSS in these bands. See also Section 7.5.3.

Recommendation ITU-R BO.789-2 [18] sets out planning objectives for S-DAB; these are closely aligned with those given for T-DAB in Recommendation ITU-R BS.774-2 [21] (see Section 7.5.3). Recommendation ITU-R BO.1130-1 [22] describes two such systems, identified as Systems A and B. System A, developed by the Eureka 147 DAB Consortium, is said to be fully developed and to meet all the objectives of Recommendation ITU-R BO.789-2 [18]; its use is recommended to Administrations wishing to implement the service soon. The development of System B is less complete; this option may be considered by Administrations which implement the service in the longer term. It may be noted that System A as described in Recommendation ITU-R BO.1130-1 [22] is virtually the same as the system described in Recommendation ITU-R BS.1114-1 [23] and recommended there for implementation for T-DAB. It is recognised that other systems meeting the agreed objectives may emerge in the future, be recommended by ITU-R and preferred by Administrations. It is recommended that organisations developing these new options should design their systems to harmonise with those already developed.

10.5 References

1 'Common functional requirements for the reception of digital multiprogramme television emissions by satellites operating in the 11/12 GHz frequency range'. Recommendation ITU-R BO.1294, ITU-R Recommendations 1997 BO Series (ITU, Geneva, 1997)

2 'MUSE system for HDTV broadcasting-satellite service'. Recommendation ITU-R BO.786, *ibid.*

3 'MAC/packet based system for HDTV broadcasting-satellite services'. Recommendation ITU-R BO.787, *ibid.*

4 'High definition television by satellite'. Report ITU-R BO.1075-2, ITU-R Reports BO Series (ITU, Geneva, 1995)

5 'Television systems'. Recommendation ITU-R BT.470-4, ITU-R Recommendations 1997 BT Series (ITU, Geneva, 1997)

6 'Standards for conventional television systems for satellite broadcasting in the channels defined by Appendix 30 of the Radio Regulations'. Recommendation ITU-R BO.650-2, ITU-R Recommendations 1997 BO Series (ITU, Geneva, 1997)

7 'Digital multi-programme emission systems for television, sound and data services for satellites operating in the 11/12 GHz frequency range'. Recommendation ITU-R BO.1211, *ibid.*

8 'Service multiplex methods for digital terrestrial television broadcasting'. Recommendation ITU-R BT.1209-1, ITU-R Recommendations 1997 BT Series (ITU, Geneva, 1997)

9 'Computer programs for planning broadcasting-satellite services in the 12 GHz band', Report ITU-R BO.812-4, ITU-R Recommendations 1997 BO Series (ITU, Geneva, 1997)

10 'Reference receiving earth station antenna patterns for planning purposes to be used in the revision of the WARC BS-77 broadcasting-satellite service plans for Regions 1 and 3'. Recommendation ITU-R BO.1213, ITU-R Recommendations 1997 BO Series (ITU, Geneva, 1997)

11 'Reference transmit earth station antenna off-axis e.i.r.p. patterns for planning purposes to be used in the revision of the Appendix 30A (Orb-88) Plans of the Radio Regulations at 14 GHz and 17 GHz in Regions 1 and 3'. Recommendation ITU-R BO.1295, *ibid.*

12 'Reference receive space station antenna patterns for planning purposes to be used for elliptical beams in the revision of Appendix 30A (Orb-88) Plans of the Radio Regulations at 14 GHz and 17 GHz in Regions 1 and 3'. Recommendation ITU-R BO.1296, *ibid.*

13 'Protection ratios to be used for planning purposes in the revision of the Appendices 30 (Orb-85) and 30A (Orb-88) Plans of the Radio Regulations in Regions 1 and 3'. Recommendation ITU-R BO.1297, *ibid.*

14 'Criteria for sharing between BSS links and other Earth-to-space or space-to-Earth links of the FSS'. Recommendation ITU-R S.1063, ITU-R Recommendations 1997 S Series (ITU, Geneva, 1997)

15 'Power flux-density values to facilitate the application of RR Article 14 for the FSS in Region 2 in relation to the BSS in the band 11.7–12.2 GHz'. Recommendation ITU-R S.1065, *ibid.*

16 'Ways of reducing the interference from the broadcasting-satellite service of one Region into the fixed-satellite service of another Region around 12 GHz'. Recommendation ITU-R S.1066, *ibid.*

17 'High-quality sound/data standards for the broadcasting-satellite service in the 12 GHz band'. Recommendation ITU-R BO.712-1, ITU-R Recommendations 1997 BO Series (ITU, Geneva, 1997)
18 'Service for digital sound broadcasting to vehicular, portable and fixed receivers for broadcasting-satellite service (sound) in the frequency range 1400–2700 MHz'. Recommendation ITU-R BO.789-2, *ibid.*
19 'Radio-frequency channel arrangements for digital radio systems operating in the range 1350–1530 MHz'. Recommendation ITU-R F.1242, ITU-R Recommendations 1997 F Series Part 1 (ITU, Geneva, 1997)
20 'Introduction of satellite and complementary terrestrial digital sound broadcasting in the WARC-92 frequency allocations'. Report ITU-R BO.2006, ITU-R Reports 1995 BO Series (ITU, Geneva, 1997)
21 'Service requirements for digital sound broadcasting to vehicular, portable and fixed receivers using terrestrial transmitters in the VHF/UHF bands'. Recommendation ITU-R BS.774-2, ITU-R Recommendations 1997 BS Series (ITU, Geneva, 1997)
22 'Systems for digital sound broadcasting to vehicular, portable and fixed receivers for broadcasting-satellite service (sound) bands in the frequency range 1400–2700 MHz'. Recommendation ITU-R BO.1130-1, ITU-R Recommendations 1997 BO Series (ITU, Geneva, 1997)
23 'System for terrestrial digital sound broadcasting to vehicular, portable and fixed receivers in the frequency range 30–3000 MHz'. Recommendation ITU-R BS.1114-1, ITU-R Recommendations 1997 BS Series (ITU, Geneva, 1997)

Chapter 11

The mobile-satellite services

11.1 The services and their frequency allocations

11.1.1 The services

The mobile-satellite service (MSS) includes the maritime mobile-satellite service (MMSS), the aeronautical mobile-satellite service (AMSS) and the land mobile-satellite service (LMSS). However, there has been a tendency for the technology and the facilities available for the various users of mobile-satellite networks to converge, and almost all the allocations are for the MSS in general.

11.1.2 Frequency allocations and sharing

The bands allocated for mobile-satellite services fall conveniently into six groups, and a seventh group contains FSS allocations which are used for feeder links for MSS satellites. These seven groups of allocations are reviewed below. In addition, some satellite constellations in low earth orbits (LEO) use direct links between satellites to connect a calling subscriber to a called subscriber or a feeder link earth station within the coverage area of another satellite. There are various ISS frequency allocations for inter-satellite links (ISLs) (see Chapter 14) although none is designated specifically for this application.

1 Bands used for commercial GSO systems

a) The commercial GSO satellite networks that are already established operate in bands with world-wide primary allocations for the MSS at 1525–1544 MHz and 1545–1559 MHz (downlinks) and 1626.5–1645.5 MHz and 1646.5–1660.5 MHz (uplinks). Frequency coordination between these systems is in accordance with the procedures of RR Resolution 46 (see RR S5.354). RR S5.351 forbids the use of these allocations for feeder links.

   These frequency bands are shared with primary allocations for the SO, the FS and the RAS in the framed part of the table of frequency allocations and with other services by footnotes to the table. The MSS uplinks are protected, to some degree, from interference from terrestrial transmitters (mode 9 in Figure 5.4) by the sharing constraint given in RR Table S21-2. There is no constraint on MSS downlink PFD to protect terrestrial receivers from emissions from satellite emissions (mode 8 in Figure 5.4).

The MSS allocation at 1660–1660.5 MHz is for uplinks from mobile earth stations that may be on land and this band is shared with the RAS which also has a primary allocation. RR S5.376A requires mobile earth stations operating in this band not to cause harmful interference to the RAS. Recommendation ITU-R M.1316 [1] provides guidance in this matter. RR Resolution 125 calls for further consideration of this sharing situation at a future WRC when experience has been gained in the use of the recommendation.

b) There has been a world-wide secondary allocation for the LMSS (uplinks) at 14.0–14.5 GHz since WARC-79. This allocation is used by transportable earth stations to get access to GSO networks of the FSS, typically for satellite news gathering (SNG); see Section 11.4. RR S22.2 applies in this band and frequency coordination is by RR Article S9.

At WRC-97 the secondary allocation for the LMSS was broadened to include the MMSS and RR Resolution 216 calls for studies of the feasibility of extending the allocation to include the AMSS also. There are primary allocations in this band for the FSS (uplinks; see item 2(a) in Section 9.1.2), the RNS, the FS and the MS and use of the band by an SNG uplink with a wide bandwidth baseband requires adherence to the constraints and coordination principles applied to the FSS in that band (see Section 9.2.1 above); other ITU-R recommendations are specific to SNG earth stations (see Section 11.4 below).

c) The bands 20.1–20.2 GHz (downlinks) and 29.9–30.0 GHz (uplinks) are world-wide primary allocations for the MSS, sharing with the FSS allocations, which are also primary and use the bands in the same directions of transmission. The bands 19.7–20.1 GHz (downlinks) and 29.5–29.9 GHz (uplinks) are also allocated for the MSS, primary in Region 2 and secondary in Regions 1 and 3. These bands are also shared with primary allocations for the FSS. RR S22.2 applies in these bands and frequency coordination is in accordance with RR Article S9.

Recommendation ITU-R S.1329 [2] is relevant. The reference in item 5 in Section 9.1.2 above to the problems that might be raised by the use of these allocations by satellites that serve both the FSS and the MSS is relevant here also. These bands are also allocated for non-GSO systems of the FSS by RR S5.484A, under the terms of RR Resolution 130; see item 6(c) in Section 9.1.2.

These bands are also shared in many countries with allocations for the FS and MS, primary at 20 GHz and secondary at 30 GHz; see RR S5.524 and S5.542. However, RR S5.524 withholds any downlink PFD constraint from MSS satellite transmitters at 20 GHz, whereas RR S5.452 applies the e.i.r.p. constraints of RR Article S21 to terrestrial transmitters at 30 GHz.

d) There is a world-wide secondary allocation of long standing for the AMS(R)S in the band 117.975–137.0 MHz (see RR S5.198), the primary allocation of this band being for the AMS(R). This allocation survives from early provision for possible aeronautical satellite system development.

2 *The government network bands*

There are three sets of MSS allocations which are used largely for government, including military, systems.

a) The bands 235–322 MHz and 335.4–399.9 MHz may be used for the MSS subject to coordination as described in Section 5.3 above; see RR S5.254. The principal allocations in these bands are for the FS and the MS.

b) The bands 7250–7375 MHz (downlinks) and 7900–8025 MHz (uplinks) are worldwide primary allocations for the MSS, subject to coordination with the sharing services both in accordance with RR Articles S9 and S11, and also as described in Section 5.3 above; see RR S5.461. RR S22.2 applies in these bands. The sharing services are the FSS, the FS and the MS.

c) The bands 20.2–21.2 GHz (downlinks) and 30–31 GHz (uplinks) are world-wide primary allocations for the MSS. These allocations are shared with FSS allocations which are also primary, systems are coordinated in accordance with RR Articles S9 and S11 and RR S22.2 is applicable. These bands are also shared in many countries with allocations for the FS and MS, primary at 20 GHz and secondary at 30 GHz. However, RR S5.524 withholds any downlink PFD constraint from MSS satellite transmitters at 20 GHz, whereas RR S5.452 applies the e.i.r.p. constraints of RR Article S21 to terrestrial transmitters at 30 GHz.

3 Bands below 1 GHz for non-GSO systems

a) Seven allocations have been made for the MSS or the LMSS between 100 MHz and 1 GHz at or since WARC-92. These allocations are all narrow and RR S5.209 limits their use to non-GSO networks, the so-called 'Little LEOs'. This part of the spectrum is heavily congested with systems of services with primary allocations. Various paragraphs of the RR require that development of the MSS does not restrain the development of these sharing services. Frequency coordination is by the procedure of RR Resolution 46; see also Section 5.2.5 above. The allocations are as follows.

- 137–138 MHz (downlinks), world-wide, partly primary, partly secondary. See also Recommendations ITU-R M.1231 [3] and ITU-R M.1232 [4]. In making assignments to satellite transmitters in this band, RR S5.208A requires Administrations to take all practicable steps to protect the operations of the RAS.
- 148.0–149.9 MHz (uplinks), world-wide, primary. Recommendation ITU-R M.1185-1 [5], incorporated into the RR by the reference in RR Appendix S5, provides a method for determining coordination distances between mobile earth stations and terrestrial stations in this band. RR S5.221, applicable in a large number of countries, requires MSS stations not to cause harmful interference to, nor claim protection from, the stations of the FS and the MS with which this band is shared,
- 149.9–150.05 MHz and 399.9–400.05 MHz (uplinks), world-wide for LMSS, primary, sharing with the RNSS. RR Resolution 715 calls for measures to facilitate the implementation of this sharing arrangement to be studied. However, these bands are to be allocated exclusively for the MSS from 2015; see RR S5.224A and S5.224B.
- 400.15–401 MHz (downlinks), world-wide, primary, shared with the MetA, the MetS and the SR (downlinks). RR S5.208A requires the protection of the RAS.
- 454–456 MHz and 459–460 MHz (uplinks), Region 2 only, MSS, primary, shared with the FS and the MS.

There is more broadly applicable information on coordination involving satellite networks using frequency bands below 1 GHz in Recommendations ITU-R M.1039-1 [6] and M.1087 [7].

b) There is a pair of secondary MSS allocations at 312–315 MHz (uplinks) and 387–390 MHz (downlinks). RR S5.255 permits these bands to be used for non-GSO systems, subject to coordination using RR Resolution 46. RR S5.208A calls for protection of the RAS in the 387–390 MHz band. In addition there are various allocations for the MSS, or the MSS except for the AMS(R)S, between 806 and 960 MHz, in accordance with RR S5.317, S5.319 and S5.320.

c) A search is in progress for more bands below 1 GHz that could be allocated for Little LEO systems. RR Resolution 728 calls for studies of the feasibility of sharing in the BS allocations between 470 and 862 MHz. RR Resolution 219, although sensitive to the need to avoid the risk of interference to signals from EPIRBs in the adjacent band (see item 6(a) below), calls for the consideration of the feasibility of an allocation for the MSS in the frequency range 401–406 MHz. RR Resolution 214 invites a more general search for possible additional bands that could be allocated for non-GSO networks below 1 GHz and calls for studies of techniques for facilitating sharing between Little LEO systems and between them and systems of other services.

d) RR Resolution 127 notes that insufficient bandwidth is available at present for feeder links for non-GSO MSS networks operating below 1 GHz and calls for the identification and study of possible new sharing allocations in bands around 1.4 GHz.

4　Non-GSO major system bands

The bands listed in Table 11.1 were allocated for the MSS at WARC-92 or subsequently. The Region 2 allocations at 1930–1970 MHz and 2120–2160 MHz are secondary, shared with primary allocations for the FS and the MS, and coordination is by the procedures of RR Articles S9 and S11. All of the other allocations in this table are primary and are to be coordinated in accordance with RR Resolution 46. These MSS allocations have substantial bandwidth and they are likely to be used for major telephony networks, using satellites in orbits which may be geostationary, low or intermediate in height. Such systems are called global mobile personal communications by satellite (GMPCS) systems. It is likely that most systems will use low orbits; these are the so-called 'Big LEO' systems. The search for possible additional bands for allocation continues. See for example RR Resolution 220, which calls for studies to be made of the feasibility of allocating some part of the range 1559–1610 MHz for MSS downlinks.

There are primary allocations for other services in all of these bands and many various and substantial sharing problems arise. The problems are too numerous for listing here, but see for example the relevant footnotes to the international table of frequency allocations. None the less, measures for facilitating the establishment of the MSS and other new services in this heavily loaded part of the spectrum have been developed. In particular, in several frequency ranges where new allocations have been made for the MSS, the radio channel plans recommended for systems of the FS in the same band have been changed to ease the introduction of MSS systems. For example, Recommendation ITU-R F.1242 [8] rearranged the frequency plan for FS systems in the band 1350–1530 MHz; see Section 6.2.1 item 3(a) above. Similarly WRC-95, through RR Resolution 716, requested Administrations to rearrange systems of the FS in the bands 1980–2010 MHz and 2170–2200 MHz, and in Region 2 in the bands 2010–2025 MHz and 2160–2170 MHz also, by implementing the revised radio

Table 11.1   Frequency allocations for big non-GSO networks

(MHz)
1492–1525, downlinks, Region 2 only.
1610–1626.5, uplinks (also downlinks 1613.8–1626.5 MHz).
1675–1710, uplinks, Region 2 only.
1930–1970, uplinks, Region 2 only, secondary.
1980–2010, uplinks, identified for use with IMT-2000.
2010–2025, uplinks, Region 2 only.
2120–2160, downlinks, Region 2 only, secondary.
2160–2170, downlinks, Region 2 only.
2170–2200, downlinks, identified for use with IMT-2000.
2483.5–2500, downlinks.
2500–2535, uplinks.
2655–2690, downlinks.

Allocations are world-wide and primary except where otherwise stated.

channel plans in Recommendation ITU-R F.1098-1 [9]; see Section 6.2.1 item 4(c) above.

5 Millimetre-wave MSS allocations
There are nine bands above 40 GHz allocated for the MSS, all world-wide and all shared with various other services. They are little used at present.

6 Bands used for distress and safety communication

a)   There are world-wide MSS allocations for the bands 121.45–121.55 MHz and 242.95–243.05 MHz, intended for the reception on satellites of distress signals transmitted by aeronautical EPIRBs at 121.5 and 243 MHz; see RR S5.199.
b)   The band 406.0–406.1 MHz is allocated, world-wide and exclusively, for MSS uplinks from maritime satellite EPIRBs. RR S5.266 and S5.267 forbid any emission that could interfere with these EPIRB signals. See also RR Resolution 205 and Recommendation ITU-R SM.1051-2 [10].
c)   The bands 1544–1545 MHz (downlinks) and 1645.5–1646.5 MHz (uplinks), adjacent to the MSS allocations listed in item 1(a) above, are reserved for distress and safety communications, including satellite EPIRBs. See RR S5.356 and S5.375. In addition there is provision for giving priority to emergency traffic in the adjacent bands; see Section 11.2 below.

7 Feeder link allocations
MSS feeder links, that is, links between MSS satellites and earth stations at fixed locations, are not normally assigned frequencies in bands allocated for the MSS. RR S1.25 and S5.351 allow this to be done but only for exceptional purposes, such as when a control link via that route is necessary for the proper functioning of the network. The feeder links that carry traffic are treated as part of the FSS; see RR S1.21.

No specific bands have been designated for feeder links for GSO MSS networks; assignments are made and coordinated in the bands that are used for uplinks and downlinks for networks of the FSS. However, this practice would not be satisfactory for most feeder links for non-GSO MSS networks, since primary allocations for non-GSO FSS networks are few and narrow. Accordingly, the use of the following world-wide allocations for the FSS is limited to such feeder links. These bands are coordinated in accordance with RR Resolution 46. See also item 8 in Section 9.1.2.

a) 5091–5150 MHz. The long term use planned for the band 5000–5150 MHz is the ARNS (see RR S5.444). The microwave landing system (MLS) is already being installed between 5000 and 5091 MHz. The band is also allocated for the ARNSS by RR S5.367. However, RR S5.444A also allocates 5091–5150 MHz for the FSS (uplinks), its use limited to non-GSO MSS feeder links in accordance with RR Resolution 114. This allocation is primary until 2010, although systems of the ARNS that cannot be accommodated below 5091 MHz will take precedence over the feeder links. From 2010 onwards the FSS allocation becomes secondary. Recommendation ITU-R S.1342 [11] considers the prospect of interference to aircraft using MLS at frequencies below 5091 MHz from feeder link earth station transmitters at frequencies above 5091 MHz. The conclusions, which are subject to further study, are that coordination may be necessary, depending on the parameters of the case, if the separation between the earth station and the MLS site is less than 450 km.

b) 5150–5250 MHz. This primary FSS (uplinks) allocation, shared with the ARNS, is limited to non-GSO MSS feeder links; see RR S5.447A. In addition RR S5.447B provides a downlink FSS allocation in the band 5150–5216 MHz, also limited to non-GSO MSS feeder links, subject to a PFD limit at the Earth's surface of $-164\,\mathrm{dB(W/m^2)}$ in any 4 kHz bandwidth. See also RR S5.447C.

c) 6700–7075 MHz. There is a primary allocation for the FSS, (uplinks and downlinks) in this band, the uplink allocation being available for FSS networks (see item 3 in Section 9.1.2) and the downlink allocation being limited by RR S5.458B to non-GSO MSS feeder links. RR S5.458C calls for coordination, as outlined in Section 5.3 above, between 7025 MHz and 7075 MHz.

d) 15.43–15.63 GHz. There is a primary allocation for the FSS in this band for uplinks and downlinks, both limited by RR S5.511A to non-GSO MSS feeder links, shared with the ARNS. Recommendation ITU-R S.1340 [12] outlines the characteristics of aeronautical systems operating in the band and estimates that coordination distances ranging between 270 and 600 km will be required when considering interference from feeder link earth stations to locations where the ARNS systems are in operation. Recommendation ITU-R S.1341 [13] finds broadly similar coordination distances for interference to feeder link earth station receivers from ARNS systems. Coordination may also be required between feeder link earth stations transmitting uplinks and other earth stations receiving downlinks in the same frequency band. See also the downlink PFD constraint applied by RR S21.16.

There are allocations for the RAS and passive sensors of the EES and the SR almost adjacent to this band, at 15.35–15.4 GHz; see RR S5.511A. Recommendation ITU-R S.1341 [13] considered the potential impact of MSS feeder links on these passive services and recommended that the protection criteria given in Recommendation ITU-R RA.769-1 [14] for this band should be taken into account when designing feeder links. The reference to Recommendation ITU-R RA.769-1 in RR S.511A has the effect of incorporating that recommendation into the RR. RR Resolution 123 of WRC-97 foresaw that the degree of protection which was desirable for the RAS might be difficult to provide and called for further studies to be done for consideration at WRC-00.

e)   19.3–19.7 GHz (downlinks) and 29.1–29.5 GHz (uplinks). There are world-wide, primary allocations in these bands for the FSS, to be used for feeder links to non-GSO satellites of the MSS, and also for links for GSO networks of the FSS (see item 5 in Section 9.1.2) operating in the same directions of transmission. See RR S5.523B, S5.523C, S5.523D and S5.523E and also RR Resolution 120. The rules as to the procedure for frequency coordination and the applicability of RR S22.2 are complex; see the regulations referenced above. Responding to RR Resolution 121, Recommendation ITU-R S.1324 [15] provides an analytical method for estimating interference levels between GSO and non-GSO links; see also Recommendation ITU-R S.1328 [16]. RR S5.541A requires means to be provided for compensating for variations in uplink transmission loss; see item 5 in Section 9.1.2.

11.2 The commercial geostationary MSS systems

Since 1997 the four bands between 1525 and 1660.5 MHz listed in Section 11.1.2 have been allocated for the MSS. Previously the bands were segmented, much of the bandwidth being allocated for either the MMSS, the AMS(R)S or the LMSS. This segmentation served an important purpose. It was thought that the reservation of a bandwidth of 1 MHz for distress and safety communication (see item 6(c) in Section 11.1.2 above) might not be an adequate provision for all of the special and urgent traffic that would arise from a major disaster at sea or in the air. The allocation of other bands for aeronautical and maritime use gave ICAO and IMO the opportunity to use their influence over the medium to ensure that adequate priority was given to emergency traffic.

Accordingly, although WRC-97 agreed that these allocations should be normally available for the MSS in general, the power to give priority to emergency traffic has been retained. See RR Articles S44 and S53 and also RR S5.353A, S5.357A and S5.362A. See also Recommendations ITU-R M.830 [17] and M.1233 [18]. These footnotes to the international table of frequency allocations have the effect of defining frequency bands within these allocations in which traffic related to maritime or aeronautical emergencies is given priority over all other uses with immediate access. RR Resolution 218 called for studies to be made of ways of adding precision to the scale of these requirements where possible and for defining how they should be implemented.

A single system using geostationary satellites, with coverage that is global except for the polar regions, has been in service in these bands since 1976. The original system was MARISAT, owned and operated by the Communications Satellite Corporation. That system was absorbed into the International Maritime Satellite Organization (INMARSAT) system in 1982. More recently a number of other satellite systems, providing more limited geographical coverage have been introduced.

The formation of INMARSAT was sponsored by the Inter-Governmental Maritime Consultative Organization, the predecessor of the IMO, and established by an agreement between the governments of the countries that expected to be major users of the system. Governments remained as parties to the INMARSAT agreement, retaining an oversight of the objectives of the business. The capital required by the operating organisation was provided by a consortium of 'designated entities', mostly PTOs, authorised by their governments to provide public access to the system. Structural changes have recently been made to the organisation, giving it a more conventional corporate structure. An affiliated company has been established to set up a non-GSO system, mainly for LMSS applications.

INMARSAT owns the geostationary satellites and manages the system. The designated entities provide access to the system for the public on land via the PSTN and the feeder link earth stations which some PTOs own. The owners of vehicles with a need to access the system, mainly ships but including aircraft, land vehicles and transportable earth stations, provide their own mobile earth stations. Feeder link earth station parameters and a range of types of mobile earth station, appropriate for the various kinds of vehicle and the facilities required, have been standardised by agreement between the various parties involved. INMARSAT provides a range of conventional telecommunication services such as telephony, facsimile, telex, low and medium speed data and packet data, plus specialised services to meet special needs, especially those of shipping.

The close relationships between the various parties involved in the INMARSAT system have made it relatively easy to achieve the level of international cooperation that a successful system requires. Until lately there has been virtually no competing system. With comprehensive participation of the various parties involved, INMARSAT has been in itself a global standardising body for mobile earth station equipment. There have been no practicable alternatives to the frequency bands around 1.6 GHz and the GSO. Finally, coordination of feeder link assignments in frequency bands already used for access to GSO FSS networks raises no new basic problems. These points are made here because the situation for non-GSO MSS systems is quite different; see Sections 11.3 and 11.4 below.

Efficient use of orbit and spectrum

The directivity of the satellite antennas and the earth station antennas of satellite networks are the chief factors enabling geostationary networks to be coordinated and operated in the same frequency band without experiencing harmful interference. Among GSO mobile-satellite services the larger ships and a small number of transportable land mobile stations have antennas with substantial directivity (see, for example, Recommendation ITU-R M.694 [19]) but the directivity of most mobile earth station antennas is small, often negligible. It is therefore particularly important that all other factors that can minimise inter-network interference are employed to the maximum extent.

Recommendation ITU-R M.1038 [20] reviews network design principles that enable spectrum and orbit to be used efficiently in GSO MSS systems. Clearly, mobile earth station antennas should be directive where feasible. But in addition, satellites should be equipped with antennas which minimise radiation outside the intended service area. The occupied bandwidth per unit of information bandwidth should be minimised except where sharing would be facilitated by broad-band signal processing or modulation techniques. Recommendation ITU-R M.1086 [21] advises on how the principles of RR Appendix S8 should be applied in the MSS situation.

National spectrum management for GSO mobile-satellite services

The predominant provider of GSO MSS service is INMARSAT, and the predominant user is shipping. This outline concentrates on this application. Governments oversee the design and inspect the condition of ships' earth station equipment, as with all ships' equipment involved with the safety of life. Part of this procedure is to verify that the equipment is type approved and properly installed. The identity of every ship with an earth station is notified to the ITU for publication in the ITU List of Ship Stations; see RR S20.8.

If a feeder link earth station is to be installed, the same considerations arise for the Administration as for an earth station which is part of a network of the FSS. The earth station should be located where coordination with terrestrial stations will not be difficult and its technical performance should meet the standards set by ITU regulations and recommendations as well as those set by INMARSAT. Feeder link frequencies must be agreed between the satellite system manager and the Administration and API must be given, followed by coordination as for FSS network assignments, notification of assignments to the ITU and registration in the MIFR.

There must be publication and coordination of assignments to ship earth stations, not individually but collectively, to secure permission, if possible, for ships to operate their earth stations within the ports and territorial waters of other countries. This permission may not be given or may be limited to part only of the MSS allocation. Thus, RR S5.374 declares that ship earth stations shall not cause harmful interference to FS stations in the bands 1631.5–1634.5 MHz and 1656.5–1660 MHz in the 44 countries listed in RR S5.359. However, an international agreement has been drawn up on the use of INMARSAT ship earth stations in territorial waters and RR Recommendation 316 urges Administrations to accede to it. It may be that countries that require to maintain the use of these bands in general for terrestrial services will, none the less, assign frequencies in other bands to stations that are vulnerable to interference from ships.

11.3 Non-GSO MSS systems

11.3.1 Little LEO systems

The MSS allocations below 1 GHz share very narrow bands with the systems of various other services, mostly terrestrial. An MSS system that can function in such allocations is likely to have a low total information throughput and the transmission system used should be robust in the presence of interference. A store-and-forward system for data packets with ARQ is a likely choice. Such a system might take several forms, two of which can be defined as follows.

A single LEO satellite in a highly inclined circular orbit, its height being in the range 1100 ± 400 km, passes within sight of every point on the Earth's surface at least once every day. It can act as a courier, receiving and storing a message when it is in sight of a sending earth station and transmitting the message to the destination earth station when its orbit has brought the latter within sight. This makes a very simple communication system, which is reliable, much faster than airmail, needing no feeder links, no infrastructure and potentially low in cost. The transmission delay is, of course, very long in comparison with conventional telecommunication systems but it is acceptable for some applications and the delay can be reduced by using several satellites in suitably phased orbits.

With a constellation of several dozens of LEO satellites in phased circular orbits, there can be a satellite within sight of any earth station in the coverage zone all of the time. A message can be transmitted at any time from a user earth station via a satellite to a feeder link earth station. The message can forwarded at the feeder link station to the addressee via the local satellite, via the PSTN or via a terrestrial link and another feeder link earth station to get access to a destination served by another satellite. This provides a message service with almost no delay but, unlike the other example, feeder links and terrestrial infrastructure are needed.

11.3.2 GMPCS systems

Non-GSO MSS systems for operation above 1 GHz are the satellite complement of cellular terrestrial networks, providing a variety of facilities for their users, most typically telephony between hand-held terminals. Most foreseen satellite systems will use large constellations of satellites in circular LEOs but smaller constellations in circular orbits at heights intermediate between a LEO and the GSO or in elliptical orbits are also proposed. Both TDMA and code division multiple access (CDMA) provided by spread spectrum modulation are proposed for use. Such systems may be virtually global in coverage, given successful frequency coordination.

It will be seen from item 4 of Section 11.1.2 that a substantial total bandwidth has been allocated for the MSS between 1 and 3 GHz. There are proposals for a considerable number of GMPCS systems. Recommendation ITU-R M.1186 [22] provides technical guidance on means for facilitating coordination between system using spread spectrum modulation. RR Resolution 215 calls for further studies of coordination techniques and criteria in the ITU-R study groups involved and urges Administrations to ensure that new systems for which they are responsible are designed to enhance efficiency of spectrum use and to facilitate coordination. RR Resolution 25 emphasises the sovereign right of each state to regulate telecommunications within the area of its jurisdiction and resolves that Administrations licensing a global satellite system shall ensure that the system can be operated only in territories in which its use has been authorised.

There is a major project in progress in ITU-R Study Group 8 to develop a text defining the IMT-2000, formerly the FPLMTS, as a terrestrial cellular system, globally standardised and providing comprehensive user facilities; see Section 8.5.4. RR S5.388 and RR Resolution 212 prepare for the use of the bands 1885–2025 MHz and 2110–2200 MHz for the system. It is foreseen that a satellite component will be integrated with IMT-2000 (see Recommendation ITU-R M.1167 [23]) using the bands 1980–2010 MHz and 2170–2200 MHz, although not necessarily to the exclusion of other systems.

Meanwhile, various other GMPCS system designs have been developed and several are, at the time of writing, in the process of entering service. Each of these systems is financed and controlled by a commercial corporation, although each is intended to provide an international, ideally a global, service and much has been done to extend commercial participation in the projects to potential user countries. The establishment of such systems involves extensive negotiation and coordination between Regulators, Administrations and the companies owning the systems; see Section 11.3.3.

11.3.3 *Spectrum management and regulation for GMPCS*

The main radio regulation elements of the action required in setting up a GMPCS are likely to be as follows.

1   The Administrator and the Regulator consider proposals for setting up systems managed from within their own jurisdiction and give provisional authority, conditional upon the use of technically acceptable users' terminals and feeder link earth stations, to proposals they consider to be:
    - adequately resourced and supported by determination to carry the project through,
    - compatible with international agreements and standards and national frequency allocations,
    - likely to be coordinatable internationally, and
    - in keeping with national objectives for service to the public, including competition.

2   The proposer of a system will invite foreign governments to allow the system to be used within their jurisdiction. Criteria similar to those listed above will probably be applied by those governments in considering their response, with the addition of consideration of proposed user tariff principles. Where the result is favourable, the proposer will seek cooperative agreements with PTOs etc. who, with their government's approval, will provide gateways into the PSTN for the GMPCS and any other necessary local participation in the system, including perhaps the provision and operation of feeder link earth stations.

3   The proposing Administration will make advance publication for the system and then coordinate with other Administrations the frequency assignments for the MSS, FSS (for feeder links) and ISS links that it proposes to make to the satellites of the proposed system.

4   The proposer will supply prototype user's terminals for type approval to the proposing Administration or to a competent agency acceptable to the Administration. This type-approval certification may also be accepted by foreign Administrations, failing which prototype terminals must qualify for type approval by foreign agencies. The proposer will ensure that supplies of user terminals, servicing facilities etc. will become available at the appropriate time.

5   The proposing Administration will coordinate the use of the MSS frequencies in all the foreign countries that have agreed to allow the system to be used and with other countries with land areas within coordination areas adjacent to the user countries.

6   The proposer will determine locations for feeder link earth stations. The proposer or local associate companies will procure the earth stations. The responsible Administrations will approve the station characteristics and license the earth stations, make advance publication, formally assign the feeder link frequencies that are to be used and coordinate them with other stations.

7   The proposer will set up the necessary satellite monitoring and telecommand earth stations and obtain licences and frequency assignments from the responsible Administrations. The Administrations will notify and, if necessary, coordinate these assignments.
8   The proposer will procure and launch the satellites.
9   Administrations will arrange for user terminals to be licensed, or will provide a class or blanket licence covering all use of them within their jurisdiction. It is important that governments allow visitors from abroad to import and re-export their terminals and use them without formality.
10 The constellation will start functioning. The ISS links, the feeder links, the telecommand and satellite monitoring links and the MSS links will all be switched on and the entry into service of all of the frequency assignments will be notified to the ITU Radiocommunication Bureau for registration in the MIFR.

Many of these elements are labour-intensive, highly complex and uncertain of outcome. In addition, there will be various issues outside radio regulation to be resolved, such as the telephone numbering scheme and the call accounting system.

It has been feared that setting up an international GMPCS system might be so difficult to bring to a satisfactory conclusion, particularly in the early years when several systems are being introduced at the same time, that good proposals would be lost. Accordingly the first meeting of the World Telecommunication Policy Forum (WTPF), held in October 1996, had as its sole business the search for a set of voluntary principles that would guide national policy makers, regulators and system operators in implementing GMPCS. The essence of the outcome was summarised in the form of a Memorandum of Understanding, which a large number of Administrations have already signed. The ITU will act as the depositary of the Memorandum of Understanding and as a registry of type-approval procedures for terminal equipments.

11.4 Satellite news gathering

SNG uses satellite facilities for transmitting signals from the scene of news, cultural or sporting events to a switching centre where they can be incorporated into broadcasting programmes. If the signal is an audio channel, an MSS system is likely to be used. Often, however, a video channel plus one or more sound programme or commentary channels will be required. Such signals need wide transmission bandwidth and this is obtained by transmitting from the transportable MSS earth station to a satellite of an FSS network, the signal being retransmitted by the satellite, received at an FSS earth station and passed on to the switching centre. See Recommendation ITU-R BT.1205 [24] for user signal quality requirements for standard definition and high definition TV transmitted by digital SNG. The 14.0–14.5 GHz uplink band is suitable for the purpose; see item 1(b) in Section 11.1.2 above. It is sometimes necessary to set up audio channels in both directions for control purposes, the downlink to the SNG earth station typically being transmitted in an FSS allocation near 12 GHz.

An SNG earth station has the functional characteristics of a mobile station. It is transportable and it is used for temporary links of short duration, set up in a hurry. Because of the circumstances in which these facilities are typically used, it is very desirable that there should be a measure of uniformity in the equipment and the operating practices that are employed. Such standards are not relevant to the core subject of this book but reference may be made to Recommendations ITU-R SNG.770-1 [25], SNG.771-1 [26], SNG.1007-1 [27], SNG.1070 [28] and SNG.1152 [29].

The satellite service provider will probably define a small number of radio channels, perhaps only one, that may be used for occasional TV uplinks from transportable earth stations. Recommendation ITU-R SNG.722-1 [30], which discusses SNG earth station characteristics in general, includes minimum standards for the more critical characteristics. In coordinating the satellite, the service provider may include earth stations meeting these standards as a typical earth station that may be located anywhere in the service area.

It is important that an Administration should ensure that transportable earth stations that operate within its jurisdiction conform to these minimum technical standards. Even so, a transportable earth station, transmitting a carrier powerful enough for a video link via a geostationary satellite, is likely to be capable of causing serious interference to distant terrestrial stations. It may be desirable to avoid a domestic interference problem by refraining from assigning for terrestrial services the frequency or frequencies made available by the satellite service provider for SNG use. Since there may not be enough time to coordinate the use of the SNG channel with a neighbouring Administration when the need arises, a good solution to the international interference problem would be for all Administrations to refrain from using the SGN channel for terrestrial services. Otherwise, it would be desirable to coordinate the activation of SNG earth stations at predictable locations.

These preparations having been made, it will usually be feasible for the Administration to provide temporary licences to use an assignment at an approved transportable earth station at a stated location on demand.

11.5 References

1 'Principles and a methodology for frequency sharing in the 1610.6–1613.8 and 1660–1660.5 MHz bands between the mobile-satellite service (Earth-to-space) and the radio astronomy service'. Recommendation ITU-R M.1316, ITU-R Recommendations 1997 M Series Part 5 (ITU, Geneva, 1997)

2 'Frequency sharing of the bands 19.7–20.2 GHz and 29.5–30.0 GHz between systems of the mobile-satellite service and systems of the fixed-satellite service'. Recommendation ITU-R S.1329, ITU-R Recommendations 1997 S Series (ITU, Geneva, 1997)

3 'Interference criteria for space-to-Earth links operating in the mobile-satellite service with non-geostationary satellites in the 137–138 MHz band'. Recommendation ITU-R M.1231, ITU-R Recommendations 1997 M Series Part 5 (ITU, Geneva, 1997)

4 'Sharing criteria for space-to-Earth links operating in the mobile-satellite service with non-geostationary satellites in the 137–138 MHz band'. Recommendation ITU-R M.1232, *ibid.*

5 'Method for determining coordination distance between ground based mobile earth stations and terrestrial stations operating in the 148.0–149.9 MHz band'. Recommendation ITU-R M.1185-1, *ibid.*

6 'Co-frequency sharing between stations in the mobile service below 1 GHz and FDMA non-geostationary-satellite orbit (non-GSO) mobile earth stations'. Recommendation ITU-R M.1039-1, *ibid.*

7 'Methods for evaluating sharing between systems in the land mobile service and spread-spectrum low-Earth orbit (LEO) systems in the mobile-satellite services (MSS) below 1 GHz'. Recommendation ITU-R M.1087, *ibid.*

8 'Radio-frequency channel arrangements for digital radio systems operating in the range 1350 to 1530 MHz'. Recommendation ITU-R F.1242, ITU-R Recommendations 1997 F Series Part 1 (ITU, Geneva, 1997)

9 'Radio-frequency channel arrangements for radio-relay systems in the 1900–2300 MHz band'. Recommendation ITU-R F.1098-1, *ibid.*

10 'Priority of identifying and eliminating radio interference in the band 406–406.1 MHz'. Recommendation ITU-R SM.1051-2, ITU-R Recommendations 1997 SM Series (ITU, Geneva, 1997)

11 'Method for determining coordination distances, in the 5 GHz band, between the international standard microwave landing system stations operating in the aeronautical radionavigation service and non-geostationary mobile satellite service stations providing feeder uplink services'. Recommendation ITU-R S.1342, ITU-R Recommendations 1997 S Series (ITU, Geneva, 1997)

12 'Sharing between feeder links for the mobile-satellite service and the aeronautical radionavigation service in the Earth-to-space direction in the band 15.4–15.7 GHz'. Recommendation ITU-R S.1340, *ibid.*

13 'Sharing between feeder links for the mobile-satellite service and the aeronautical radionavigation service in the space-to-Earth direction in the band 15.4–15.7 GHz and the protection of the radio astronomy service in the band 15.35–15.4 GHz'. Recommendation ITU-R S.1341, *ibid.*

14 'Protection criteria used for radioastronomical measurements'. Recommendation ITU-R RA.769-1, ITU-R Recommendations 1997 RA Series (ITU, Geneva, 1997)

15 'Analytical method for estimating interference between non-geostationary mobile-satellite feeder links and geostationary fixed-satellite networks operating co-frequency and codirectionally'. Recommendation ITU-R S.1324, ITU-R Recommendations 1997 S Series (ITU, Geneva, 1997)

16 'Satellite system characteristics to be considered in frequency sharing analyses between geostationary-satellite orbit (GSO) and non-GSO satellite systems in the fixed-satellite service (FSS) including feeder links for the mobile-satellite service'. Recommendation ITU-R S.1328, *ibid.*

17 'Operational procedures for mobile-satellite networks or systems in the bands 1530–1544 MHz and 1626.5–1645.5 MHz which are used for distress and safety purposes as specified for GMDSS'. Recommendation ITU-R M.830, ITU-R Recommendations 1997 M Series Part 5 (ITU, Geneva, 1997)

18 'Technical considerations for sharing satellite network resources between the mobile-satellite service (MSS) (other than the aeronautical mobile-satellite (R) service (AMS(R)S)) and AMS(R)S'. Recommendation ITU-R M.1233, *ibid.*

19 'Reference radiation pattern for ship earth station antennas'. Recommendation ITU-R M.694, ITU-R Recommendations 1997 M Series Part 5 (ITU, Geneva, 1997)

20 'Efficient use of the geostationary-satellite orbit and spectrum in the 1–3 GHz range by mobile-satellite systems'. Recommendation ITU-R M.1038, *ibid.*

21 'Determination of the need for coordination between geostationary mobile satellite networks sharing the same frequency band'. Recommendation ITU-R M.1086, *ibid.*

22 'Technical considerations for the coordination between mobile satellite service (MSS) networks using code division multiple access (CDMA) and other spread spectrum techniques in the 1–3 GHz band'. Recommendation ITU-R M.1186, *ibid.*

23 'Framework for the satellite component of Future Public Land Mobile Telecommunication Systems'. Recommendation ITU-R M.1167, ITU-R Recommendations 1997 M Series Part 2 (ITU, Geneva, 1997)

24 'User requirements for the quality of baseband SDTV and HDTV signals when transmitted by digital Satellite News Gathering (SNG)'. Recommendation ITU-R BT.1205, ITU-R Recommendations 1997 BT Series (ITU, Geneva, 1997)

25 'Uniform operational procedures for Satellite News Gathering (SNG)'. Recommendation ITU-R SNG.770-1, ITU-R Recommendations 1997 SNG Series (ITU, Geneva, 1997)

26 'Auxiliary coordination satellite circuits for SNG terminals'. Recommendation ITU-R SNG.771-1, *ibid.*

27 'Uniform technical standards (digital) for Satellite News Gathering (SNG)'. Recommendation ITU-R SNG.1007-1, *ibid.*

28 'An automatic transmitter identification system (ATIS) for analogue-modulation transmissions for Satellite News Gathering and outside broadcasts'. Recommendation ITU-R SNG.1070, *ibid.*

29 'Use of digital transmission techniques for Satellite News Gathering (SNG) sound'. Recommendation ITU-R SNG.1152, *ibid.*

30 'Uniform technical standards (analogue) for Satellite News Gathering (SNG)'. Recommendation ITU-R SNG.722-1, *ibid.*

Chapter 12

The amateur services

Spectrum is allocated to the AmS and the AmSS for the use of authorised radio amateurs for self-training and technical investigations; see RR S1.56 and S1.57. Recommendation ITU-R M.1042 [1] refers to the contribution that radio amateurs and their stations can make to the restoration of communications in the event of natural disasters and Recommendation ITU-R M.1043 [2] discusses how Administrations can develop this resource. Recommendation ITU-R M.1041 [3] reviews the directions in which amateur radio could most desirably develop in the future.

Frequency allocations and sharing

The following frequency bands are allocated for the AmS and the AmSS in the international table of frequency allocations, all allocations being primary except where secondary status is indicated. Some of these allocations are shared with other services. It should be noted, however, as with any service, that an Administration may refrain from implementing such an allocation within its own jurisdiction. Conversely, any Administration may make a national allocation in favour of the amateur services in a band that has no international amateur allocation, subject to any necessary coordination and agreement with other Administrations with services that might be affected by interference. Some but not all of these national variations are indicated in footnotes to the international table.

1 The 1800 kHz band

a)   In Region 1, 1830–1850 kHz is an exclusive allocation for the AmS; 1810–1830 kHz is also allocated exclusively for the AmS in many Region 1 countries, subject to the consultations indicated in RR S5.100. RR S5.96 makes provision for additional AmS allocations in a part of the frequency range 1715–2000 kHz to be made in a number of Region 1 countries, amateur transmitter power being limited to 10 W and subject to consultation with neighbouring countries to avoid harmful interference to other services.

b)   In Region 2 there is an exclusive AmS allocation at 1800–1850 kHz and the band 1850–2000 kHz is also allocated for the AmS, sharing with the FS, the MS (excluding the AMS), the RLS and the RNS.

c)   In Region 3 the band 1800–2000 kHz is allocated for the AmS, shared with the FS, the MS (excluding the AMS) and the RNS, but RR S5.97 requires that none of these services cause harmful interference to LORAN systems (see Section 15.2 item 5) that have frequency assignments at 1850 and 1950 kHz.

2   The HF/VHF group (subject to withdrawal during natural disasters)
Eight frequency bands between 3500 kHz and 148 MHz are normally in heavy use by radio amateurs. However, Administrations may authorise the use of these bands by other stations for international communication in times of natural disaster; see RR S5.120. The bands are as follows;

a)   In Region 1 there is an AmS allocation at 3500–3800 kHz, sharing with the FS and the MS (except the AMS). In Region 2 the corresponding AmS allocation at 3500–3750 kHz is exclusive in most countries and is shared with the FS and the MS (except the AMS) at 3750–4000 kHz. In Region 3 the AmS shares 3500–3900 kHz with the FS and the MS.

b)   The band 7000–7100 kHz is shared by the AmS and the AmSS and with no other service in almost all countries world-wide. Since there are allocations for the BS in adjacent bands, RR Resolution 641 emphasises that there should be no broadcasting in this band. In Region 2 there is also an exclusive allocation for the AmS at 7100–7300 kHz but this band is allocated for the BS in Regions 1 and 3. However, RR S5.142 denies the AmS in Region 2 protection from interference from the BS in Regions 1 and 3. WARC-92 recognised that the Regional differences in the use of these bands was unsatisfactory and called for attention to be given to the matter as soon as the opportunity arises; see RR Recommendation 718.

c)   There is a world-wide secondary allocation for the AmS at 10.10–10.15 MHz, shared with the FS.

d)   The following bands are allocated for the AmS and some are shared with the AmSS. With the one exception indicated below, these bands are shared with virtually no other service:

> world-wide; 14.0–14.25 MHz (AmS and AmSS);
> world-wide; 14.25–14.35 MHz (AmS);
> world-wide; 18.068 -18.168 MHz (AmS and AmSS);
> world-wide; 21.0–21.45 MHz (AmS and AmSS);
> world-wide; 24.89–24.99 MHz (AmS and AmSS);
> world-wide; 144–146 MHz (AmS and AmSS);
> Region 2; 146–148 MHz (AmS).
> Region 3; 146–148 MHz (AmS, shared with the FS and the MS).

3   The 28 MHz band
The band 28.0–29.7 MHz is allocated for the AmS and the AmSS and no other service, world-wide.

4   The 50 MHz band
The band 50–54 MHz is allocated for the AmS in Regions 2 and 3 and exclusive in almost all countries.

5 The 220 MHz band
The band 220–225 MHz is allocated for the AmS in Region 2 only, sharing with the FS and the MS.

6 The 430 MHz band
In Region 1 there is a primary allocation for the AmS at 430–440 MHz, sharing with the RLS. In Regions 2 and 3 the AmS allocation is secondary, the RLS being primary. The band 435–438 MHz is also allocated for the AmSS (see RR S5.282) provided that no harmful interference is caused to services with allocations in the international table. These bands are shared, in whole or in part, with various other services in many countries. The band 433.05–434.79 MHz is also used for ISM; see RR S5.138 and S5.280.

7 The 902 MHz band
There is a secondary allocation for the AmS in the band 902–928 MHz in Region 2 only, sharing with the FS (primary) and the RLS (primary in USA). The band is also used for ISM; see RR S5.150.

8 The secondary microwave allocations
There are secondary allocations for the AmS in five bands between 1 and 11 GHz, as follows;

a) 1260–1300 MHz, world-wide, sharing with primary allocations for the RLS and various other services.
b) 2300–2450 MHz, world-wide, sharing with primary allocations for the FS, the MS, the RLS, the BSS (USA and India) etc. The band 2400–2450 MHz is used for ISM; see RR S5.150.
c) 3300–3500 MHz in Regions 2 and 3 and in some countries in Region 1 (see RR S5.431), sharing with primary allocations for the RLS, the FS, the FSS (downlinks) etc.
d) 5650–5850 MHz world-wide, plus 5850–5925 MHz in Region 2 only, sharing with primary allocations for the RLS, the FSS (uplinks), the FS and the MS.
e) 10.0–10.5 GHz world-wide, sharing with primary allocations for the FS, the MS and the RLS.

Within these bands there are also the following secondary allocations for the AmSS; 1260–1270 MHz (uplinks only), 2400–2450 MHz, 3400–3410 MHz (Regions 2 and 3 only), 5650–5670 MHz (uplinks only) and 10.45–10.50 GHz.

9 The millimetre-wave primary allocations
There are primary world-wide allocations for the AmS and the AmSS, shared with almost no other services, at 24.0–24.05 GHz, 47.0–47.2 GHz, 75.5–76 GHz, 142–144 GHz and 248–250 GHz.

10 The millimetre-wave secondary allocations
There are secondary world-wide allocations for the AmS at 24.05–24.25 GHz, and for the AmS and AmSS at 76–81 GHz, 144–149 GHz and 241–248 GHz. In each case the RLS has a primary allocation.

Recommendation ITU-R M.1044 [4] considers how the problems of interference arising in shared amateur and amateur-satellite bands may best be tackled.

Licensing radio amateurs

In other services, the owner or operator of a radio station is licensed to operate that specific station using frequencies assigned to that station. A call sign or some other identifying signal is assigned to the station. However, in the amateur services a licence is granted, and a call sign is assigned, to a person who is thereby authorised to operate any amateur radio station coming within the terms of the licence using any frequency in any band which the Administration authorises that amateur to use. RR S25.6 requires Administrations to ensure that amateur licences are awarded only to persons who have demonstrated the operational and technical abilities judged necessary for the operation of an amateur station. RR S25.5 requires licensed amateurs to be competent in sending Morse by hand and receiving Morse by ear, although this requirement may be waived for a licence limited to operation at frequencies above 30 MHz.

Spectrum management for the AmS

RR S25.8 applies all of the general rules of the ITU to amateur stations where appropriate, making special mention of the need for emissions to be as stable in frequency and as free from spurious emissions as the state of technical development for such stations permits.

Administrations, and in some bands the RR, set constraints on the transmitter power and types of emission that may be used by amateur stations to enable the spectrum available to be shared fairly between amateurs and with other services to which bands may also be allocated; see RR S25.7 and, for example, S5.96.

RR S25.1 recognises an Administrations's right to forbid its radio amateurs to use their stations to communicate with amateur stations in another country. RR S25.2 requires trans-national communications between amateur stations to be made in plain language and to be limited to technical matters and remarks of a trivial personal nature. RR S25.3 and S25.4 emphasise that amateur radio stations shall not be used for transmitting international communications on behalf of third parties unless the Administrations concerned have come to some other arrangement. An Administration may include constraints on the content of signals between the amateur stations within its own jurisdiction in the terms of its amateur licences.

However, in general, Administrations do not attempt to manage the spectrum made available for the AmS. Frequencies are not assigned to terrestrial amateur stations, nor are the frequencies that they use notified to the ITU for registration in the MIFR. In principle amateurs are free to use any frequency in bands made available by the terms of their licence. In practice the orderly use of spectrum is brought about by management provided by national amateur radio societies, coordinated by the International Amateur Radio Union (IARU).

Spectrum management for the AmSS

The rules and practices set out above for the AmS apply also where appropriate for the AmSS but additional provisions are made necessary by the particular nature of satellite communication and the amateur service.

In principle the frequencies emitted by amateur satellites and earth stations are subject to coordination and registration in accordance with the procedure of RR Articles S9 and S11. The frequencies must therefore be formally assigned by an

Administration. Since Administrations do not accept responsibility for interference entering amateur systems, the complex procedure of notification and coordination that is found necessary, for example for networks of the FSS is not required for the AmSS. Coordination between a proposed new AmSS satellite and other stations of the AmSS and the AmS is carried out informally by the amateur radio societies. However, RR Resolution 642 sets out a simple procedure which provides for the need to coordinate with stations of other services.

In addition to this coordination procedure, RR S25.11 requires AmSS satellites to be equipped with devices for controlling satellite emissions in the event that harmful interference arises. Also, the Administrations authorising an AmSS satellite are required to ensure that enough satellite telecommand earth stations have been established before the satellite is launched, to ensure adequate control of interference.

12.1 References

1 'Disaster communications in the amateur and amateur-satellite services'. Recommendation ITU-R M.1042, ITU-R Recommendations 1997 M Series Part 6 (ITU, Geneva, 1997)
2 'Use of the amateur and amateur-satellite services in developing countries'. Recommendation ITU-R M.1043, *ibid.*
3 'Future amateur radio systems (FARS)'. Recommendation ITU-R M.1041, *ibid.*
4 'Frequency sharing criteria in the amateur and amateur-satellite services'. Recommendation ITU-R M.1044, *ibid.*

Chapter 13

The science radio services

13.1 The services

Much of the use made of radio by the science services is similar in kind to the communication facilities which most other radio services provide; information available at one location is transmitted over a radio link to another location where it is required. However, in the science services the information to be transmitted has often been acquired directly by observations of nature made, not by the senses of human observers, but by sensors that function at frequencies within the radio spectrum that are also allocated for the service.

These sensors may be passive or active. Passive sensors take the form of sensitive radiometers which measure the strength of the natural radiation within chosen frequency limits emanating, for example, from the surface of the Earth or from within its atmosphere. Active sensors are, in essence, radars which analyse radiation returned by a target that has been illuminated by transmissions at chosen frequencies; usually the target is also the Earth's surface or its atmosphere. Many of the activities of the space research (SR) and Earth exploration-satellite (EES) services, including the meteorological-satellite service (MetS), are of this kind, typically using passive or active sensors carried by satellites in low Earth orbits. A special case is the radio astronomy service (RAS) which observes emissions of natural origin arriving from beyond the Earth's atmosphere; all RAS allocations are used passively.

The SR, the EES and the MetS also require to transmit large amounts of data between spacecraft and the Earth and from one spacecraft to another. Frequency allocations have been made for space–Earth transmission, although a data link between a spacecraft and an earth station at a fixed location can also be treated as a feeder link and can be assigned a frequency in an FSS allocation; see RR S1.21. These services also use ISLs for their data and here again there are two options for frequency assignments; the bands allocated for the data links for these services can be used (see RR S1.51, S1.52 and S1.55), as can the ISS allocations.

There are three other science-related radio services. The meteorological aids service (MetA) is used for links to platforms, airborne or sea-borne, which gather meteorological data. The standard frequency and time signal service (TFS) and the corresponding satellite service (TFSS) are used for comparison of time and frequency standards and for the dissemination of these standards.

Finally, it is convenient to include the space operation service (SO) in this group. The SO is used for telecommand and telemetry links between Earth and satellites and spacecraft in sub-orbital trajectories. However, the total bandwidth allocated for the SO is quite limited and the number of satellites in operation is continually rising. Accordingly, while telecommand and telemetry links for satellites of, for example, the FSS may be assigned frequencies in SO allocations for use during the launch phase, these links are usually transferred to assignments in the bands in which the satellite will normally operate once the operational orbit has been attained. This is in accordance with the formal definition of the SO; see RR S1.23.

RR S5.341 advises that a programme of passive research, searching for intentional emissions of extra-terrestrial origin, also known as the Search for Extra-terrestrial Intelligence (SETI) is in progress in some countries in the frequency ranges 1400–1727 MHz, 101–120 GHz and 197–220 GHz. This activity is akin to radio astronomy but it is not treated as part of the RAS. RR Recommendation 702 recommends that the desirability of making some provision in the Radio Regulations to facilitate this research should be considered by a future WRC. However, there is no provision for it at present.

13.2 The use of passive sensors

13.2.1 Protection of passive sensors from interference

Radio astronomy observatories and the passive sensors used by other science services are extremely sensitive to interference from active radio services, that is, those that use radio transmitters. Various means have been used in the construction of the international table of frequency allocations to minimise in-band interference (that is, interference from transmitters operating in the frequency bands allocated for these passive services) in the bands in which the passive services have primary allocations, as follows.

1   Grossly unsuitable band sharing arrangements, such as the RAS sharing with satellite downlinks and passive sensors used for Earth exploration sharing with terrestrial broadcasting, are not made.
2   RR S5.149 lists many of the shared primary allocations for the RAS and urges Administrations to take all practical steps to protect the RAS from interference in these bands.
3   For a few frequency bands shared by passive services and active services there are specific sharing constraints on the active services to give a measure of protection to the passive services or a requirement that the active services do not cause harmful interference to the passive services.
4   Many of the more important frequency allocations for the passive services are not shared in the international table with an allocation for any active service. It should, however, be noted that an Administration is free to assign frequencies to stations of active services in such a band unless those stations are identified as a source of harmful interference, and a distant RAS observatory may suffer significant interference from signals which are too weak to permit identification of the source.

5   RR S5.340 lists seven bands, allocated for passive services, in which the operation of any transmitter is prohibited. In eight more bands, the operation of active services is limited to specific cases, most of which would be relatively harmless to passive services.

These measures help to solve the problem of interference to passive sensors of the SR, the EES and the MetS in most, though not all, of their allocations. The RAS also benefits in the same ways, although the extremely high sensitivity of its observatories ensure that their performance may be degraded by low levels of in-band interference from untraceable sources which may be of no consequence to the other passive services.

The RAS is the only passive service that observes at a fixed location. If the operator wishes to observe at a given frequency, the responsible Administration is asked for the frequency to be assigned to the observatory. If the Administration is willing and able to do so, other radio stations within its jurisdiction will not be given assignments that would interfere with the assignment to the observatory and the assignment may be notified to the ITU for registration in the MIFR. If the frequency is within a primary RAS allocation, the Administrations of neighbouring countries will probably be helpful in suppressing in-band interference. If there should be in-band interference from abroad the procedures and conventions outlined in Section 4.5 apply. There is general advice on the problem in Recommendations ITU-R RA.1031-1 [1] and RA.1272 [2]. See also RR Article S29.

However, the following points should be noted for a radio astronomy observatory.

- It is much more sensitive to interference than the passive sensors used for Earth exploration. The levels of interference that are harmful to radio astronomy in the various parts of the radio spectrum are given in Recommendation ITU-R RA.769-1 [3]. This recommendation is incorporated into the Radio Regulations by the reference in RR S29.12.
- Its antenna is directed towards the sky, in line of sight of an increasingly numerous population of artificial satellites,
- it may observe very distant astronomic objects, some of them in other galaxies and consequently having a Doppler-shifted noise spectrum.

These circumstances affect the vulnerability of the RAS to interference in several ways.

Thus, an observatory may suffer severe interference from the out-of-band emission products of a transmitter operating in a band adjacent to an RAS allocation. Recommendation ITU-R RA.517-2 [4] identifies some situations of particular concern. Recommendation ITU-R RA.1237 [5] recommends that all practicable steps should be taken to reduce spectral components from such emissions which fall outside their allocated band. RR S4.5 requires the frequencies assigned for emissions to be separated from the limits of the band to a degree sufficient to ensure that harmful interference does not occur in an adjacent band. However, RR S4.6 rules that protection must be provided for the RAS only to the extent that the stations of other services provide for one another. Thus the protection available from the application of the Radio Regulations, which may be adequate when the interference is from a transmitter operating at an assignment within the band allocated for the RAS, is likely to fall far short of what is required for the RAS if the interference is from an emission assigned a frequency in an adjacent band.

Also an observatory may suffer severe interference from spurious emissions of transmitters operating at any frequency. The ITU seeks to limit interference from spurious emissions only by applying mandatory limits to the spurious emission power as a function of the power of the emission as a whole; see RR Appendix S3 and Section 4.2.3 above. Furthermore, no numerical value has been set so far on the limit for satellite transmitters, nor for systems using digital modulation. The RAS is particularly vulnerable to interference from harmonic radiation and other spurious emissions from satellite transmitters, especially those from geostationary satellites; see Recommendation ITU-R RA.611-2 [6]. See also RR S5.413 and S5.402.

Finally, if the Doppler shift suffered by a line emission from a distant source is big enough to move the signal out of the band allocated for the RAS, as it may well be, the observatory has no remedy at all under ITU regulations if it experiences interference in receiving it.

13.2.2 Radio astronomy

Radio astronomers explore the cosmos by observing the radio noise reaching the Earth from space. Observations are of two basic kinds. Line measurements observe the distribution in the frequency domain of the noise power flux received from an astronomical source in the vicinity of spectral lines of particular interest. Continuum measurements measure and compare the total noise power fluxes received within a substantial bandwidth at different points in the radio spectrum.

There are 23 frequency allocations for the RAS that are intended for continuum measurements; they are listed in Table 13.1. Some allocations are secondary and some are Regional but most are primary and world-wide, as shown in the table. The bandwidth of some of these allocations is considerably less than would be necessary for accurate measurements.

At least 60 spectral lines within the frequency range allocated by the ITU are of major interest to astronomers. There is a list of the line frequencies of greatest importance in Recommendation ITU-R RA.314-8 [7]. Some of these lines lie within bands allocated for the RAS for continuum measurements. In addition there are frequency allocations for the RAS at the frequencies of many of the other spectral lines, primary in many cases. These allocations for line observations are narrow in bandwidth. There are other lines which are observed by astronomers without the benefit of a frequency allocation, although attention may be drawn to these observations by an advisory footnote to the international table of frequency allocations; see for example RR S5.458A.

Interference and the RAS

Progress in radio astronomy depends largely on making precise measurements of noise fluxes which are extremely weak. Equipment is available which enables accurate astronomical measurements to be made, in the absence of interference from radio transmitters, at levels that are far below the noise level in radio receivers that are used for communications systems; see for example Recommendation ITU-R RA.769-1 [3]. These criteria show that radio astronomy is extremely vulnerable to interference from radio transmitters. Administrative means for dealing with interference are

Table 13.1   Frequency allocations for RAS continuum measure-ments (allocations are primary unless otherwise indicated)

MHz	GHz
13.36–13.41	10.6–10.68
25.55–25.67[3]	10.68–10.7[2]
37.50–38.25 (secondary)	15.35–15.4[2]
73–74.6 (Region 2 only)	22.21–22.5
150.05–153.0 (Region 1 only)	23.6–24.0[1]
406.1–410	31.3–31.5[1]
608–614 (Region 2 only)	31.5–31.8[2]
1400–1427[1]	42.5–43.5
1660-1670	86–92[2]
2655–2690 (secondary)	105–116[2]
2690–2700[2]	164–168[3]
4800–4990 (secondary)	182–185[2]
4990–5000	217–231[1]
	265–275

Notes. [1]All emissions prohibited in this band (RR S5.340).
[2]Most emissions, although with specified exceptions (see RR S5.340), prohibited in this band.
[3]All international allocations in this band are passive.

available (see Section 13.2.1 above, also item 1 in Section 13.2.3 below) but they fall far short of what astronomers would desire. But see also items 2 and 4 in Section 13.2.3.

Radio astronomy in the shielded zone of the Moon

It is thought that the Earth's atmosphere masks phenomena that would otherwise be observable by radio astronomy. Furthermore, the problem of interference to the RAS is unlikely to be totally eliminated; indeed, interference may become worse as the use of radio becomes more intense. It is foreseen that radio astronomy observatories will be set up in the area of the Moon's surface that is never illuminated by radio emissions from the Earth's surface or by satellites in typical Earth orbits. Accordingly, RR S22.22 to S22.25 outline a code of practice designed to keep the shielded zone of the Moon free from interference over most of the radio spectrum. See also Recommendation ITU-R RA.479-4 [8].

13.2.3 Passive sensors for the SR and the EES

Satellite-borne passive sensors operating in a substantial number of microwave and millimetre-wave frequency bands are used in the SR and the EES, including the MetS, to observe the Earth and its atmosphere. Recommendation ITU-R SA.515-3 [9] lists the purposes which the various frequency bands serve. Table 13.2 lists the

Table 13.2   Frequency allocations for passive sensors on satellites of the SR and/or the EES (GHz) (allocations are primary unless otherwise indicated)

1.37–1.4 (secondary)	21.2–21.4	116–126
1.4–1.427[1]	22.21–22.5	140.69–140.98[2]
2.64–2.655 (secondary)	23.6–24.0[1]	150–151
2.67–2.69 (secondary)	31.3–31.5[1]	156–158
2.69–2.7[2]	31.5–31.8[2]	164–168[3]
4.95–4.99 (secondary)	48.94–49.04[2]	174.5–176.5
4.99–5.00 (secondary)	50.2–50.4[2]	182–185[1]
10.6–10.68	52.6–54.25[1]	200–202
10.68–10.7[2]	54.25–59.3	217–231[1]
15.2–15.35 (secondary)	86–92[1]	235–238
15.35–15.4[1]	100–102	250-252[3]
18.6–18.8[4]	105-116[1]	

Notes.   [1]All emissions prohibited in this band (see RR S5.340).
[2]Most emissions, although with specified exceptions (see RR S5.340), prohibited in this band.
[3]All international allocations in this band are passive.
[4]Primary in Region 2 and secondary in Regions 1 and 3.

present international allocations. The band 4990–5000 MHz is allocated for the passive sensors of the SR only but all the other bands are allocated for passive sensors of both the SR and the EES and, by inference, the MetS. With the exceptions indicated in Table 13.2 these allocations are primary and world-wide.

In addition to the frequency bands listed in Table 13.2, RR S5.438 recognised the access of passive sensors of the SR and EES to the band 4200–4400 MHz on a secondary allocation basis but offers no protection from interference from the radio altimeters that operate in this band.

Many of these allocations are shared with the RAS. Some bands are shared with services that use transmitters but many are covered by the prohibition of emissions, or of the emissions that would be damaging to the operation of passive sensors, contained in RR S5.340; see Table 13.2. However, potential interference problems have been identified in some of these bands and solutions have been agreed, or are being sought, in some of them. Thus;

1   *The 10.6–10.68 GHz band*
This band is shared by passive sensors with primary allocations for the FS and the MS (except the AMS). To protect the passive services, RR S5.482 imposes power constraints on emissions of the active services that are more severe than those of RR Article S21.

2   *The 15.35–15.4 GHz band*
WRC-97 allocated the band 15.43–15.63 GHz for the FSS, to be used for feeder uplinks and downlinks for non-GSO MSS networks but concern was expressed about the possible effects of these emissions on passive sensors below 15.4 GHz. RR Resolution 123 calls for studies of this matter. See also item 7(d) in Section 11.1.

3   *The 18.6–18.8 GHz band*

See RR Recommendation 706. The passive service allocations are primary in Region 2 and secondary in Regions 1 and 3, the band being shared with primary allocations for the FS, the MS and the FSS (downlinks) world-wide. Note that RR S4.7 qualifies the degree of protection provided for EES and SR passive sensors against out-of-band emissions in the same way that RR S4.6 serves the RAS. Studies reported in Recommendation ITU-R F.761 [10] found some risk of interference to passive sensors from transmitters of the FS and RR S5.522 urges the use of power levels lower than the limits imposed on the FS by RR Article S21 in order to protect the passive sensors. Also a risk has been perceived that energy from satellite transmitters of the FSS would degrade the accuracy of passive sensors after reflection from the ground has led to RR S5.523, which requests that the downlink PFD should be as small as is practicable. See also item 8(b) in Section 6.2.1 and item 5 in Section 9.1.2.

4   *The 22.21–22.5 GHz band*

This band is shared by the passive services with the FS and the MS (except the AMS). RR S5.532 rules that the passive services shall not impose constraints on the FS and the MS.

Sharing studies have been made in the bands between 50 and 60 GHz, where there are extensive allocations for passive sensors. There are allocations for the FS, the MS and the ISS in this frequency range, although they are not in heavy use at present. Recommendation ITU-R SA.1259 [11] shows that sharing between passive sensors and the FS is only likely to be feasible at frequencies where absorption of emissions from stations on the Earth's surface by oxygen in the atmosphere is substantial. Recommendation ITU-R SA.1279 [12] concludes that sharing between passive sensors and ISLs serving non-GSO satellites is not feasible but sharing with ISLs serving geostationary satellites may be feasible, subject to constraints.

13.3 Active sensors for the SR and the EES

There are 12 allocations for satellite-borne active sensors or radars; see Table 13.3. This range of frequencies enables observations to be made of a variety of phenomena; see Recommendation ITU-R SA.577-5 [13].

Most of the allocations for active sensors are for both the SR and the EES, although a few cover only one of the services concerned. All are world-wide. All are shared with

Table 13.3 Frequency allocations for active sensors on satellites of the SR and/or the EES (GHz) (allocations are primary unless otherwise indicated)

1.215–1.300	9.5–9.8	24.05–24.25[1]
3.1–3.3[1]	9.975–10.025[1]	35.5–36.0
5.25–5.46	13.25–13.75	78–79
8.55–8.65	17.2-17.3	94.0–94.1

Note.   [1]Secondary.

the RLS and some are shared with one or other of the radionavigation services. Most of the active sensor allocations are primary, although footnotes to the frequency allocation table detract from the primary status of some by requiring that active sensors do not constrain the future expansion of the sharing services or cause harmful interference to radionavigation services.

Recommendation ITU-R SA.1280 [14] advises on the choice of active sensor emission characteristics to enhance compatibility with terrestrial radars. RR Resolutions 724 and 725 call for further study of sharing around 5350 MHz, Recommendation ITU-R SA.1281 [15] recommends a constraint on the power of active sensors in the 13.4–13.75 GHz band to protect the RLS and the RNS.

Recommendation ITU-R SA.1282 [16] recommends that active sensors and wind profiler radars of the RLS around 1260 MHz should not operate in the same frequency bands. Recommendation ITU-R SA.1261 [17] finds the band 94.0–94.1 GHz suitable for active sensors used as cloud radars but recommends that the band should not be used for other active sensor applications. However, in general, sharing between active sensors and the RLS does not raise major interference problems; see Recommendation ITU-R SA.516-1 [18].

Among the objectives set out by WRC-95 in its RR Resolution 712 is a new allocation for satellite-borne active sensors with a bandwidth of 3.5 MHz between 420 and 470 MHz. This is required to enable assessments to be made of forest cover and the rate of forest degradation in tropical and temperate regions, as requested by the United Nations Conference on Environment and Development (Rio de Janeiro, 1992). Recommendation ITU-R SA.1260 [19] is the first outcome of this search; sharing problems were found with existing allocations, perhaps most readily solvable between 430 and 440 MHz. RR Resolution 727 of WRC-97 increased the estimate of the bandwidth required to 6 MHz and called for the search to be continued, and in particular to define the criteria that would make sharing feasible.

13.4 Communication bands for the science services

13.4.1 *The EES, the MetS, the SR and the SO*

The various needs of the EES, the MetS, the SR and the SO for spectrum for the transmission of data are closely akin and it is convenient to consider their allocations together. They are reviewed here under six items.

1 The HF bands

There are nine allocations for the SR distributed over the HF range. They are all worldwide, secondary and very narrow, a few kilohertz wide at most.

2 Bands between 100 MHz and 1 GHz

All the four services have allocations in this frequency range, about 25 in all. Primary allocations for the SO, the MetS and the SR (all for downlinks) all share 137–138 MHz with allocations for the MSS (downlinks) and other services. The MetS is being phased out of this band; see RR S5.203. There is a primary world-wide allocation for SR downlinks at 143.6–143.65 MHz, plus secondary allocations at 138–143.6 MHz in Regions 2 and 3. There is a primary allocation for SO uplinks at 148–149.9 MHz (see RR S5.218) and one for SO downlinks, mostly secondary, at 267–273 MHz. There is a cluster of narrow allocations for the SO, the MetS, the SR and the EES between

400.15 MHz and 403 MHz. The use of the SR primary allocation at 410–420 MHz is limited by RR S5.268 to extra-vehicular activities in space in the vicinity of a spacecraft. The band 449.75–450.25 MHz is allocated for uplinks of the SO and the SR; see RR S5.286. Finally, 460–470 MHz is a secondary allocation for downlinks of the EES and the MetS.

All of these allocations are shared with other services, which in many cases have superior allocation status, and the science services must coordinate with the other allocated services as described in Section 5.3 in some bands. Recommendation ITU-R SA.1258 [20] recommends constraints to facilitate sharing between the science services at 401–403 MHz. Recommendation ITU-R SA.1236 [21] recommends constraints on SR emissions at 410–420 MHz to avoid causing unacceptable interference to systems of the FS and the MS. RR Resolution 723 seeks the allocation, by a future WRC, of up to 3 MHz of bandwidth between 100 MHz and 1 GHz for telecommand links of the SR and the SO.

3 The SR deep space allocations

Space more than 2×10^6 km from the Earth is defined as 'deep space' in RR S15.177. A group of SR allocations is reserved for communication between Earth and stations in deep space. The optimum frequency for links such as these, that involve exceptional transmission loss and quite limited transmitter power for the return link, depends on atmospheric conditions on Earth, the direction of transmission and the characteristics of the antennas used; see Recommendations ITU-R SA.1012 [22] and SA.1013 [23]. The current allocations are shown in Table 13.4.

Some of these allocations are primary, some are secondary and all are shared. Recommendation ITU-R SA.1016 [24] discusses sharing with deep space links. In practice, interference is not usually a major problem, despite the very low power level of Earth-bound signals from distant spacecraft. The exceptionally high e.i.r.p. of the earth station emissions makes uplinks relatively immune from interference and the downlink frequency bands are not shared with services likely to cause interference. Finally, the earth stations used for deep space communication are few and their locations have been selected with the objective of minimising interference with terrestrial stations.

4 Allocations for data relay satellite systems

Most satellite-borne passive and active sensors are carried by satellites in LEO and in some circumstances it is convenient to transfer the data that they collect to geostationary data relay satellites via ISLs for onward transmission to the ground. Recommendation ITU-R SA.1019 [25] recommends the use of assignments in the vicinity of 2.1, 14.0 and 24 GHz for ISLs, depending upon the data transmission rate required.

Table 13.4   Frequency allocations for communication with deep space

2110–2120 MHz (uplinks)	12.75–13.25 GHz (downlinks)
2290–2300 MHz (downlinks)	16.6–17.1 GHz (uplinks)
5650–5725 MHz	31.8–32.3 GHz (downlinks)
7145–7190 MHz (uplinks)	34.2–34.7 GHz (uplinks)
8400–8450 MHz (downlinks)	

The bands 2025–2110 MHz and 2200–2290 MHz are primary allocations for ISLs of the SO, the SR and the EES; in addition, these two bands are allocated for uplinks and downlinks respectively, of these services, both bands also being shared with primary allocations for the FS and the MS. This complex sharing situation in a heavily used range of the spectrum gives rise to a need for measures to limit interference and substantial progress has been made in providing them, as follows. See also item 4 in Section 6.2.1.

a)   Recommendation ITU-R SA.1273 [26] recommends downlink PFD limits at the Earth's surface from satellites of the SR, the SO and the EES in these bands; the limits are expressed for a sampling bandwidth of 1 MHz but they are otherwise closely aligned with those in RR Article S21, Table S21-4.

b)   Recommendation ITU-R SA.1274 [27] recommends a level of interference power which data relay satellite links should be designed to tolerate from FS systems.

c)   Recommendations ITU-R F.1248 [28] and F.1247 [29] recommend severe limits on the power of tropospheric scatter transmitters and line-of-sight transmitters of the FS in these bands. Both recommendations recommend particularly severe limits for stations with beams directed towards points in the GSO which Recommendation ITU-R SA.1275 [30] indicates to be preferred locations for data relay satellites.

d)   Recommendations ITU-R F.382-7 [31] and F.1098-1 [32] recommend radio frequency channel plans for line-of-sight FS systems which have been devised to enable the FS to be developed in ways that are harmonised with the allocations for science services in this frequency range.

e)   Recommendation ITU-R SA.1154 [33] offers various measures designed to facilitate sharing between the science services and the MS in these bands. One of the points made is that these bands should not be used for the kinds of mobile system which it describes as 'high density' systems, such as a cellular radiotelephone system like IMT-2000; see Section 8.5.4 above. RR S5.391 gives regulatory force to this point.

f)   RR Recommendation 622 recommends that these various measures should be taken into account by Administrations as soon as is practicable.

There is, at present, no band allocated for ISLs near 14 GHz. The allocation for the ISS at 25.25–27.5 GHz is virtually reserved for ISLs of the science services.

5   The remaining allocations between 1 and 10 GHz
With the one exception identified below, these allocations are world-wide.

a) The SO has primary world-wide allocations at 1427–1429 MHz (uplinks), shared with the FS and MS (excluding the AMS), and at 1525–1535 MHz (downlinks), shared with the FS and MSS (downlinks).

b)   The main activity of the EES in this part of the spectrum is the delivery to the ground of raw data from satellite sensors and the distribution to users of processed data. There is a secondary allocation for EES at 1525–1535 MHz, sharing with primary allocations for the SO and the MSS (both for downlinks) and other secondary allocations. RR S5.289 allows EES downlinks to use the band 1690–1710 MHz on condition that they cause no harmful interference to the services for which it is allocated. Finally, there is a primary allocation for EES downlinks at 8025–8400 MHz, subject in Regions 1 and 3 to a relatively severe constraint on the spectral PFD at the ground, Recommendation ITU-R SA.1277 [34] not having been implemented by WRC-97; see RR S5.462A. There is the possibility that this constraint will be modified soon; see RR Resolution 124 and a note in Section 6.2.3 above.

c) The MetS also downlinks data, raw and processed, in this frequency range. The band 1670–1710 MHz is a primary allocation for MetS downlinks, shared with various other services, most notably the MSS; Recommendation ITU-R SA.1158-1 [35] proposes a possible basis for sharing the band between these services. Much of this band is also shared with the MetA; see item 5 in Section 13.4.2 below.

There are three more primary allocations for the MetS. At present the 7450–7550 MHz allocation, which is for downlinks and is shared with primary allocations for FSS downlinks, the FS and the MS (excluding the AMS), is used by both GSO and non-GSO MetS satellites but RR S5.461A limits this allocation, in future, to GSO MetS systems. New non-GSO MetS satellites are to be assigned downlink frequencies in a newly allocated band at 7750–7850 MHz, also shared with the FS and the MS (excluding the AMS); see RR S5.461B. Finally, the band 8175–8215 MHz is allocated for MetS uplinks, shared with primary allocations for FSS uplinks, EES downlinks, the FS and the MS.

d) The SR has three allocations in this frequency range. There is a secondary allocation at 5250–5255 MHz. The primary allocation for uplinks at 7190–7235 MHz, shared with the FS and the MS, is subject to coordination as discussed in Section 5.3 above. The primary allocation for downlinks at 8450–8500 MHz is shared with the FS and the MS (excluding the AMS).

6 The allocations above 10 GHz.
These allocations, listed in Table 13.5, are all world-wide but most of them are involved in a complex pattern of sharing with other services.

13.4.2 The meteorological aids service

Typical stations of the MetA are of three kinds. Radiosondes are flown on free-floating balloons to measure meteorological conditions in the upper atmosphere. Other sensors are carried by hydrological buoys at sea, where data on marine conditions are collected. The third kind of station is located on the ground, typically at a permanent site, and used to command data to be transmitted from radiosondes and buoys, and to receive it.

The frequency allocations for the MetA are as follows;

1   2025–2045 kHz. This allocation is secondary, for Region 1 only and limited by RR S5.104 to oceanographic buoys.
2   27.5–28.0 MHz. This allocation is primary, shared with the FS and MS which are also primary.
3   153–154 MHz. This allocation is secondary, for Region 1 only, shared with the FS and the MS, both of which are primary.
4   400.15–406 MHz. This is a world-wide primary allocation, half of which is not shared with any other primary allocation.

Table 13.5   Frequency allocations for science services above 10 GHz

---

13.4–14.3 GHz, secondary, for the SR. See Note (i) below.

14.4–14.47 GHz, secondary, for SR downlinks.

14.5–15.35 GHz, secondary, for the SR.

18.1–18.3 GHz, primary, for MetS downlinks; see RR para. S5 519 and item 5 in Section 9.1.2 above.

25.5–27.0 GHz, secondary, for EES downlinks. See Note (ii) below.

28.5–30.0 GHz, secondary, for EES uplinks. See Note (iii) below.

31.0–31.3 GHz, secondary, for the SR.

34.7–35.2 GHz, secondary, for the SR.

37.0–38.0 GHz, primary, for SR downlinks. See Note (iv) below.

37.5–40.5 GHz, secondary, for EES downlinks. See Note (iv) below.

40.0–40.5 GHz, primary, for SR and EES uplinks.

65.0–66.0 GHz, primary, for the SR and EES.

74.0–84.0 GHz, secondary for SR downlinks.

Note (i) The band 13.75–14.0 GHz was allocated, with primary status, for the FSS (uplinks) at WARC-92 and this has had an impact on the use made of the band by the SR; see RR S5.503 and S5.503A. Recommendation ITU-R S.1069 [36] provides constraints and guidelines for the phased discontinuation of some of the former SR uses of this band.

(ii) The band 25.5–27.0 GHz is also allocated for the ISS and the MS, with primary status, plus a secondary allocation for the TFSS. Recommendation ITU-R SA.1278 [37] offers a basis for use of the band by the EES. Note also the constraint on the e.i.r.p. of FS and MS transmitting stations with antennas directed within 1.5° of the GSO, applied wherever practicable by RR S21.2.

(iii) The band 29.95–30.0 GHz may be used for ISLs of the EES on a secondary basis; see RR S5.543. See also RR S22.4, which puts an explicit limit on the interference that these ISLs may cause to a GSO network of the FSS.

(iv) All or part of the band 37–40 GHz is also allocated on a primary basis for the FS, the MS, the FSS (downlinks) and the MSS (downlinks). RR Resolution 133 calls for study to be given to criteria for sharing this band between these various services; see item 14 in Section 6.2.1 and item 9(b) in Section 9.1.2.

5   1668.4–1700 MHz. This allocation is also world-wide and primary. It is, however, shared with various other primary services, the RAS, the FS, the MS (except AMS), the MetS (downlinks) and the MSS (uplinks). A sharing constraint limiting the PFD from MetS downlinks is imposed by RR Table S21-4 to protect the MetA.

6   35.2–36.0 GHz. This allocation is world-wide and primary, shared with the RLS.

In addition, meteorologists use ground-based radars to map the distribution of cloud and rain. Aircraft have weather radars for similar purposes. These radars are of the RLS, or perhaps the RNS in the case of airborne radars. However, it should be noted that RR S5.423 and S5.452 authorise RNS allocations at 2700–2900 MHz and 5600–5650 MHz to be used by ground-based meteorological radars on equal terms with other radionavigation radars, and RR S5.475 provides similarly for both ground-based and airborne radars in the 9300–9500 MHz band.

13.4.3 The standard frequency and time signal services

The terrestrial service

There are frequency allocations for the TFS, each a few kilohertz wide, centred on 20, 2500 and 5000 kHz and on 10, 15, 20 and 25 MHz. These allocations are primary and world-wide. The transmission of standard frequency and time signals is also authorised in Region 3 in allocations to the FS in narrow bands centred on 4000 and 8000 kHz and 16 MHz; see RR S5.126, S5.144 and S5.153. There are a number of other footnotes which authorise TFS emissions on a limited geographical basis elsewhere in the HF frequency range.

Standardising laboratories in many countries use these allocations to make time and frequency standards available to the public. Many aim to make the signals available day and night and this requires the use of several of these frequencies to overcome the diurnal variations of ionospheric propagation. There is interference between the TFS emissions from different stations, and to reduce this interference, some transmissions are made at the centre frequency of these allocated bands and others are displaced from the centre by a small, accurately known amount. There is a list of current TFS emissions using the allocations listed above in Table 1 of Recommendation ITU-R TF.768-3 [38]. In order that the limited spectrum that is available should be put to the best use, Administrations authorising new TFS emissions are required to coordinate the assignments first with the ITU Radiocommunication Bureau and the Chairman of ITU-R Study Group 7; see Resolution ITU-R 28 [39] and RR S26.1 to S26.6.

Table 2 of Recommendation ITU-R TF 768-3 [38] lists standard time and frequency signals transmitted in other bands. In addition to these TFS transmissions, there are many emissions of other services, above all navigation systems and broadcasting transmissions, which can be used to provide time and/or frequency signals of great precision. Some navigation stations that can be used in this way are listed in Table 3 of the recommendation.

The satellite service

The main application for the TFSS would be the comparison of the time and frequency standards maintained at the various standardising laboratories. Recommendation ITU-R TF.1011-1 [40] compares the various media available for standards comparison, including the use of satellite links. The following frequency bands are available for the TFSS.

1   400.05–400.15 MHz; this is a primary allocation.
2   4200–4204 MHz (downlinks) and 6425–6429 MHz (uplinks): subject to coordination as described in Section 5.3 above; see RR S5.440.
3   13.4–14.0 GHz (uplinks), 20.2–21.2 GHz (downlinks), 25.25–27.0 GHz (uplinks), and 30.0–31.3 GHz (downlinks): these allocations are all secondary.

13.5 Managing spectrum for the science services

The responsibilities of an Administration and the procedures that are used in assigning, coordinating where necessary and registering in the MIFR a frequency for use in the science services are basically the same as for other services and as outlined in

Chapters 4 and 5 above. Section 13.4 above makes reference to the special coordination requirement for TFS transmissions called for by Resolution ITU-R 28. There is relevant guidance as to interference criteria etc. in many ITU-R Recommendations of Series RA and Series SA. See in particular Recommendation ITU-R SA.1027-2 [41]. Particular mention should perhaps be made of the need for the Administration to take particular care to protect a radio astronomy observatory from other stations using frequencies that have been assigned to the observatory.

It might seem that the choice of frequencies for assignment to stations of the science services and coordination with the assignments made to other science service stations would place a particularly heavy burden on the Administration but this is not necessarily so. There is close international collaboration in such matters between and within the various agencies, national and international, involved in the scientific use of radio, relieving the Administration of much detailed coordination, and neither conflicts of interest nor competition are likely to arise at a national level.

13.6 References

1 'Protection of the radioastronomy service in frequency bands shared with other services'. Recommendation ITU-R RA.1031-1, ITU-R Recommendations 1997 RA Series (ITU, Geneva, 1997)

2 'Protection of radio astronomy measurements above 60 GHz from ground based interference'. Recommendation ITU-R RA.1272, *ibid.*

3 'Protection criteria used for radioastronomical measurements'. Recommendation ITU-R RA.769-1, *ibid.*

4 'Protection of the radioastronomy service from transmitters in adjacent bands'. Recommendation ITU-R RA.517-2, *ibid.*

5 'Protection of the radio astronomy service from unwanted emissions resulting from applications of wideband digital modulation'. Recommendation ITU-R RA.1237, *ibid.*

6 'Protection of the radioastronomy service from spurious emissions'. Recommendation ITU-R RA.611-2, *ibid.*

7 'Preferred frequency bands for radioastronomical measurements'. Recommendation ITU-R RA.314-8, *ibid.*

8 'Protection of frequencies for radioastronomical measurements in the shielded zone of the Moon'. Recommendation ITU-R RA.479-4, *ibid.*

9 'Frequency bands and bandwidths used for satellite passive sensing'. Recommendation ITU-R SA.515-3, ITU-R Recommendations 1997 SA Series (ITU, Geneva, 1997)

10 'Frequency sharing between the fixed service and passive sensors in the band 18.6–18.8 GHz'. Recommendation ITU-R F.761, ITU-R Recommendations 1997 F Series Part 2 (ITU, Geneva, 1997)

11 'Feasibility of sharing between spaceborne passive sensors and the fixed service from 50 to 60 GHz'. Recommendation ITU-R SA.1259, ITU-R Recommendations 1997 SA Series (ITU, Geneva, 1997)

12 'Spectrum sharing between spaceborne passive sensors and inter-satellite links in the range 50.2–59.3 GHz'. Recommendation ITU-R SA.1279, *ibid.*

13 'Preferred frequencies and necessary bandwidths for spaceborne active remote sensors'. Recommendation ITU-R SA.577-5, *ibid.*

14 'Selection of active spaceborne sensor emission characteristics to mitigate the potential for interference to terrestrial radars operating in frequency band 1–10 GHz'. Recommendation ITU-R SA.1280, *ibid.*

15 'Protection of stations in the radiolocation service from emissions from active spaceborne sensors in the band 13.4–13.75 GHz'. Recommendation ITU-R SA.1281, *ibid.*

16 'Feasibility of sharing between wind profiler radars and active spaceborne sensors in the vicinity of 1260 MHz'. Recommendation ITU-R SA.1282, *ibid.*

17 'Feasibility of sharing between spaceborne cloud radars and other services in the range of 92–95 GHz'. Recommendation ITU-R SA.1261, *ibid.*

18 'Feasibility of sharing between active sensors used on Earth exploration and meteorological satellites and the radiolocation service'. Recommendation ITU-R SA.516-1, *ibid.*

19 'Feasibility of sharing between active spaceborne sensors and other services in the vicinity of 410–470 MHz'. Recommendation ITU-R SA.1260, *ibid.*

20 'Sharing of the frequency band 401–403 MHz between the meteorological satellite service, Earth exploration-satellite service and meteorological aids service'. Recommendation ITU-R SA.1258, *ibid.*

21 'Frequency sharing between space research service extra-vehicular activity (EVA) links and fixed and mobile service links in the 410–420 MHz band'. Recommendation ITU-R SA.1236, *ibid.*

22 'Preferred frequency bands for deep-space research in the 1–40 GHz range'. Recommendation ITU-R SA.1012, *ibid.*

23 'Preferred frequency bands for deep-space research in the 40–120 GHz range'. Recommendation ITU-R SA.1013, *ibid.*

24 'Sharing conditions relating to deep-space research'. Recommendation ITU-R SA.1016, *ibid.*

25 'Preferred frequency bands and transmission directions for data relay satellite systems'. Recommendation ITU-R SA.1019, *ibid.*

26 'Power flux-density levels from the space research, space operation and Earth exploration-satellite services at the surface of the Earth required to protect the fixed service in the bands 2015–2110 MHz and 2200–2290 MHz'. Recommendation ITU-R SA.1273, *ibid.*

27 'Criteria for data relay satellite networks to facilitate sharing with systems in the fixed service in the bands 2025–2110 MHz and 2200–2290 MHz'. Recommendation ITU-R SA.1274, *ibid.*

28 'Limiting interference to satellites in the space science services from the emissions of trans-horizon radio relay systems in the bands 2025–2110 MHz and 2200–2290 MHz'. Recommendation ITU-R F.1248, ITU-R Recommendations 1997 F Series Part 2 (ITU, Geneva, 1997)

29 'Technical and operational characteristics of systems in the fixed service to facilitate sharing with the space research, space operation and Earth-exploration services operating in the bands 2025–2110 MHz and 2200–2290 MHz'. Recommendation ITU-R F.1247, *ibid.*

30 'Orbital locations of data relay satellites to be protected from the emissions of fixed service systems operating in the band 2200–2290 MHz'. Recommendation ITU-R SA.1275, ITU-R Recommendations 1997 SA Series (ITU, Geneva, 1997)

31 'Radio-frequency channel arrangements for radio-relay systems operating in the 2 and 4 GHz bands'. Recommendation ITU-R F.382-7, ITU-R Recommendations 1997 F Series Part 1 (ITU, Geneva, 1997)

32 'Radio-frequency channel arrangements for radio-relay systems in the 1900–2300 MHz band'. Recommendation ITU-R F1098-1, *ibid.*

33 'Provisions to protect the space research (SR), space operations (SO) and Earth-exploration satellite services (EES) and to facilitate sharing with the mobile service in the 2025–2110 MHz and 2200–2290 MHz bands'. Recommendation ITU-R SA.1154, ITU-R Recommendations 1997 SA Series (ITU, Geneva, 1997)

34 'Sharing in the 8025–8400 MHz frequency band between the Earth exploration-satellite service and the fixed, fixed-satellite, meteorological-satellite and mobile services in Regions 1, 2 and 3'. Recommendation ITU-R SA.1277, *ibid.*

35 'Sharing in the 1675–1710 MHz band between the meteorological-satellite service (space-to-Earth) and the mobile-satellite service (Earth-to-space)'. Recommendation ITU-R SA.1158-1, *ibid.*

36 'Compatibility between the fixed-satellite service and the space science services in the band 13.75–14 GHz'. Recommendation ITU-R S.1069, ITU-R Recommendations 1997 S Series. (ITU, Geneva, 1997)

37 'Feasibility of sharing between the Earth exploration-satellite service (space-to-Earth) and the fixed, inter-satellite and mobile services in the band 25.5–27.0 GHz'. Recommendation ITU-R SA.1278, ITU-R Recommendations 1997 SA Series (ITU, Geneva, 1997)
38 'Standard frequencies and time signals'. Recommendation ITU-R TF.768-3, ITU-R Recommendations 1997 TF Series (ITU, Geneva, 1997)
39 'Standard frequency and time signal emissions. Resolution ITU-R 28, *ibid*.
40 'Systems, techniques and services for time and frequency transfer'. Recommendation ITU-R TF.1011-1, *ibid*.
41 'Sharing and coordination criteria for space-to-Earth data transmission systems in the Earth exploration-satellite and meteorological-satellite services using satellites in low earth orbit'. Recommendation ITU-R SA.1027-2, ITU-R Recommendations 1997 SA Series (ITU, Geneva, 1997)

Chapter 14

The inter-satellite service

It will be seen, in particular from item 4 in Section 13.4.1 and Note (iii) in Table 13.5, that the SR and EES use certain of their allocations for links between spacecraft. However, in addition, a number of millimetre-wave bands have been allocated for the inter-satellite service (ISS) to provide for other links between Earth satellites. All of these ISS allocations are primary and all but one (that is, 25.25–27.5 GHz, see item 3 below) are available for assignments for any satellite-using service.

All of these ISS allocations are shared with primary allocations for other services and the potential exists for strong interference between ISLs and the stations of sharing services. Much depends on three factors; the orbits of the satellites to be linked, the e.i.r.p. of the terrestrial emissions and the absorption which takes place due to molecular resonances in the gases of the atmosphere, which is closely dependent on the frequency (see the Appendix, Section A2.2.4). Where there can be interference between an ISL and an aircraft station, the height of the aircraft above the ground may have a very significant effect on the amount of the absorption that occurs.

If an ISL is used to connect a geostationary satellite and a spacecraft at a greater distance from the Earth, RR S22.3 provides a constraint to limit interference to other geostationary satellite networks. This constraint, regardless of the frequency band, requires the boresight axis of the ISL antenna at the geostationary satellite to be pointed at least 15° away from any point on the GSO. Other constraints have been imposed to limit interference in specific bands; see below.

Several surveys of the feasibility of sharing between the ISS and other services have been carried out. Thus, Recommendation ITU-R S.1327 [1] is a general survey, covering various potential sharing situations between 50.2 and 71 GHz. Recommendations ITU-R SA.1279 [2] and S.1339 [3] consider sharing between the ISS and passive sensors of the SR and the EES in this spectrum range in more detail. Recommendation ITU-R S.1326 [4], responding to a resolution of WRC-95, provides constraints that would facilitate sharing between the ISS and geostationary satellite uplinks of the FSS at 50.4–51.4 GHz, although this band is not, at present, allocated for the ISS.

The ISS allocations, and the sharing constraints and limitations of use which have been agreed to facilitate sharing and to limit the impact of the development of one service upon another, are as follows.

1   *22.55–23.55 GHz.* No allocations for the RAS share this band with the ISS but RR S5.149 advises that radio astronomers make spectral line observations in the sub-bands 22.81–22.86 GHz and 23.07–23.12 GHz and urges Administrations to protect these sub-bands when making ISS assignments in this band.

2   *24.45–24.75 GHz.* In Regions 2 and 3 the ISS shares this band with a primary alloca-tion for the RNS, and RR S5.533 denies the ISS protection from harmful interfer-ence from airport surface detection equipment of the RNS.

3   *25.25–27.5 GHz.* Use of this allocation is virtually limited by RR S5.536 to ISLs for geostationary data relay satellites of the SR and the EES. This band is shared with the FS and MS and an arrangement to minimise the exposure of the ISS to terrestrial interference is set out in Recommendation ITU-R SA.1278 [5]. See also item 13 in Section 6.2.1 and Note (ii) in Table 13.5 above. The sub-band 27.0–27.5 GHz is shared with the FSS in Regions 2 and 3 but RR S5.537 rules that RR S22.2 does not apply to the ISS if interference to the FSS should occur.

4   *32.0–33.0 GHz.* This band is allocated for the ISS, shared with the RNS and the FS and, between 32.0 and 32.3 GHz, with downlinks of the SR (deep space). Recom-mendation ITU-R S.1151 [6], responding to RR Recommendation 707, advises on sharing between the RNS and the ISS, in response to which RR S5.548 calls for all necessary care to be taken in the design of systems to avoid interference to the RNS, a safety service. However, the allocation for the FS in this band is new, made by WRC-97, and intended for use for high density applications (see RR S5.547 and Section 6.2.4), in consequence of which it will be necessary to consider again the conditions under which these various services may share this band; see RR Resolutions 126 and 726.

5   *54.25–58.2 GHz.* The ISS shares the various parts of this frequency band with the FS, the MS and passive sensors of the EES and the SR. The terms under which the ISS may operate are set out in RR S5.556A and S5.558A. RR S5.558 requires use of the band by the AMS to not cause harmful interference to the ISS.

6   *59.0–71.0 GHz.* Between 59.0 and 59.3 GHz, RR S5.556A applies constraints to the ISS. In the band 59.3–71.0 GHz there are no constraints on the ISS; the band is shared with primary allocations for various other services, but RR S5.553, S5.558 and S5.559 protect the ISS from the sharing services most likely to cause interfer-ence.

7   *116–134 GHz, 170–182 GHz and 185–190 GHz.* These bands are also allocated for the ISS, sharing with various other services with primary allocations. However, there are no constraints on the ISS in any of these bands and RR S5.558 and S5.559 protect the ISS from the only sharing services (namely the RLS and the AMS) that are likely to cause harmful interference to the ISS. However, RR S5.149 draws attention to the fact that the RAS makes spectral line observations in the bands 174.42–175.02 GHz, 177.0–177.4 GHz, 178.2–178.6 GHz and 181.0–181.46 GHz and urges Administrations to take all practicable steps to protect the RAS from harmful interference when assigning frequencies to ISLs.

Frequency assignments for links in the ISS are notified, coordinated and registed by the same basic procedures as are used for satellite uplinks and down-links; see RR Articles S9 and S11 and Section 5.2 above.

14.1 References

1 'Requirements and suitable bands for operation of the inter-satellite service within the range 50.2–71 GHz'. Recommendation ITU-R S.1327, ITU-R Recommendations 1997 S Series (ITU, Geneva, 1997)

2 'Spectrum sharing between spaceborne passive sensors and inter-satellite links in the range 50.2–59.3 GHz'. Recommendation ITU-R SA.1279, ITU-R Recommendations 1997 SA Series (ITU, Geneva, 1997)

3 'Feasibility of sharing between spaceborne passive sensors of the Earth exploration-satellite service and inter-satellite links of geostationary-satellite networks in the range 50–65 GHz'. Recommendation ITU-R S.1339, ITU-R Recommendations 1997 S Series (ITU, Geneva, 1997)

4 'Feasibility of sharing between the inter-satellite service and the fixed-satellite service in the frequency band 50.4–51.4 GHz'. Recommendation ITU-R S.1326, *ibid.*

5 'Feasibility of sharing between the Earth exploration-satellite service (space-to-Earth) and the fixed, inter-satellite and mobile services in the band 25.5–27.0 GHz'. Recommendation ITU-R SA.1278, ITU-R Recommendations 1997 SA Series (ITU, Geneva, 1997)

6 'Sharing between the inter-satellite service involving geostationary satellites in the fixed-satellite service and the radionavigation service at 33 GHz'. Recommendation ITU-R S.1151. *ibid.*

Chapter 15

Radionavigation and radiolocation

15.1 Introduction

Radiodetermination is defined in RR S1.9 as

> the determination of the position, velocity and/or other characteristics of an object, or the obtaining of information relating to these parameters, by means of the propagation properties of radio waves.

There are no frequency allocations for the radiodetermination service (RDS) itself but there are many for its specialised subdivisions. Many of the stations of this family of services are categorised with the radionavigation service (RNS) which is further subdivided between the maritime and the aeronautical radionavigation services (MRNS and ARNS). Systems of the RNS are used for the navigation of ships or aircraft. These systems provide safety services which, by definition, involve the safeguarding of human life and property and must therefore be given a high degree of protection from interference; see RR S1.59 and S15.8. The other RDS systems are categorised with the radiolocation service (RLS).

In addition to these terrestrial services, there is the radiodetermination-satellite service (RDSS), itself subdivided into the radionavigation-satellite and radiolocation-satellite services (RNSS and RLSS). Finally, the RNSS is further subdivided into maritime and aeronautical services (MRNSS and ARNSS).

The radionavigation services

Frequency bands are allocated for terrestrial radionavigation in all parts of the radio spectrum from 9 kHz, upwards to 265 GHz, except in the HF range where radio propagation conditions are unfavourable. Many of these bands are allocated to the MRNS or the ARNS. There are several allocations for the RNSS above 1 GHz. As would be appropriate for a safety service, the RNS and RNSS allocations are primary and many are world-wide and exclusive. It has not yet been found necessary to allocate spectrum specifically for the MRNSS or the ARNSS.

The systems of these services take various forms. Below about 500 MHz a typical system consists of one or several stations at fixed, known locations transmitting continuous wave emissions which, received on ships or aircraft, enable the mobile station to determine its position or the heading it requires to reach its destination or help aircraft to land safely. Above about 500 MHz the typical system is a pulsed primary radar. Some radars are at fixed locations and are used, for example, for air traffic control or for the oversight of the movement of ships in congested waters and aircraft on the ground at airports. Other radars are mobile, carried by aircraft or ships, as an aid to navigation and the avoidance of collision. In most cases the use to be made of a frequency band has been internationally agreed, often for a specific system or kind of system, through the ITU, the IMO or the ICAO; see Section 15.2 below.

The radiolocation services

Most allocations for the RLS are above 1 GHz. The allocations for the RLSS are few and above 24 GHz. Some RLS systems are short range distance-measuring systems, typically narrow-band devices using a continuous wave emission. The many secondary allocations for the RLS, often with limited geographical coverage, made by footnotes to the international frequency allocation table, tend to be used for these. Other RLS allocations, however, usually primary and world-wide or Region-wide and in some cases exclusive, are used mainly for military radars, fixed or mobile and the management of these bands is not in the public domain.

Pulse radars and interference

The occupied bandwidth of a pulse radar is wide, typically tens of megahertz. Most radars, using no frequency determining device other than a cavity magnetron, have an operating frequency that, compared with other kinds of radio system, is ill-defined and of poor stability. Spot frequencies are not usually assigned to such radars; instead stations may be authorised to operate in a specified frequency band. The magnetrons that the radar uses will have been tested to confirm that they will operate within the allocated band, leaving a substantial margin at each end of the band to avoid interference to stations operating in adjacent bands; see RR Appendix S2, footnote 33. Radiation of powerful spurious emissions from radars, including harmonic radiation, often occurs.

Pulse radars, whether of the RNS or the RLS, are not very sensitive to interference from other pulse radars. However, communication systems, especially digital systems, are susceptible to interference from pulse radars and interference from a radar assigned to operate in an adjacent frequency band often arises. Powerful spurious emissions may raise problems for any radio system. Consequently, most RNS and RLS allocations are exclusive although sometimes the two services share a frequency band. Where the RDS and a conventional communication service share a frequency band, in-band interference problems are usually avoided or evaded by the implementation of sharing constraints or by giving secondary status to one or other of the allocations.

Recommendation ITU-R F.1190 [1] offers protection criteria for interference to digital radio relay systems from pulse radars. Recommendation ITU-R F.1097 [2] discusses means for mitigating the impact of radar interference on radio relay systems, whether by modification of the radio relay station or the radar station. Recommendation ITU-R M.1179 [3] discusses in some detail why interference arises in particular cases and offers suggestions for its suppression. Recommendation ITU-R M.1177-1 [4] recommends techniques for measuring spurious emission levels at radar stations. Finally, Recommendation ITU-R M.1314 [5] suggests ways of reducing the level of spurious emissions radiated by pulse radars.

15.2 International management of radiodetermination allocations

As indicated above, RLS bands and some RNS bands are used in accordance with national requirements, under national management. Cross-frontier interference is cleared by whatever means are available. Most RNS bands, however, are used for internationally agreed purposes and with an element of international spectrum management; these main applications are reviewed, item by item, in this Section.

1 The OMEGA system
The band 9–14 kHz, a world-wide exclusive primary allocation for the RNS, is used for the OMEGA global position-fixing system.

2 LF medium-range position-fixing systems
There are primary allocations for the RNS or the MRNS world-wide between 70 and 130 kHz and these are used for various medium-range position-fixing systems such as LORAN-C and Decca Navigator. Continuous wave systems in specified sub-bands are given protection from pulsed RNS systems by RR S5.60. RR S5.62 urges Administrations responsible for RNS systems operating between 90 and 110 kHz to coordinate with one another to avoid harmful interference.

In part, these allocations are shared with the MMS and/or the FS. Coordination is made mandatory in Region 2 by RR S5.61 for all services in the bands 70–90 kHz and 110–130 kHz. RR Resolution 705 and Recommendation ITU-R M.589-2 [6] provide sharing criteria for the protection of LORAN-C and Decca Navigator. RR Resolution 706 proposes the deletion of the secondary allocation for the FS in the band 90–110 kHz but this had not been implemented at the time of writing.

3 Radiobeacons between 130 and 535 kHz
A basic radiobeacon signal consists of a carrier wave, modulated for a part of each minute by an identification signal and transmitted from a known location. Ships and aircraft receive these emissions using a direction-finding receiver, enabling a course to be set or an approximate position to be fixed, the latter by measurements of the directions of several beacon stations. The frequency bands 283.5–315 kHz (Region 1) or 285–325 kHz (Regions 2 and 3) are used, largely or wholly, for maritime radiobeacons, and there is some use of higher frequencies for the purpose, up to 335 kHz. Aeronautical radiobeacons are to be found between 160 and 535 kHz, the main concentration being at 315–405 kHz and 505–535 kHz.

Frequency plans for maritime radiobeacons at 283.5–315 kHz in the European Maritime Area and for aeronautical radiobeacons at 415–435 kHz and 505–526.5 kHz in Region 1 were drawn up at planning conferences in 1985; see References [7] and [8], respectively. (The European Maritime Area is defined in RR S5.15; simplifying somewhat, it is bounded by longitudes 32°W and 43°E and by latitudes 30°N and 72°N.) General regulations on the operation of radiobeacons are to be found in RR S28.18–S28.23. These regulations include a requirement that the radiated power of a beacon should be enough, but not more, to set up a field strength level given in RR Appendix

S12 at the limit of its service range. In determining whether interference from another beacon is to be regarded as acceptable, it is to be assumed that the given wanted field strength is, in fact, achieved.

Radiobeacons of this kind have been in use for many years but there are prospects that the present usage for maritime purposes will evolve to provide additional facilities. RR S5.73 permits the transmission by radiobeacon stations at 285–325 kHz (283.5–325 kHz in Region 1) of supplementary material, provided that the prime function of the beacon is not degraded thereby. It is contemplated that this opportunity will be used in two ways; to enable beacon emissions to evolve into hyperbolic position-fixing systems (see Recommendation ITU-R M.631-1 [9]) and to provide channels for the distribution in real time of information on variations in the propagation delay of the signals of differential radionavigation systems (see RR Resolution 602 and Recommendations ITU-R M.1178 [10] and M.823-2 [11]).

4 Maritime direction-finding service
On request, some coast stations will measure and report the bearing of a ship transmitting at 410 kHz; see RR S28.12. RR S5.76 protects this frequency from harmful interference from other radionavigation systems operating in the same frequency band. Direction-finding facilities may also be provided at 2182 kHz, 156.8 MHz and 156.525 MHz; see RR S28.13 and S28.14.

5 MF position-fixing systems
Several frequency bands around 1900 kHz are allocated for the RNS and are used, for example, for LORAN; see RR S5.97. See also various footnotes to the international table of frequency allocations, such as RR S5.92, indicating other RNS activity.

6 Wind profiler radars
New secondary allocations were made at WRC-97 for the RLS in a number of countries in the bands 46–68 MHz and 470–494 MHz, their use being limited to wind profiler radars. RR Resolution 217 urges that, in addition to these new allocations, wind profiler radars should be set up in the bands 440–450 MHz, 904–928 MHz (Region 2 only), 1270–1295 MHz and 1300–1375 MHz but not in other bands. Where neither 470–494 MHz nor 440–450 MHz is available for this purpose, the Resolution suggests the use of 420–435 MHz or 438–440 MHz instead. The use of the band 400.15–406 MHz for wind profiler radars is causing interference to the safety service in the band 406–406.1 MHz and WRC-97 urged Administrations to cease permitting the use of this band for radars. Recommended characteristics for these radars are to be found in Recommendations ITU-R M.1226 [12], M.1085-1 [13] and M.1227 [14].

7 Instrument Landing System
Three frequency bands allocated exclusively world-wide for the ARNS are used for the Instrument Landing System (ILS). The band 74.8–75.2 MHz is used for marker beacons with a nominal frequency of 75.0 MHz. The band 328.6–335.4 MHz is used for glide path signals with a nominal frequency of 332 MHz. Frequencies in the band 108–111.975 MHz are assigned in accordance with an ICAO plan for ILS localiser emissions. This last band is also used for aeronautical VHF omnidirectional range (VOR) beacons.

The allocation for the ARNS of relatively wide bands for narrow-band emissions centred on 75 MHz and 332 MHz arises in part from the use by some airlines of aircraft receivers which are seriously deficient in selectivity; see RR S5.180. However, the ILS is being phased out, to be replaced eventually by the Microwave Landing System (MLS) (see item 14 below). RR S5.181, S5.197 and S5.259 foreshadow the reallocation of these bands, with appropriate safeguards, for the MS.

In Section 7.5.3 there is an account of interference affecting aircraft receivers in the bands just above 108 MHz. This interference arises by various mechanisms, the ultimate source being broadcasting transmissions in the band just below 108 MHz. Chapter 7 refers to the measures that are being adopted to limit this interference.

8 Radionavigation aids for ships and aircraft in distress
Several special purpose radionavigation systems are used by aircraft and ships, complementing specialised communication systems, specifically as aids to rescue in the event of disaster. These navigation and communication systems are reviewed together in Section 8.5 above.

9 Global satellite navigation systems at 150 and 400 MHz
The RNSS allocations in the bands 149.9–150.05 MHz and 399.9–400.05 MHz are used for the Transit system. An allocation to permit sharing of these bands with uplinks of non-GSO networks of the LMSS was added by WRC-95. RR Resolution 715 calls for study of the measures that will be necessary to facilitate this sharing. However, the Transit system is approaching the end of its working life and it is to be superseded by other satellite navigation systems operating in other bands. The allocation for the RNSS is to cease in 2015; see RR S5.224B.

10 The ARNS at 960–1215 MHz
The world-wide, exclusive, primary allocation for the ARNS at 960–1215 MHz is used for various aids for air navigation and RR S5.328 reserves the future use of the band for such purposes.

11 Global satellite navigation systems at 1250 and 1580 MHz
The RNSS allocations at 1215–1260 MHz and 1559–1610 MHz are used for position-fixing systems such as the Global Positioning System (GPS) and GLONASS. The 1215–1260 MHz allocation is constrained to some degree by RR S5.329 and S5.331, which require that the satellite service cause no harmful interference to RNS systems in a number of countries. This allocation is also shared with primary allocations for active sensors of the SR and the EES but RR S5.332 and S5.335 have the effect of reducing the status of these allocations relative to that of the RNSS.

12 RNS radars between 1 and 10 GHz
A number of bands between 1 and 10 GHz are allocated for the RNS, the ARNS or the MRNS and used for radar and associated systems. All of these allocations are world-wide and primary. All are shared with the RLS although in most bands the RLS allocation is secondary.

a)   Five bands are allocated for the ARNS. The use of 1300–1350 MHz, 2700–2900 MHz and 9000–9200 MHz is limited by RR S5.337 to ground-based radars and associated airborne transponders. The use of 5350–5470 MHz is limited by RR S5.449 to airborne radars and associated airborne beacons. The use of 8750–8850 MHz is limited by RR S5.470 to airborne Doppler navigation aids operating with a centre frequency of 8800 MHz.

b)   Two bands that are allocated for the MRNS, at 8850–9000 MHz and 9200–9225 GHz, are used mainly for shore-based radars. The other two MRNS allocations, at 5470–5650 MHz and 9225–9300 MHz are used mainly for ship-borne radars.

c)   In addition, the bands 2900–3100 MHz and 9300–9500 MHz are allocated for the RNS. Aeronautical use of these bands is limited to ground-based radars plus, at 9 GHz, airborne weather radars and ground-based aeronautical beacons; see RR S5.426 and S5.475. The main use of these bands is for ship-borne radars. The band 9500–9800 MHz is allocated for the RNS and the RLS with equal status. This band is also shared with primary allocations for active sensors of the SR and the EES but RR S5.476A has the effect of making the status of these allocations inferior to that of the RNS and the RLS.

Recommendation ITU-R M.629 [15] identifies issues of policy arising in the use made of these frequency bands for maritime purposes. See also RR Recommendation 605. The key issues arise from the use of various devices to enhance the operation of primary radar systems. The simplest of these are radar target enhancers; Recommendation ITU-R M.1176 [16] describes the use of an amplifier to increase the radar response from buoys or small craft. A radar beacon, or racon, is somewhat more complicated, transmitting a code letter in response to the reception of a radar signal; see Recommendation ITU-R M.824-2 [17]. Ship-borne interrogator transponders provide identification of the responding ship; see Recommendation ITU-R M.630 [18].

13 Radio altimeters
Radio altimeters operate in various bands, typically allocations for the RLS. However, the allocation for the ARNS at 4200–4400 MHz is reserved for the exclusive use of altimeters.

14 The microwave landing system
The MLS is being deployed in the band 5000–5150 MHz at the time of writing, but there is an arrangement, terminating in 2010, whereby the sub-band 5091–5150 MHz may be used for feeder links for LEO systems of the MSS while it is not required for the MLS. See item 7(a) in Section 11.1.2 above.

15 Above 10 GHz
Allocations for the various RNS services below 10 GHz are exclusive or, if shared, are shared either with allocations that have secondary status or with the RLS; because of these circumstances severe problems of spectrum management do not arise. In recent years, however, it has been necessary to provide a number of new allocations of wide bandwidth in the millimetre-wave range for other services. Some of these new allocations are shared with the RNS or the RLS. Sharing constraints which would permit the radionavigation services to grow into these bands have been agreed in some cases; in other cases sharing arrangements are under development.

a)   The band 13.75–14.0 GHz is allocated for the RLS. RR S5.501 adds a primary allocation for the RNS in a number of countries. The band has also been allocated on a primary basis for FSS uplinks. Sharing constraints on all three services are given in RR S5.502. See also Recommendation ITU-R S.1068 [19] and item 2(a) in Section 9.1.2. above.

b) The band 14.0–14.3 is allocated for the RNS and for FSS uplinks. RR S5.504 requires stations of the RNS to protect satellite receivers of the FSS; see also Recommendation ITU-R M.496-3 [20].

c) The band 15.4–15.7 GHz is allocated for the ARNS and the band 15.43–15.63 GHz is allocated for the FSS, limited to feeder links with either direction of transmission serving non-GSO satellites of the MSS. See RR S5.511A, S5.511C, S5.511D and item 7(d) in Section 11.1.2 above.

d) In Regions 2 and 3 the band 24.25–24.65 GHz is allocated for the RNS, sharing with the ISS between 24.45 and 24.65 GHz. In Region 3 there are also allocations for the FS and the MS. All of these allocations are primary. However, RR S5.533 does not permit the ISS to claim protection from harmful interference from airport surface protection radar systems of the RNS.

e) The band 32.0–33.0 GHz is allocated for the RNS, shared with the ISS and, between 32.0 and 32.3 GHz, with downlinks of the SR (deep space); see RR Recommendation 707. Recommendation ITU-R S.1151 [21] advises on sharing between the RNS and the ISS, the latter being limited in this case to links between geostationary satellites of the FSS. However, WRC-97 made an additional allocation for the FS which includes this band, to be used for high density applications (see Section 6.2.4). In consequence of this FS allocation, it will be necessary to consider again the conditions under which these various services may share this band; see RR Resolution 726 and Section 14 item 4 above.

f) The bands 59–64 GHz and 126–134 GHz are allocated for the ISS, sharing with the RLS, use for airborne radars being subject to not causing harmful interference to the ISS; see RR S5.559. RR Recommendation 710 calls for studies to be made to determine criteria for this sharing arrangement.

15.3 National spectrum management

Administrations assign frequencies to RNS and RLS stations at fixed locations, typically in accordance with international plans or after consultation with other Administrations or international bodies such as IMO or ICAO that may be affected. These assignments are notified to the Radiocommunication Bureau of the ITU in accordance with RR Article S11 for entry in the MIFR, although notification may be dispensed with if the assignment is not capable of causing harmful interference to the services of another country. Assignments to satellites and earth stations at fixed locations belonging to the RNSS and RLSS are notified, coordinated and registered in accordance with RR Articles S9 and S11.

Information on assignments made to stations at fixed locations that may be of use in the navigation of ships, and details of the facilities that these stations provide, are listed in a document [22] published by the ITU, based on information supplied by Administrations; see RR S20.16. This publication is revised as required.

Administrations assign frequencies to mobile and portable RDS stations. For radar systems operating above 2.45 GHz which have poor frequency determination, the assignment often takes the form of authority to operate with an emission that is maintained wholly within an allocated frequency band. These assignments are not notified to the ITU for inclusion in the MIFR.

15.4 References

1 'Protection criteria for digital radio-relay systems to ensure compatibility with radar systems in the radiodetermination service'. Recommendation ITU-R F.1190, ITU-R Recommendations 1997 F Series Part 1 (ITU, Geneva, 1997)

2 'Interference mitigation options to enhance compatibility between radar systems and digital radio-relay systems'. Recommendation ITU-R F.1097, *ibid.*

3 'Procedures for determining the interference coupling mechanisms and mitigation options for systems operating in bands adjacent to and in harmonic relationship with radar stations in the radiodetermination service'. Recommendation ITU-R M.1179, ITU-R Recommendations 1997 M Series Part 4 (ITU, Geneva, 1997)

4 'Techniques for measurement of spurious emissions of radar systems'. Recommendation ITU-R M.1177-1, *ibid.*

5 'Reduction of spurious emissions of radar systems operating in the 3 GHz and 5 GHz bands'. Recommendation ITU-R M.1314, *ibid.*

6 'Interference to radionavigation services from other services in the frequency bands between 70 kHz and 130 kHz'. Recommendation ITU-R M.589-2, *ibid.*

7 'Final Acts of the Regional Administrative Radio Conference for the planning of the maritime radionavigation service (Radiobeacons) in the European Maritime Area', Geneva, 1985 (ITU, Geneva, 1986)

8 'Final Acts of the Regional Administrative Radio Conference for the planning of the MF maritime mobile and aeronautical radionavigation services (Region 1)', Geneva, 1985 (ITU, Geneva, 1986)

9 'Use of hyperbolic maritime radionavigation systems in the band 283.5–315 kHz'. Recommendation ITU-R M.631-1, ITU-R Recommendations 1997 M Series Part 4 (ITU, Geneva, 1997)

10 Use of the maritime radionavigation band (283.5–315 kHz in Region 1 and 285–325 kHz in Regions 2 and 3)'. Recommendation ITU-R M.1178, *ibid.*

11 'Technical characteristics of differential transmissions for global navigation satellite systems from maritime radio beacons in the frequency band 238.5–315 kHz in Region 1 and 285–325 kHz in Regions 2 and 3'. Recommendation ITU-R M.823-2, *ibid.*

12 'Technical and operational characteristics of wind profiler radars in bands in the vicinity of 50 MHz'. Recommendation ITU-R M.1226, *ibid.*

13 'Technical and operational characteristics of wind profiler radars in bands in the vicinity of 400 MHz'. Recommendation ITU-R M.1085-1, *ibid.*

14 'Technical and operational characteristics of wind profiler radars in bands in the vicinity of 1000 MHz'. Recommendation ITU-R M.1227, *ibid.*

15 'Use for the radionavigation service of the frequency bands 2900–3100 MHz, 5470–5650 MHz, 9200–9300 MHz, 9300–9500 MHz and 9500–9800 MHz'. Recommendation ITU-R M.629, *ibid.*

16 'Technical parameters of radar target enhancers'. Recommendation ITU-R M.1176, *ibid.*

17 'Technical parameters of radar beacons (Racons)'. Recommendation ITU-R M.824-2, *ibid.*

18 'Main characteristics of two frequency shipborne interrogator transponders (SIT)'. Recommendation ITU-R M.630, *ibid.*

19 'Fixed-satellite and radiolocation/radionavigation services sharing in the band 13.75–14 GHz'. Recommendation ITU-R S.1068, ITU-R Recommendations 1997 Series S (ITU, Geneva, 1997)

20 'Limits of power flux-density of radionavigation transmitters to protect space station receivers of the fixed-satellite service in the 14 GHz band'. Recommendation ITU-R M.496-3, ITU-R Recommendations 1997 M Series Part 4 (ITU, Geneva, 1997)

21 'Sharing between the inter-satellite service involving geostationary satellites in the fixed-satellite service and the radionavigation service at 33 GHz'. Recommendation ITU-R S.1151, ITU-R Recommendations 1997 S Series (ITU, Geneva, 1997)

22 'List of radiodetermination and special service stations'. List VI, published by the ITU with recapitulative supplements

Radio propagation and radio noise

A1 Introduction

When a spectrum manager is making a frequency assignment it may be necessary to estimate the minimum transmitter power level that is consistent with the achievement of specified link performance objectives. It is seldom feasible to use direct measurements for this purpose. Similarly, the impact on the reception of a wanted signal of interference from known sources may have to be estimated. For point-to-area services such as broadcasting or land mobile, the power level required to set up a specified signal strength anywhere within a specified service area may have to be estimated. Accurate estimates enable the best use to be made of the spectrum.

Accurate estimates require knowledge of the local topography and of various characteristics of the radio stations involved, including the gain, polarisation characteristics and effective height of the antennas, the modulation parameters of the emissions involved, the bandwidth of the receivers and the power of interfering transmitters. In the absence of such knowledge it will be necessary to make assumptions about them, and standards recommended by ITU-R are often used for this purpose. But above all it is necessary to be able to estimate the transmission loss arising from propagation between the transmitting antennas and the receiving antennas. It may also be necessary to estimate radio noise levels.

The transmission loss depends mainly on the distances involved, the frequency and the propagation mode or modes arising, but the loss is also dependent to a significant degree on environmental factors which vary from place to place. Noise levels also depend on the locality. It is preferable for estimates of link performance to be based on statistically dependable data gathered in the same region as the transmission path in question. Local data are particularly desirable on the electrical parameters of the ground at low frequencies, on man-made noise at VHF and UHF, on absorption due to rainfall on space–Earth links above about 15 GHz and on the prevalence of tropospheric ducts. However, statistically dependable local data may not exist. Furthermore, estimates of the level of interference from a transmitting station in another country may be needed when Administrations are seeking to resolve interference problems, and estimating techniques and local data that one administration considers to be reliable may be rejected by the other.

There is an extensive literature on radio wave propagation and a number of textbooks on particular aspects of the subject; see for example References [1–7]. Hall *et al.* [8] provides an excellent general review. These books provide valuable insights into the phenomena involved but they do not meet the spectrum manager's requirement for universal acceptability and some do not provide a basis for quantitative assessments of the impact of propagation degradations on systems.

Consequently, one of the more important aspects of the work of the ITU-R study groups and of their predecessors in the CCIR has been to bring together the available geophysical and environmental information bearing on radio propagation and to present it so that it can be used for any location. This material can be used in relatively simple ways to forecast radio propagation conditions anywhere with an accuracy sufficient for most administrative purposes. It is accepted within its limitations by all ITU Members. The basic material and methods for its use are to be found in the recommendations of ITU-R Study Group 3 (Series P, Parts 1 and 2). Some of this material has also been re-processed and presented elsewhere, for example in the ITU Radio Regulations and in frequency assignment plan agreements, to facilitate particular applications. An introduction to these resources follows.

Radio signals propagated by five regular modes are used for terrestrial communication systems, namely the tropospheric wave, the ground wave, the ionospheric wave, tropospheric scatter and meteor-burst propagation. Propagation involving spacecraft and satellites can conveniently be considered as a special case of the tropospheric wave. The main phenomena associated with these five modes are outlined in Sections A2 to A6 below. Which of these modes is operative on any particular link will depend on the carrier frequency, the length of the link and various other factors. Different modes may be operative for a link simultaneously or at different times.

These five regular modes are, of course, also active in propagating interference but other propagation modes which are active for some but not most of the time must also be taken into account in a careful assessment of potential sources of interference and as an aid in identifying the source when interference arises; these irregular modes are outlined in Section A7. Section A8 reviews the incidence of radio noise. And Section A9 outlines means for estimating the impact of interference and noise on wanted emissions.

A2 Propagation by the tropospheric wave mode

A2.1 Propagation in free space

In free space, regardless of frequency, a wave is propagated in a straight line and at a uniform velocity from a transmitting antenna to a receiving antenna without loss of power, other than the inverse distance squared attenuation that is due to the expansion of a spherical wavefront. If the power radiated from the transmitting antenna is p_t and the power delivered into a correctly matched load at the receiving antenna is p_r, both antennas being assumed to be isotropic and matched in polarisation characteristics, then

$$10 \log p_t/p_r = 20 \log 4\pi d/\lambda \ \text{dB} \qquad \text{(A.1a)}$$

$$= 20 \log 4\pi df/c \ \text{dB} \qquad \text{(A.1b)}$$

$$= L_{bf} \ \text{dB} \qquad \text{(A.1c)}$$

where d is the distance between the antennas (m),
$\quad \lambda$ is the wavelength (m),
$\quad f$ is the frequency (Hz), and
$\quad c$ is the velocity of an electromagnetic wave in free space $\approx 3 \times 10^8$ (m/s).

L_{bf} is called the free space basic transmission loss. In more convenient units,

$$L_{bf} \approx 32.4 + 20 \log F + 20 \log D \ \text{dB} \qquad \text{(A.ld)}$$

where F is the frequency (MHz), and D is the distance between antennas (km).

For some purposes it is convenient to express the power of a wave at a distance from a transmitting antenna in the form of Φ, the power flux density (PFD). If an isotropic antenna is radiating p_t watts, then

$$\Phi = 10 \log p_t - 10 \log 4\pi d^2 \ \text{dB}(\text{W/m}^2) \qquad \text{(A.2)}$$

Thus, the power received by a correctly matched isotropic receiving antenna situated where the PFD equals Φ dB(W/m^2) is given by

$$10 \log p_r = \Phi + 10 \log \lambda^2/4\pi \ \text{dB(W)} \qquad \text{(A.3)}$$

$\lambda^2/4\pi$ can be regarded as the capture cross-section from which an isotropic antenna extracts the power flux; $4\pi d^2$ is sometimes called the 'wave spreading factor'.

Equations A.1 to A.3 express in logarithmic terms the power of the wave at a distance but, depending on the type of receiving antenna that is to be used, it is sometimes more convenient to consider the field strength, e, instead. Thus, using convenient units,

$$e = 173P^{1/2}/D \ \text{mV/m} \qquad \text{(A.4)}$$

where P, the power radiated by an isotropic transmitting antenna and D, the distance, are expressed in (kW) and (km).

The relationships given above apply exactly to links between antennas in space, allowance being made for antenna gain in the relevant directions, provided that the path of the wave from one antenna to the other does not pass close to the Earth. If the wave does pass close to the Earth there may be additional losses due to the atmosphere. These relationships apply approximately to propagation at the higher frequencies within the atmosphere but a number of atmospheric and environmental phenomena may absorb, deflect, delay or distort the wave. The more significant of these phenomena are described briefly in Section A2.2. Some propagation phenomena arise from the physical state and chemical constitution of the atmosphere itself; an ITU-R handbook [9] provides a broad review of 'radiometeorology'. This modified form of free space propagation, called the tropospheric wave, is the normal mode for terrestrial links at frequencies above about 30 MHz.

Links not wholly in free space

A link between an earth station and a spacecraft, typically an artificial Earth satellite, is in free space for part of its length and in the atmosphere for the remainder. There are two ITU-R handbooks [10, 11] which provide a broad review of propagation between spacecraft and Earth.

The length of space–Earth links

The distance (d) between an earth station and a geostationary satellite (neglecting trivial inaccuracies arising from the height of the earth station above sea level and departures of the shape of the Earth from that of a true sphere) can be calculated from

$$d = (R_E^2 + r_g^2 - 2R_E r_g \cos C)^{-1/2} \text{ km} \tag{A.5}$$

where R_E is the mean equatorial radius of the Earth $= 6378$ km,
r_g is the radius of the geostationary orbit $= 42\,174$ km, and
C is the angle at the Earth's centre subtended by the earth station and the satellite.

$$C = \text{arc } \cos(\cos A \cdot \cos B) \text{ degrees} \tag{A.6}$$

where A is the difference between the longitude of the satellite and the longitude of the earth station, and B is the latitude of the earth station

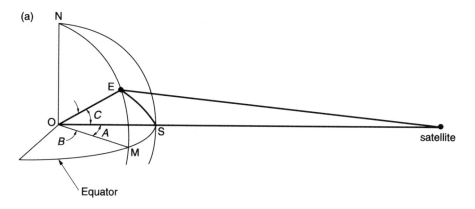

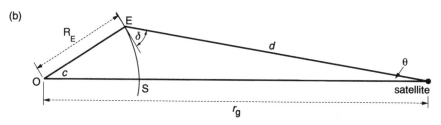

Figure A.1   The geometry of a link between an earth station (E) and a geostationary satellite: (a) shows the centre of the Earth (O), the North Pole (N) and the sub-satellite point (S) in a three-dimensional sketch; (b) is two-dimensional in the plane of OES

See Figure A.1. The error that would arise from applying these equations to a satellite that was only approximately geostationary would be very small.

The corresponding calculation, for any given moment in time, for a satellite in a highly elliptical orbit or an orbit that is circular but has an orbital period substantially differing from that of a geostationary satellite is much more complex and demands a knowledge of the full orbital parameters of the satellite. Furthermore, the distance between the satellite and the earth station will be continually varying. However, the maximum and minimum distances can readily be calculated for a satellite in a circular orbit having an angle of inclination of the orbital plane which is greater than the latitude of the earth station. The minimum distance occurs when the satellite is directly over the earth station, thus

$$d_{min} = h \, \text{km} \tag{A.7}$$

and the maximum distance occurs when the satellite crosses the horizon as seen from the earth station. Thus, disregarding ray bending in the troposphere,

$$d_{max} = (h^2 + 2R_E h)^{1/2} \, \text{km} \tag{A.8}$$

where h is the height of the satellite orbit above the surface of the Earth (km). See Figure A.2.

A2.2 Tropospheric phenomena

A2.2.1 Refraction

The refractive index of the air (n) is a function of the atmospheric pressure, the water vapour density and the temperature. Under normal conditions the refractive index decreases exponentially with increasing height above ground and it can be represented with sufficient accuracy by an expression of the form

$$n(h) = 1 + a \times \exp(-bh) \tag{A.9}$$

where $n(h)$ is the refractive index at height h above sea level, and a and b are parameters that depend on the climate.

There are specimen climatic data in Recommendation ITU-R P.453-6 [12]. A standard reference atmosphere for refraction, recommended by ITU-R, is given by

$$n(h) = 1 + 315 \times 10^{-6} \times \exp(-h/7.35) \tag{A.10}$$

h being expressed in kilometres. With the refractive index varying with height in this way, the rays of a wave which is initially propagated horizontally will follow a curved path, approximately an arc of a circle.

It is often necessary to trace the curved route followed by a ray over the curved surface of the Earth, typically to determine whether the direct path between a transmitting antenna and a receiving antenna is likely to be blocked by an obstruction. To facilitate this process it is usual to produce a height profile of the ground between the two stations, distorted by assuming that the radius of the Earth is not R_E, its true value (≈ 6371 km), but kR_E, where k is a function of $n(h)$ appropriate to the climate of the locality and the height of the stations above sea level. A value for k equal to 4/3 is sufficiently accurate for many locations and is widely used; see also Recommendation ITU-R P.834-2 [13]. For this distorted profile, and with the standard reference atmosphere conditions, the ray path can be assumed to be straight.

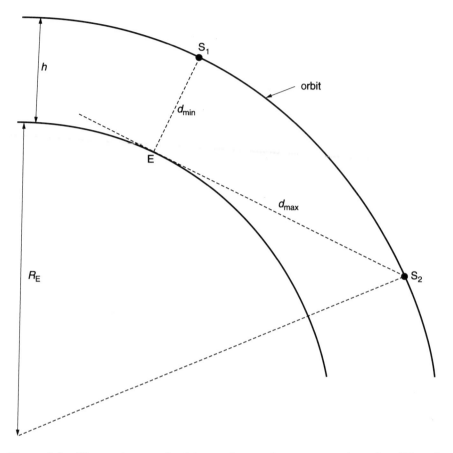

Figure A.2   The maximum and minimum distances between an earth station (E) and a satellite (S) that may pass over the earth station in its circular orbit

Thus Figure A.3 shows the height profile between antennas at T and R assuming the true Earth radius (Figure A.3a) and a false Earth radius appropriate to normal local climate conditions (Figure A.3b). The path N of the ray that links the two antennas under normal conditions is curved in Figure A.3a but the corresponding path N' on Figure A.3b is straight. Ray path H (and H'), having greater curvature than N (and N'), may be followed at times when climatic conditions lead to stronger refraction. The contrary occurs with ray path L (and L') when refraction is weak. Note that ray path Z (and Z'), which encounters an obstruction at high ground before the ray can reach R, may be operative under conditions of very weak refraction, possibly leading to failure of the link; see Section A2.2.2. Line A in Figure A.3a is a straight line between the two antennas, showing that this link would be grossly obstructed in the total absence of atmospheric refraction.

This distorted profile technique lends itself to computer application wherever sufficiently accurate and detailed relief maps in digital database form are available.

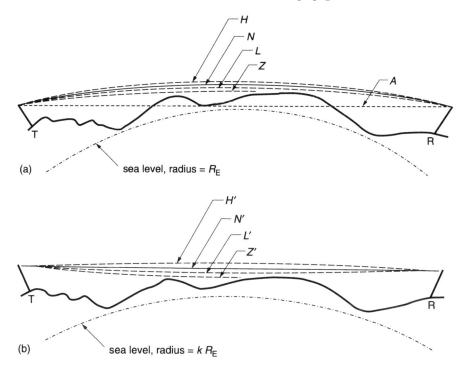

Figure A.3 *A height profile between fixed radio stations, T and R, drawn for a true Earth radius* R_E *(a) and a false Earth radius* kR_E *(b). (Not to scale; the height dimension is greatly exaggerated.) For a key to the various rays, see the text*

Ray bending

Curvature of the ray path due to this mechanism shows itself in other ways. It is obvious that, for maximum effective antenna gain, the angle of elevation of the principal axis of the antenna (the boresight axis) will be higher than that of a straight line between the antennas. This ray bending is most evident at an earth station communicating with a satellite and it is illustrated in Figure A.4. However, the effect is not large; it is not likely to exceed 0.15° when the true angle of elevation is 10° or more, rising to about 0.5° for an angle of elevation of 1°. See Recommendations ITU-R P.834-2 [13] and F.1333 [14].

Antenna gain reduction due to ray bending

Another effect that arises from ray bending, in particular when the boresight direction is at an elevated angle, as at an earth station, is illustrated in Figure A.5. There is a broadening of the beam, and therefore a reduction of the gain, of a high-gain antenna which is directed close to the horizon. In the figure, the ray OC leaves the antenna at a somewhat higher angle of elevation than ray OA and consequently OC is bent rather less than OA. Ray OD, radiated at a lower angle, is bent more than either of the others. Thus, in the far field, the beam behaves as if it is diverging, not from its true point of origin at O but from O′, above and in front of O. Consequently, not only is the angle of elevation of the beam as a whole less than that of the geometric or

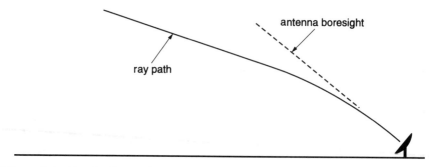

Figure A.4 Ray bending in the troposphere affecting the angle of arrival at an antenna. The
effect, as shown, is greatly exaggerated

'boresight' axis of the antenna OB (to the extent of the angle *a* in the figure) but the
beamwidth is greater and so the gain of the antenna is less than the physical parameters
of the antenna would indicate. Here again, the impact on gain is greatest when the
angle of elevation is low; for a high gain antenna the reduction of gain would be less
than 0.2 dB at an elevation of 5°, rising to about 1 dB at 1°.

Tropospheric duct formation

Circumstances arise when d*n*/d*h*, the rate of change with height of the refractive index,
has a large negative value over a small height range. When this happens a tropospheric
duct may be formed. This duct may trap a wave which is being propagated at an angle
close to the horizontal, sometimes preventing a wanted wave from reaching its
intended destination. The probability that ducting will occur depends on the terrain
beneath the ray path and the climate. Situations in which there is a considerable risk
of link failure due to duct formation must be avoided, say by careful selection of the
route of a radio relay chain or counteracted by such devices as antenna height diversity.

Propagation through the duct itself is a low loss mode, but the condition is sporadic
and so it is not to be relied upon for regular communication systems. On the other
hand, a duct may transmit strong intermittent interference over long distances; see
Section A7.6.

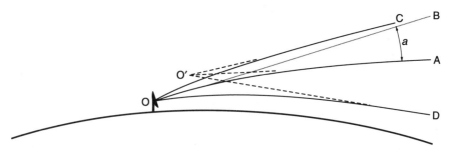

Figure A.5 Beam broadening due to ray bending in the troposphere. The effect, as shown, is
greatly exaggerated

Tropospheric scintillation
The refraction phenomena discussed above arise from an atmospheric structure in which the refractive index varies smoothly in the vertical direction and is uniform in the horizontal direction. This model is broadly valid for most of the time but the troposphere is turbulent and a wave passing through it encounters frequent and rapid, usually small, changes in refractive index. In the SHF frequency range this leads to some incoherence in the wavefront, causing an effective loss of antenna gain; the gain loss is small and even for a high gain antenna it is unlikely to exceed 0.5 dB.

Of greater significance is the small-scale multipath propagation which this non-uniformity of the medium causes, shown by rapid amplitude and phase variations of signals, called tropospheric scintillation. The effect is not important for terrestrial systems below 40 GHz but it may degrade the performance of a satellite-Earth link, particularly those above 10 GHz having a low angle of elevation at the earth station. Recommendation ITU-R P.618-5 [15] provides a method for estimating the magnitude of the effect.

A2.2.2 The effects of solid obstructions

Ground reflection
The signal at a receiving antenna in sight of the transmitting antenna usually consists of the direct ray accompanied by at least one indirect ray, usually reflected from the ground. Other indirect rays may be reflected from other points on the ground or from buildings close to either of the antennas, but for the present purpose it will be assumed that one ground-reflected ray is the strongest and has a substantial specular component.

For a low-gain base station antenna the strength of the ground-reflected ray is likely to be comparable with the direct ray at all angles of elevation, setting up a stable wave-interference pattern in combination with the direct ray. It is usually convenient to treat the radiation pattern that the antenna would have in free space, as modified by the specular component of the ground-reflected wave, as the radiation pattern of the antenna as a whole. If the height of the antenna above the ground is not large compared with the wavelength, the pattern will be stable and simple; with greater height it will be more complex. If it is necessary to optimise the performance of the system in a specific angle of elevation, this can be done, to some degree, by choice of the height of the antenna.

For a high-gain antenna operating at an angle of elevation which is not very close to the horizontal, the gain of the antenna in the direction of a potential point of ground reflection will be small and the ground-reflected ray can be neglected. This condition is present, for example, at most FSS earth stations. However, for most MSS systems the antenna gain at the mobile earth station is quite small and low angles of satellite elevation may arise in some parts of the service area. In this latter case the ground-reflected or sea-reflected ray may be strong, causing severe multipath fading. Methods for estimating the characteristics of this fading are to be found in Recommendations ITU-R P.680-2 [16], P.682-1 [17] and P.681-3 [18].

Diversity reception
Line of-sight radio relay stations have high gain antennas but the angular separation between the direct ray and the ground reflected ray will be quite small if the distance

between stations is large; see Figure A.6. Here again there may not be a big difference between the strengths of the direct ray and the strongest ground reflected ray, especially if reflection occurs at a water surface. Multipath fading may be deep. It may also be frequency-selective. Other fading problems may also arise. For example, Figure A.3 shows a ray path (Z) that would be non-optical under conditions of very low tropospheric refraction, at which time the direct ray might be sustained, up to a point, by a fading signal which is diffracted at the edge of the obstacle; see below.

It may be possible to reduce fading arising from multipath propagation or diffraction to a tolerable level by a space diversity configuration, typically by means of twin receiving channels fed from vertically spaced antennas or by frequency diversity, perhaps augmented by adaptive equalisation of the received signal; see Recommendations ITU-R F.1093-1 [19] and ITU-R F.752-1 [20]. There may indeed be circumstances in which it is economical to use radio station sites which are non-optical even under normal refraction conditions, using space diversity to overcome diffraction fading. Methods for designing such links for acceptable performance at least cost are set out in Recommendation ITU-R P.530-7 [21].

Diffraction at obstacles

An obstacle encountered by a wave may not wholly exclude the wave from the space behind the obstacle; some of the energy of the wave is usually diffracted at the edge of the obstacle into that space. The reduction of the signal level imposed by the obstacle is usually large compared with the signal level that could be expected in the space in the absence of the obstacle, but the scale of the loss depends mainly on two factors. The first is the angle, θ in Figure A.7, through which the wave has to be diffracted in order to reach the receiving antenna. The second is the shape of the vertical profile of the obstacle along a line joining the two antennas. An idealised knife-edge profile, as shown in Figure A.7, causes least loss; it also lends itself most readily to analysis. Where a knife-edge does not represent the actual profile sufficiently well, other idealisations, such as a smooth cylinder or a sequence of knife-edges or cylinders, may enable a more realistic estimate to be made.

Diffraction occurs over the spherical surface of smooth, 'level' ground, augmenting to a small extent the signal due to regular tropospheric refraction over a long propagation path. However, this diffracted signal component may well be smaller than that which would arise if there were a solid obstruction, a hill perhaps, disrupting the smooth surface below the path.

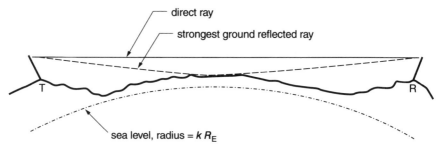

Figure A.6   The direct ray and the strongest ground-reflected ray of a long line-of-sight link. (Not to scale; the height dimension is greatly exaggerated.)

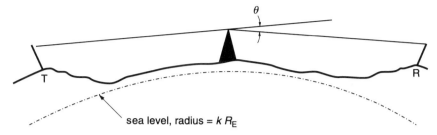

Figure A.7 *A long link served by a wave diffracted through an angle of θ at an obstruction which is idealised as a knife edge*

Recommendations ITU-R P.526-5 [22] and ITU-R P.452-8 [23] provide information on methods for estimating the loss, additional to the free space loss and despite diffraction, due to an obstacle, or the curved surface of smooth ground.

Screening by obstacles
In favourable circumstances, diffraction may allow a radio link to be operated reliably by the tropospheric wave mode even though the receiving antenna is never within line-of-sight of the transmitting antenna. Nevertheless, the transmission loss due to a big obstacle, natural or man-made, is large. This may sometimes be used for screening a receiving station from a source of interference. Such screening is particularly important in frequency bands shared by the FS and the FSS; see Recommendations ITU-R P.526-5 [22] and ITU-R P.452-8 [23].

Point-to-area propagation for mobile services and broadcasting
The methods outlined above for estimating the effect of tropospheric refraction, ground reflection and diffraction are applicable to specific ray paths. These methods are not suitable for use in the context of terrestrial point-to-multipoint services such as broadcasting and the mobile services. This is especially true in urban service areas, where great numbers of stations, many of them in motion and with antennas close to the ground, may be immersed in a environment containing many high buildings. For these services it is necessary to adopt a statistical approach to the determination of system parameters that would provide an acceptably high probability of satisfactory performance.

Recommendation ITU-R P.370-7 [24], provides a series of curves of the field strength estimated to be exceeded at 50% of locations for various percentages of the time for an effective radiated power of 1 kW as a function of distance and the effective height of the transmitting antenna. The frequency range covered is 30–1000 MHz. It is assumed that the receiving antenna will be at a height of 10 m above the local ground level and there is provision for various general climatic categories and various assumptions about the nature of the terrain. These data are intended for use for broadcasting and, with the authority of Recommendation ITU-R P.616 [25] for the maritime mobile service. For broadcasting between 1 and 3 GHz, see Recommendation ITU-R P.1146 [26].

For the aeronautical mobile and radionavigation services, Recommendation ITU-R P.528-2 [27] provides material corresponding to that of Recommendation ITU-R

P.370-7 [24] in a form more suitable for aeronautical use and covering the frequency ranges of concern between 125 and 15 500 MHz. Recommendations ITU-R P.529-2 [28], P.1145 [29] and P.1146 [26] provide similarly for the land mobile service.

There is, however, a prospect for the future of reducing the range of the uncertainties of forecasting propagation conditions, especially in urban environments, by using the detailed topographical information that is already becoming available in some countries. Such topographical databases may map not only the features of the natural terrain but also the nature of the man-made environment and even the location and dimensions of the larger buildings. Recommendation ITU-R P.1058-1 [30] seeks to encourage this development and to standardise the parameters of topographical surveys.

Attenuation by vegetation

The effect of vegetation on propagation to and from radio stations in shrubby and forested areas is not well characterised, but it clearly depends greatly on the frequency.

Below 2 MHz foliage has little direct effect on propagation, although there may be an indirect effect due to the influence of vegetation on the amount of moisture in the subsoil and therefore on the ground conductivity.

Between 2 and 30 MHz there is some absorption of the ground wave by vegetation, rising to about 0.01 dB/m at 30 MHz, the effect being rather worse for vertical than for horizontal polarisation. However, while this absorption would become substantial if the link were long, the lossy path close to the ground may sometimes be bypassed by reflection in the ionosphere.

Above 30 MHz absorption in forest continues to rise with frequency and it is appreciably more severe when the trees are in leaf; see Recommendation ITU-R P.833-1 [31]. In the SHF frequency range the loss may be as high as 1 dB/m.

Propagation in buildings and tunnels

Radio waves may penetrate into a building unless the building is constructed specifically to exclude them. Buildings of conventional construction are almost transparent at the lower frequencies. At VHF and UHF waves permeate through the building, typically passing through windows, doorways and lightly constructed partitions and suffering multiple reflections. The extent of the propagation loss in this latter mode is of concern, for example, in planning paging and cordless telephone systems and the results of various measurement programmes are to be found in CCIR Report 567-4 [32]; most of these measurements relate to frequencies around 900 MHz. For a more general study of propagation within buildings, see Recommendation ITU-R P.1238 [33].

At VLF and LF there is some significant penetration of solid rock by radio waves; see Recommendation ITU-R P.527-3 [34]. At higher frequencies, more suitable for mobile communication in mines and tunnels, penetration through rock is negligible. However, tunnel walls can to some degree support propagation by a waveguide mode, provided that the frequency used is above the equivalent waveguide cutoff frequency; see CCIR Report 880-2 [35]. Losses are severe at sharp corners in a tunnel, or when the presence of a vehicle seriously disturbs the behaviour of the tunnel as a waveguide. Less lossy propagation can be achieved by the use of a leaky feeder system or its equivalent; see Recommendation ITU-R M.1075 [36].

A2.2.3 The effects of rain and other airborne particles

Absorption by rain, snow and hail

Dry snow and hail do not absorb energy from a radio wave. At 3 GHz there is some absorption by very heavy rain. The effect becomes more severe at higher frequencies. At 15 GHz, absorption in a tropical rainstorm exceeds 10 dB/km. At still higher frequencies, absorption exceeding 10 dB/km can be expected occasionally in any rainy climate, The peak rainfall rate is likely to be the main factor to be considered, even in regions with a temperate climate, in choosing the frequency band for use for long links of the FS which carry channels that must compete in performance with systems transmitted by cable or optical fibre. Consideration of the effect of heavy rain is even more necessary for FSS systems. Wet snow may also cause severe absorption, sometimes greater than rain of the equivalent total water content.

Rain and hail also scatter energy from a wave. This has little effect on the level of the wave at the intended receiving station but it may cause sporadic interference to reach another receiving station where interference which does not arise via the regular propagation modes; see Section A7.8.

Estimation of absorption by rain

Problems arise in estimating the effects due to rainfall at earth stations of systems of the FSS which are used to provide telephone and data channels for the international tele-communications network but a sufficiently accurate estimate may be of vital commercial importance. Radio relay systems of the FS, used for the same purpose, raise rather similar, though less complex, problems. The methods used for predicting rain absorption on links between earth stations and satellites are outlined below. A corresponding method for predicting rain absorption on line-of-sight radio relay links is to be found in Recommendation ITU-R P.530-7 [21]. These methods can be simplified and used, where necessary, for other kinds of system.

Performance objectives for satellite links

The performance objective, for digital channels carried by an FSS system providing circuits for the ISDN, is specified in terms of the BER to be attained for all but 10, 2 and 0.03% of the worst month in an average year, disregarding loss of communication due to equipment failures. There are corresponding agreed objectives for satellite links used for other purposes. It can be assumed for the present purpose that these tolerated degradations are due to rain in the vicinity of the earth stations, the more important being usually the receiving earth station and that the short-term (0.03% of worst month) specification will be the most critical.

It is therefore necessary to make a realistic estimate of the peak absorption statistics or the peak rainfall rate. However, weather is changeable from year to year and in many countries the climate varies considerably from place to place. Within relatively small areas there may also be sub-climates, caused in particular by the interaction of the prevailing wind and the local hills and mountains. It is desirable for estimates to take these factors into account.

Prediction using local records

There may be an earth station already in operation at or near the site proposed for a new installation and records of the carrier level of a tracking beacon received from a satellite may have been kept for a period of some years. If so, an analysis of these records would provide the best basis for forecasting the absorption that will be exceeded at the new installation for 0.03% of the worst month of an average year. If the frequency band to be used differs from the band to which the records relate, the existing data can be scaled to the new frequency by means of the coefficients given in Recommendation ITU-R P.838 [37]. If the angle of elevation at which the data was recorded is different from the angle to be used by the new installation, a simple adjustment can take this into account.

In the absence of local absorption records it may be feasible to base predictions of absorption on rainfall records. If the rainfall rate has been recorded in the vicinity of the proposed earth station for several years using a rain gauge with a rapid (one minute) response time, this provides an adequate basis for forecasting absorption. Methods for converting the rainfall rate records exceeded for 0.03% of the worst month of the year to an absorption forecast can readily be divised using the guidance of Recommendation ITU-R P.618-5 [15].

Prediction using meteorological data

If sufficiently reliable local records of absorption or rainfall rate measured with a rapid response time recorder are not available, it is necessary to base a prediction of absorption on meteorological data, although these are mostly derived from measurements made by rain gauges having a slow response time, which leaves peak rainfall rates uncertain. An additional disadvantage is that the meteorological statistics are usually analysed on an annual basis whereas channel performance objectives are specified on a worst-month basis.

Maps in Recommendation ITU-R P.837-1 [38] divide the world into areas, categorised into 15 rain climatic zones, A to Q, ranging from polar regions and the driest deserts (Zone A) where rain is very rare, to tropical rainforest regions (Zones P and Q). For each zone there is an estimate of the rainfall rate in millimetres per hour that will be exceeded for various small percentages of the year. Also, since heavy precipitation usually begins high in the troposphere as hail, which does not absorb energy from the wave, Recommendation ITU-R P.839 [39] provides estimates of the height above sea level of the $0°C$ isotherm. This recommendation enables an estimate to be made of the length of the slant path of the ray at the required angle of elevation that is exposed to rain.

Recommendation ITU-R P.618-5 [15] provides detailed instructions for using the estimate of rainfall exceeded for 0.01% of the year and the estimate of the length of the path exposed to rain for predicting the absorption that will occur for various small percentages of an average year. Figure A.8, produced using these instructions, shows estimates of absorption occurring at angles of elevation of 10° and 30° for 0.1% of an average year in rain climate zones E (temperate) and P (tropical rainforest) as a function of frequency. The available advice for converting these 'average year' predictions to 'worst month' predictions is to be found in Recommendation ITU-R P.581-2 [40] and CCIR Report 723-3 [41].

Space diversity

In very heavy rain it is usually found that the rainfall rate is highest in small rain cells, only a few kilometres in diameter, which are embedded in an extensive area of lighter rain. Thus, for links, whether space–Earth or terrestrial, which may be exposed to rain over several tens of kilometres, it can be assumed that the whole of that path seldom if ever experiences the maximum rainfall rate at any one time. On the other

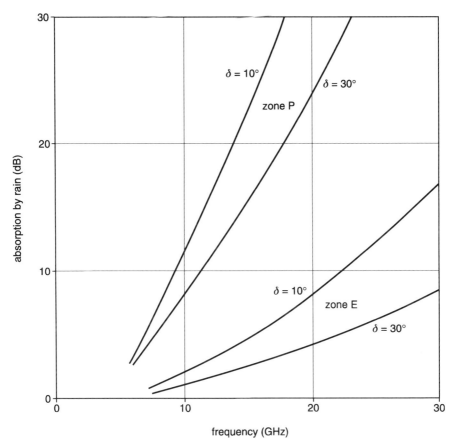

Figure A.8   *Predicted absorption by rain on a space–Earth path for 0.1% of an average year for earth stations operating at angles of elevation (δ) of 10° and 30° in rain climate zones E and P, using the method and data of Recommendations ITU-R P.618-5 [15] and P.837-1 [38], respectively*

hand, the link will suffer that maximum rainfall rate over some part of its length more frequently than the rainfall statistics of any single place would indicate. A procedure for taking this into account is incorporated into the method for predicting rain absorption in Recommendations ITU-R P.618-5 [15] and P.530-7 [21], referred to above.

However, a further consequence of this cellular rainstorm structure is that the most severe effects of rain absorption could usually be avoided by using pairs of radio stations, a few kilometres apart and operated in space diversity. There is an analysis of the potential benefit to be had in this way in the FSS in Recommendation ITU-R P.618-5 [15]. However, this solution is economically unattractive at present. Instead, where absorption is unacceptably high at one frequency, a lower frequency is used. Where this solution is not viable for the FS, the higher frequencies are used only for the shorter links.

Absorption by cloud, mist and fog

Droplets of water suspended in the air in the form of cloud, mist or fog may cause some small amount of absorption at the highest frequencies which are currently in substantial use. See Recommendation ITU-R P.840-2 [42] for data. At frequencies over 100 GHz, considerable loss is likely to arise in thick fog.

Depolarisation by rain and by airborne ice crystals

Raindrops are not perfectly spherical and their asymmetry is not random; the orientation of raindrops tends to become aligned in falling through the air and the direction of the orientation relative to radio station antennas is affected by the direction and velocity of the wind. This raindrop asymmetry may change the state of the polarisation of a radio wave, reducing the cross-polar discrimination of an antenna which is receiving the wave. This may be a source of serious interference in an FS or FSS system which re-uses the spectrum by means of dual polarisation. A second atmospheric mechanism which causes depolarisation at times arises from asymmetrical ice crystals in clouds above the 0°C isotherm when their asymmetry becomes aligned, probably by the electrostatic potential distribution in the cloud.

There is a description of these phenomena in CCIR Report 722-3 [43] and procedures for estimating the severity of their impact on radio waves in the context of the FS and the FSS are set out in Recommendations ITU-R P.530-7 [21] and P.618-5 [15], respectively.

A2.2.4 *Absorption by atmospheric gases*

In certain specific frequency bands, all above 20 GHz, there is substantial absorption of energy from radio waves due to molecular resonances in water vapour and the permanent gases in the atmosphere. Unlike the absorption due to rain considered in Section A2.2.3, the absorption due to the permanent gases is always present, and in some bands it is very severe; see Recommendation ITU-R P.676-3 [44]. Figure A.9 shows the specific attenuation (that is, the attenuation per kilometre) at sea level under standard atmospheric conditions; attenuation is substantially less at high altitudes.

There is a weak absorption band centred on 22.3 GHz due to a molecular resonance in water vapour. Very strong water vapour absorption bands are centred on 183.3 and 323.8 GHz. The absorption due to water vapour in the gaps between these lines is substantial and rises with frequency. As would be expected, the amount of this absorption depends on the absolute humidity of the atmosphere.

There is an absorption band due to oxygen centred on 60 GHz and a relatively weak line at 118.7 GHz. At sea level, attenuation in the 60 GHz band is very strong and it extends relatively uniformly over a bandwidth of several gigahertz. At high altitudes this absorption band is resolved into a series of narrow lines, attenuation remaining strong at the peaks but becoming relatively weak in the troughs between the lines; see Figure 7 of Recommendation ITU-R P.676-3 [44].

There are many absorption lines above 350 GHz due to water vapour, oxygen and the other atmospheric gases. Consequently there is high specific attenuation within the atmosphere for almost all frequencies between 275 GHz (the highest frequency allocated so far to a service in the international table of frequency allocations) and the infra-red band. Data that enable the specific attenuation for water vapour and oxygen to be estimated for horizontal and slant paths at frequencies below 1000 GHz are to be found in Recommendations ITU-R P.836-1 [45] and P.835-2 [46].

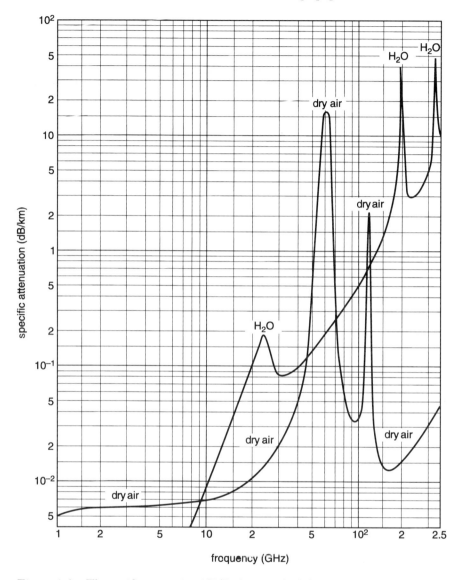

Figure A.9   *The specific attenuation (dB/km) at sea level due to absorption by atmospheric gases. Atmospheric pressure = 1013 hPa. Temperature = 15 °C. Water vapour = 7.5 g/m³ (Recommendation ITU-R P.676-3 [44])*

A2.2.5 Effects of the ionosphere on transmitted waves

Above a height of about 60 km the Earth's atmosphere is ionised, mainly by solar radiation. This ionised region is the ionosphere. The lower frequency radio waves from terrestrial transmitters are reflected back to Earth in the lower reaches of the ionosphere; see Section A4 below. Radio waves at higher frequencies pass through the ionosphere, suffering various distortions and losses; these phenomena are outlined, for example, in Recommendation ITU-R P.531-4 [47]. Two ITU-R handbooks [10, 48] review these phenomena in more detail. The more important of these phenomena are outlined below in terms of transmission in the space-to-Earth direction, typically from a satellite transmitter to a receiving earth station, but the effects are reciprocal.

Ionospheric scintillation
Irregularities in the electron density in the ionosphere cause variations in the amplitude, phase, polarisation and angle-of-arrival of a wave received at an earth station. These phenomena, called ionospheric scintillation, are commonest between sunset and local midnight in the equatorial zone. At high latitudes they are at a peak at night and around noon. Fluctuations in the amplitude of received signals, above and below the level obtained when scintillation is not present, are greatest at the lower frequencies, at times of high solar activity ('sunspot maximum') and for receiving earth stations operating at a low angle of elevation. The relationship between the amplitude of the fluctuations and the frequency is of the order of $f^{-3/2}$.

At sunspot maximum the maximum peak-to-peak fluctuation may exceed 20 dB at 1 GHz and 4 dB at 10 GHz and the period of the fluctuations is likely to range from a fraction of a second to 10 s. Significant scintillation can be expected to occur at tropical locations for 2–5% of the time at sunspot maximum, falling away to a negligible incidence at sunspot minimum.

Absorption in the ionosphere
Significant absorption in the ionosphere is limited to the polar regions and the VHF and UHF frequency ranges. Two phenomena are involved, auroral absorption and polar cap absorption.

Auroral absorption occurs during periods when the electron flux reaching the Earth from the Sun is particularly high. Its effects are observed over a latitude range of 10° to 20° relative to the latitude of maximum occurrence of visible aurorae. The duration of each occurrence ranges from a few minutes up to a few hours. Measurements made at 127 MHz show an absorption of 1 or 2 dB for up to 5% of the time at an angle of elevation of 5°. Absorption falls with an increasing angle of elevation and increasing frequency (proportionate to f^{-2}), being negligible at and above 500 MHz.

Polar cap absorption is also associated with solar abnormalities, such as solar flares. The area affected is around whichever of the Earth's magnetic poles is sunlit at the time and extending 24° of geomagnetic latitude from that pole. (There is a world map of geomagnetic latitude at Figure 2 of Recommendation ITU-R P.1239 [49].) There may be perhaps a dozen such events in a year at sunspot maximum but fewer in other years. Each event lasts several days. The severity of the absorption is of the same order as for auroral absorption.

Transmission time
A radio wave travels more slowly in a medium containing free ions than in free space; thus the transmission time varies with the total electron content (TEC) of the path along which the wave travels. This is a matter of concern in the design of precise radio-navigation systems operating at VHF and UHF. The effect is greatest, of the order of 20 ns, in tropical regions and at sunspot maximum and it declines with frequency, being proportionate to f^{-2}.

Rotation of the plane of linear polarisation

The presence of the Earth's magnetic field in an ionised medium causes the plane of polarisation of a linearly polarised wave passing through it to be rotated, to an extent which is a function of the TEC along the path of the wave and an inverse function of the square of the frequency. This phenomenon is called Faraday rotation. At VHF the plane of polarisation is rotated a total of many complete turns in a single passage through the ionosphere, and variations of the TEC from moment to moment ensure that the plane of polarisation of a wave arriving at a receiving earth station is continually changing. At 3 GHz the total variation of the angle of the plane of polarisation on arrival at the receiving antenna may be some tens of degrees; this is a very significant amount and it is usual to use circular polarisation to avoid depolarisation in C band systems. The effect becomes negligible for most purposes above 10 GHz.

A3 Ground wave propagation

If a wave of relatively low frequency is radiated horizontally from an antenna which is close to the ground, any horizontally polarised component of the wave suffers severe attenuation due to losses in the ground, and any vertically polarised component tends to follow a curved path, above and to some extent also below the curved surface of the Earth; see Figure A.10. Energy radiated at higher angles of elevation does not join this so-called ground wave, although some of it may return to Earth after reflection in the ionosphere.

Thus a radio system designed to take advantage of ground wave propagation for communication beyond the local horizon will be equipped with a transmitting antenna which radiates with vertical polarisation and with its greatest gain in the

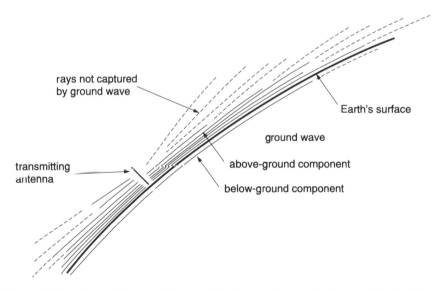

Figure A.10   *Rays of the ground wave set up by an antenna radiating a vertically polarised wave*

horizontal plane, typically a short vertical monopole. Such an antenna, radiating 1 kW, sets up a field strength of 300 mV/m on the ground at a distance of 1 km, subject to any adjustment necessary to make allowance for the near-field effect. At greater distances the field strength falls away relatively slowly, although faster than the inverse distance rate.

The curvature of the path of the wave results from a combination of three causes. The least important of these causes is diffraction of the wave over the curved surface of the Earth. Rather more significant is atmospheric refraction, due to the decline with increasing height of the refractive index of the air. But the main cause arises from the penetration of part of the wave into the ground underlying the propagation path. Depending on the electrical parameters (permeability, conductivity and permittivity) of the ground, the velocity of the wave below ground is less than the velocity of the wave above the ground, and coupling between the above-ground and below-ground components causes the path of the former to curve downwards.

Prediction of ground wave field strength

The permeability of soil and subsoil is usually close to the permeability of space, so this parameter can be disregarded. However, the conductivity and the permittivity of the ground depend on local geological factors and on how much water is present, in the ground or in the form of open water. The depth of penetration of the wave into the ground, and hence the depth of the geological strata that influence the process, varies mainly with frequency and ground conductivity, ranging from some tens or hundreds of metres at VLF (except in sea water) to less than one metre at VHF (except in fresh water ice or very dry ground); see Recommendation ITU-R P.527-3 [34]. Recommendation ITU-R P.832-1 [50] is an atlas of maps of ground conductivity covering a substantial proportion of the world.

Given the frequency and the data on ground conductivity and making standard assumptions as to the effects of diffraction and refraction, the program GRWAVE can be used to compute the ground wave field strength at a distance from an antenna radiating a known power. The program is available from the ITU. Recommendation ITU-R P.368-7 [51] shows an extensive series of curves of calculated field strength for a receiving antenna at or near the ground as a function of the distance from a transmitting antenna radiating 1 kW at frequencies between 10 kHz and 30 MHz. Figure A.11 shows a representative sample of curves taken from that recommendation. It may be seen that the field strength at distances greater than a few kilometres declines rapidly as the frequency is increased or the ground conductivity is reduced. A more comprehensive series of curves, including field strength predictions for receiving antennas which are not close to the ground, is to be found in an ITU handbook [52].

The phase of the ground wave

The phase difference (Φ) between the carrier of an emission at a transmitting antenna and the same emission at a receiving antenna after ground wave propagation can be represented by

$$\Phi = 2\pi n d/\lambda + \Phi_s \text{ radians} \tag{A.11}$$

where n is the refractive index of the atmosphere at the Earth's surface, d is the distance between the antennas, and λ is the wavelength in the same units as d.

Φ_s is an adjustment factor, called the 'secondary phase', which allows for departures from completely regular transmission. At frequencies below about 3 MHz, the value

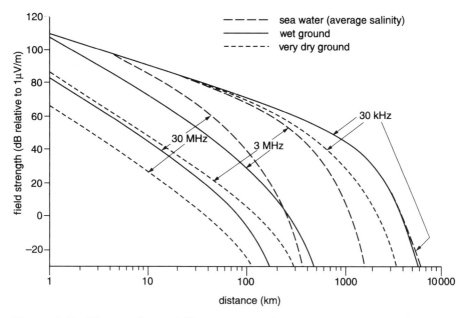

Figure A.11 *The ground wave field strength at a distance from a short vertical monopole radiating 1 kW at 30 kHz, 3 MHz and 30 MHz over various types of terrain (after Recommendation ITU-R P.368-7 [51])*

of Φ_s is small for transmission over sea. Over land, Φ_s is affected by the minerals lying below the surface. There are means for estimating the scale of this effect, outlined in CCIR Report 716-3 [53].

A4 Ionospheric wave propagation

A4.1 The ionosphere and its behaviour

Virtually all of the energy of ultra-violet and X-ray radiation from the Sun which reaches the Earth is absorbed in the process of ionising the upper atmosphere. On the illuminated side of the Earth this ionisation starts about 60 km above the ground, rises to a maximum intensity at a height of about 300 km and then tapers off slowly with still greater height as the atmospheric pressure declines. This ionised outer region of the atmosphere is called the ionosphere. Ions recombine quickly in the lower parts of the ionosphere, where the density of the atmosphere is relatively high. Thus, little ionisation persists below about 90 km at night, when few new ions are being generated. The rate of recombination is much less at higher levels and substantial ionisation continues through the night above 200 km. See Figure A.12. Several strata can be identified within the ionosphere because of their distinctive behaviour; the principal ones, indicated in Figure A.12, are called the D, E and F regions, the F region being divisible into F1 and F2 by day.

Radio waves at the lower frequencies, radiated from the ground above the horizontal, may be reflected back to Earth by the ionosphere. This ionospheric wave, usually

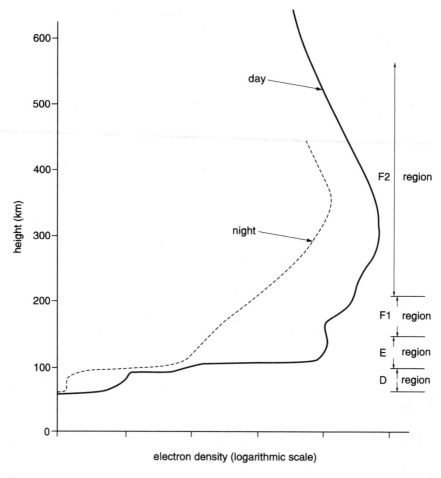

Figure A.12   Sketch of typical distribution of ion density with height in the ionosphere, by day and night

called the 'sky wave', provides the chief propagation medium for long distance links at frequencies below about 30 MHz.

Disturbances in the ionosphere
In addition to the solar radiation which is the source of the regular ionisation, high energy particles enter the Earth's atmosphere from space, mostly protons and electrons from the Sun but some of cosmic origin, including heavier positive ions. These charged particles come under the influence of the Earth's magnetic field and most of them are trapped in the Van Allen belts, above the level of maximum ionisation in the ionosphere. Such particles occasionally disturb the ionosphere, especially in the auroral zones, when a solar disturbance has temporarily increased the flux of solar particles; see also Section A2.2.5. Cosmic rays may break through to the lower levels

of the ionosphere, where they maintain a residue of ionisation in the D region throughout the night. At times enhanced X-ray radiation arising from a large solar flare may cause a large temporary increase in the density of ionisation in the D region of the ionosphere, causing a sudden ionospheric disturbance which may disrupt communication by the ionospheric wave for some tens of minutes. Often a sudden ionospheric disturbance precedes an increased flux of solar particles.

Events of these kinds are of concern to operators of systems making use of ionospheric propagation and Recommendation ITU-R P.313-8 [54] outlines a scheme for expediting the flow of information on ionospheric disturbances, so that their operational impact can be minimised. However, these are not a matter of basic concern for spectrum managers.

Radio waves in a model ionised layer
The presence of the Earth's magnetic field makes the interaction of radio waves and the ionosphere theoretically complex, but the simplified explanation which follows, which disregards the effect of the magnetic field, is sufficient for the identification of the basic phenomena of the effect of the ionosphere on radio waves.

The electric vector of a radio wave interacts with the ions. The ions oscillate to and fro, taking energy from the wave at some phases of the cycle and returning the energy at other phases. This causes a phase advance of the electrical vector, increasing the phase velocity of the wave, the effect increasing with ion density and declining with frequency. The group velocity of the wave is correspondingly reduced. There is a net loss of energy from the wave when ions, while in motion due to the presence of the wave, recombine to form neutral atoms. This loss of energy increases with the recombination rate, and thus with atmospheric pressure. Absorption varies inversely with frequency.

Thus, Figure A.13 shows a model ionised layer with a parabolic vertical distribution of ionisation. A wave is transmitted from T, on Earth, and rays A, B, C and D are incident upon the lower surface of the ionised layer at various angles. Ray A enters the layer obliquely from below with an angle of incidence i_A. Each wavefront finds itself in a medium with a transverse ion density gradient, the density being higher on the side further from the Earth's surface; a greater phase advance therefore occurs on that side, causing the wavefront to swing somewhat towards the Earth. If, as is assumed for ray A, the frequency is not too high, the wave will have swung through the horizontal without penetrating the layer as far as h_m, the level of maximum ionisation and it will then continue to swing, by now downwards, until it leaves the lower side of the layer and regains a straight path Ray A may then be received on the ground at R.

However, another part of the wavefront, entering the ionised layer at a smaller angle of incidence, must swing through a larger angle before it reaches a horizontal path. Thus ray B may reach h_m without having become horizontal. It will then swing, not further towards the horizontal but upwards and out through the top of the layer. There will be a third part of the wave, represented by ray C, between A and B, which attains a horizontal path at height h_m and then, notionally, follows the horizontal line of maximum ionisation. Let the angle of incidence on the layer of ray C be i_C. The frequency of the wave, f_C, is called the critical frequency for angle i_C. At that angle of incidence, a wave at a higher frequency would penetrate the layer and at any lower frequency it would be reflected back to Earth. In either case there may be significant absorption of energy from the wave.

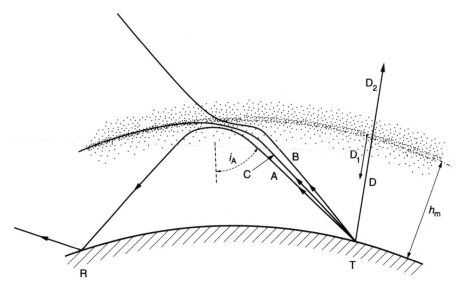

Figure A.13   Reflection from, or penetration of, a model ionised layer by radio waves entering it from below

For ray D in Figure A.13, propagated vertically up into the ionised layer, the phase velocity will increase and the group velocity will fall as height h_m is approached, but when that level has been passed the ray will penetrate the layer and emerge at the top as ray D_2. However, if the frequency of the wave were progressively reduced, a point would be reached at which the group velocity had fallen to zero by the time that h_m was reached; with any further reduction of frequency the ray would reverse its direction and emerge from the lower side of the layer as ray D_1. The frequency at which this occurs is called the vertical incidence critical frequency, f_0. The relationship between f_c and f_0 is given by

$$f_c = f_0 \sec i_C \qquad (A.12)$$

After reflection from a model ionised layer, a ray emerges at the same angle to the vertical as the angle of incidence, as if it had been reflected at a sudden discontinuity at h', called the effective height; see Figure A.14. Furthermore, the transmission time of the wave in its journey within the layer at a group velocity lower than c, the velocity in free space, will be the same as it would have been if reflection had at h' and the velocity had been equal to c throughout. Finally, since there is no variation of ion density in the horizontal plane in this model ionised layer, the ray will not deviate to right or left on reflection; it will continue along the great circle path that it was following between the transmitting antenna and the point at which it entered the ionised layer.

A wave in the ionosphere
The above account is simplistic. Among the other factors that have to be taken into account in some circumstances are the following.

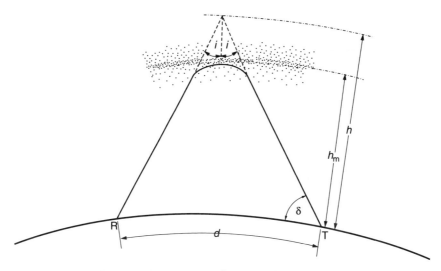

Figure A.14   The effective height of reflection, h', in a model ionised layer

1   It will be clear from Figure A.12 that the variation of ion density with height is not generally parabolic. In particular a wave may pass through a level of maximum ionisation because its frequency exceeds f_c for the effective angle of incidence, before moving up to a greater ion density at a higher level which reflects the wave.

2   There is a natural resonance in the motion of charged particles under the simultaneous influence of the Earth's magnetic field and a radio wave. For an electron the frequency of this resonance, called the 'gyrofrequency', ranges from about 0.8 MHz near the Equator to about 1.7 MHz near the poles.

3   The presence of the Earth's magnetic field causes the wave to split into two components, usually elliptically polarised in contrary senses, called the ordinary and the extraordinary waves. As a result, the polarisation after transmission through the ionosphere is usually random and the amplitude is usually varying due to wave interference between several components of the wave at the receiving antenna. This variation is one of the phenomena that are called fading.

4   The degree and distribution of ionisation varies according to the time of day, geographically, seasonally and from year to year. Some phenomena, in particular those related to the F2 region, vary significantly from day to day. These variations, whether regular or sporadic, can have a major impact on the effect of the ionosphere on the propagation of radio waves.

However, the necessary data and methods are available for taking these factors into account to produce predictions of propagation conditions with an accuracy that is acceptable for most purposes. These methods are outlined in Sections A4.2 and A4.3.

Solar activity indices
The diurnal, geographical and seasonal variations in the condition of the ionosphere at any location arise mainly from differences in the angle of elevation of the Sun at that location. If these variations are disregarded, a year-to-year variation remains due to

the behaviour of the Sun itself. The Sun is a variable star and the intensity of its ultra-violet and X-ray radiation varies with a period of about 11 years, rising from a base level, passing through a maximum and returning to the base level again before the commencement of a new cycle. The period is somewhat erratic and the range of radiation intensity from maximum to minimum varies substantially from cycle to cycle. It is helpful to use records of past solar activity in predicting future ionospheric conditions.

Direct measurement of the ultra-violet or X-ray radiation from the Sun is not possible from the ground, and observations made from artificial satellites do not have a long history. However, another of the Sun's characteristics that varies with this 11-year cycle is the display of sunspots. There have been systematic observations of sunspots since 1749 and records of the years of maximum and minimum sunspot activity go back to 1610. This long history is a valuable foundation for the prediction of future activity. An index of sunspot activity, developed by Rudolf Wolf (1816–93), calculated from the number, size and disposition of the spots that are visible each day and averaged monthly, is called the 'sunspot number' and has the symbol R. There is a wide scatter of the month-to-month values of R which is not closely indicative of the general trend of solar activity, but it is found that the 12-month running average of R, which has the symbol R_{12}, represents well the solar activity at the middle of that 12-month period.

A second solar characteristic, measurable on Earth, which provides a general measure of solar activity is the intensity of solar radio noise at a frequency that suffers little attenuation in the atmosphere. The solar noise flux at around 3 GHz has been measured at various observatories since 1947. The monthly average of this parameter has the symbol Φ. It is found that Φ has a considerably smaller month-to-month scatter than R. The 12-month running average of Φ has the symbol Φ_{12}.

In addition to these indirect indices, various studies are in progress seeking to devise a reliable basis for forecasting future ionospheric conditions from measurements of current and past ionospheric critical frequencies. The index known as *IG* is the best-established of these. For further information on R, Φ, *IG* and the other ionospherically derived indices, see Recommendation ITU-R P.371-7 [55]. The same recommendation recommends that predictions of R_{12} should be used for all predictions of ionospheric parameters for dates more than 12 months ahead of the most recent evaluation of that index. For shorter-term predictions of conditions in the F2 region, the 12-month running averages of R, Φ and *IG* are considered to be equally valid. For shorter-term predictions of conditions in the E and F1 regions, the 12-month running average of Φ is preferred.

A4.2 Ionospheric wave propagation below about 2 MHz

The pattern of propagation

The pattern of radio propagation below about 2 MHz is illustrated in Figure A.15. Thus, at 100 kHz (Figure A.15a) the energy radiated by the transmitting antenna at an angle too high to be captured by the ground wave is reflected downwards at a low level in the ionosphere, the D region by day and the E region by night. Much of the energy of this sky wave is then reflected at the Earth's surface up to the ionosphere again, and there may be multiple reflections between the ionosphere and the ground until the wave peters out, attenuated by distance and by losses on reflection. Loss on

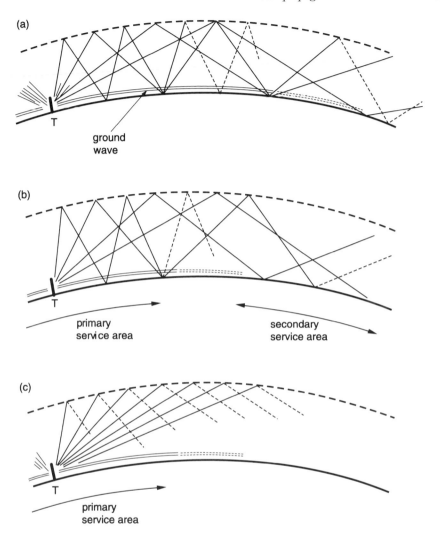

Figure A.15   *The pattern of propagation at 100 kHz and 1 MHz, showing rays of the ground wave and the sky wave: (a) represents 100 kHz propagation by day and night; (b) represents 1 MHz propagation by night; (c) shows 1 MHz daytime propagation*

reflection in the ionosphere is greater by day than by night. At any point on the ground within reach of the transmission, the signal that can be received is likely to be the vector sum of the ground wave and several rays of the sky wave. Close to the transmitter the ground wave will be dominant and there will be little amplitude variation (fading) of the signal due to wave interference. At distances where the ground wave is not dominant, fading may be deep.

At 1.0 MHz by night (Figure A.15b) the pattern of propagation is similar, with the sky wave being reflected from the E Region. At higher frequencies, above perhaps 1.5 MHz, the high angle radiation from the transmitting antenna may penetrate the E region at night under some circumstances, but it will then be reflected in the F region. However, at this frequency the range of the ground wave is considerably less than that of the night-time sky wave and three concentric zones of reception quality can be identified. The ground wave dominates close to the transmitter and there is little fading. This frequency range is used mainly for MF broadcasting using amplitude modulation and this first zone is the primary service area. Outside this zone there is another zone where the ground wave and the sky wave have comparable amplitudes and fading may be severe; this zone is of little value for AM broadcasting. In the third zone the ground wave is negligible and fading may not be unacceptably deep if circumstances cause one ray of the sky wave to predominate; this is the secondary service area in MF broadcasting.

At 1.0 MHz by day (Figure A.15c) most of the energy radiated by the transmitting antenna which is not captured by the ground wave enters the ionosphere and is absorbed in the D region. Daytime absorption is particularly strong in summer. Thus, in MF broadcasting by day there is no secondary service area and the power of the sky wave is too low to cause fading in the outer part of the primary service area. This changeover in the daytime pattern of propagation, from strong absorption of the sky wave at the higher frequencies (Figure A.15c) to low absorption at the lower frequencies (Figure A.15a) takes place gradually between about 400 and about 700 kHz.

Estimating field strengths

It is particularly important to be able to estimate the field strength of the sky wave at great distances from the transmitter, because some services, and in particular AM broadcasting, are susceptible to low levels of interference. Unfortunately, it is difficult to do this analytically, because much depends on the losses that occur on reflection from the ionosphere and the ground, and these losses involve many factors, especially the geomagnetic parameters.

Until recently it has been necessary to rely mainly on empirical prediction methods, especially to meet the needs of the broadcasting service operating between 150 and 1705 kHz. Field strengths had been measured at night at long distances from a large number of transmitting stations, the results being normalised for a standard transmitter power and analysed statistically to correlate systematic differences between stations with identifiable physical causes on the ground and in the ionosphere. However, a method is now available for computing the strength of the night-time sky wave in this frequency range; see Recommendation ITU-R P.1147 [56].

A different method is preferred for use at frequencies below about 500 kHz and there is a further option for use below about 30 kHz. These methods can be used to calculate the sum of the ground wave and the sky wave. The 'wave hop' method combines an analytical approach to calculating the field strength from a distant transmitter with the use of empirically obtained data where necessary. The strength and phase of the various components, ground wave and sky wave, of the signal are calculated and summed vectorially. Recommendation ITU-R P.684-1 [57] recommends the use of this method for frequencies between 60 and 500 kHz and for lower frequencies down to 16 kHz for distances less than 1000 km. Annex 2 to the recommendation reviews the extent to which the application of this method has been verified; at present the verification process is incomplete.

The wave-hop method would demand the summation of an impractically large series of propagation modes if used for frequencies below 30 kHz and distances greater than 1000 km. However, between 30 kHz and 10 kHz, the lowest frequency range in substantial use for radio systems, the wavelength is between 10 km and 30 km. These wavelengths are not small when compared with the height above ground of the bottom of the ionosphere. At such frequencies it is valid, within limits, to treat the ground–ionosphere space as a waveguide, and for distances beyond 1000 km the number of modes that need to be considered in predicting field strengths is manageably small. A procedure for applying the waveguide concept below 30 kHz for distances greater than 1000 km is also given in Recommendation ITU-R P.684-1 [57].

Phase shift variations

It may be noted that Φ_s the secondary phase factor discussed briefly at the end of Section A3, which is small at the lower frequencies when the ground wave predominates, is large and unstable where the sky wave is strong. This factor may be very significant, for example, for LF and VLF radionavigation systems and may be a limiting factor for broadcasting systems using digital modulation; see Recommendations ITU-R P.1321 [58] and BS.1349 [59].

A4.3 Ionospheric wave propagation between 2 and 30 MHz

The characteristics of the regions of the ionosphere

The D region is between 60 and 90 km above the ground. The intensity of the ionisation over any geographical location, being largely due to solar radiation, varies as a function of a solar activity index and the angle between the Sun and the zenith at that location, with some additional ionisation, very variable in intensity, in the auroral zones. The ion density is never great enough to reflect waves above 2 MHz, but it causes absorption which varies as an inverse function of frequency, called non-deviative absorption.

The E and F1 regions are centred about 130 km and 190 km above the ground, respectively. The ion densities of both are stable and predictable by day, closely following functions of the Sun's zenithal angle and a solar activity index, although ionisation levels in the F1 region are influenced to some extent by geomagnetic latitude. Ionisation densities by day are substantial, sufficient to provide critical frequencies at vertical incidence of several megahertz, rising towards 20 MHz at oblique incidence. There is a limited amount of absorption on reflection, called deviative absorption, which tends to increase with frequency. Some ionisation persists in both regions at night but that in the E region is not dense enough to reflect waves above 2 MHz and it is more satisfactory to consider the F1 and the F2 regions as a single whole at night, forming the F region, its characteristics dominated by those of the F2 region.

The E region is subject to another form of ionisation, occurring quite frequently at some locations and more intense than the regular E region ionisation but sporadic. This sporadic E ionisation, denoted by the symbol E_s, is not usable for regular services but it is often a source of long distance interference; see Section A7.4.

The F2 region, including the Fl region at night, extends over a much greater height range than the other regions, the maximum level of ionisation being as high as 400 km at some times. Vertical incidence critical frequencies are higher than those of the other regions (excluding E_s ionisation) typically in the range 3 to 10 MHz and often higher. At oblique incidence, under extreme conditions, frequencies exceeding 50 MHz are regularly reflected in daytime. As with the other regions, the most important influence on the level of ionisation at a location in the F2 region is the degree of solar illumination and the solar activity index, although the geomagnetic field also plays a significant role.

However, whereas it is feasible to predict accurately the ion density of the E and Fl regions by computation from basic geophysical parameters this is not feasible at present for the F2 region or the F region at night. The long lifetime of free ions in the low density atmosphere several hundreds of kilometres above the ground ensures that the ion density at any place does not follow closely the daily variations of solar illumination. In consequence, ionisation established by day declines but remains quite strong during the night. Furthermore, fast winds that blow in the F region move large volumes of ionised air from place to place and from height to height, rarifying the gas in some places, concentrating it in others, with corresponding effects on the ion density. Three consequences that flow from this are as follows.

1   The F region ion density shows a much greater variability, from day to day and from hour to hour, than the E and Fl regions.
2   The average geographical distribution of F region ionisation is much less regular than that of the regions in which it is related closely to the zenith angle of the Sun. See for example, Figure A.16, which shows a sample of the world-wide distribution of the maximum usable frequencies for a single hop 4000 km long via the F2 region, constructed from the records of ionosphere soundings in the month of May over many years and normalised for sunspot minimum ($R_{12} = 0$) conditions. See the CCIR atlas of ionospheric characteristics [60].
3   It is necessary at present to base predictions of F region characteristics on normalised measurements of vertical incidence critical frequency and effective height made over the years at a world-wide network of ionosphere sounding observatories.

Single-hop and multi-hop propagation
Equation (A.12) and the geometry of Figure A.14 enable values of sec i to be calculated as a function of the length (d) of a single hop of a ray for a given value of effective height, h'. Figure A.17 shows this relationship for effective heights of 150, 250 and 450 km, which can be taken as typical values of h' for critical frequencies in the E, Fl and F2 regions respectively, an allowance being made for refraction in the troposphere by assuming that the radius of the Earth is 4/3 times the true value. Figure A.17 also gives an indication of the angles of elevation (δ) of rays at the radio stations.

Thus, Figure A.18 illustrates a link between transmitting station T and receiving station R propagated via E region reflection. If the distance between T and R is 2000 km and $h'E$ is assumed to be 150 km, Figure A.17 shows that sec i would be about 5.25. Then, if the vertical incidence critical frequency for the E region midway between T and R is assumed to be $f_c E$, the maximum usable frequency (MUF) for this link will be $5.25 \times f_c E$. The expression sec i is called the MUF Factor for this

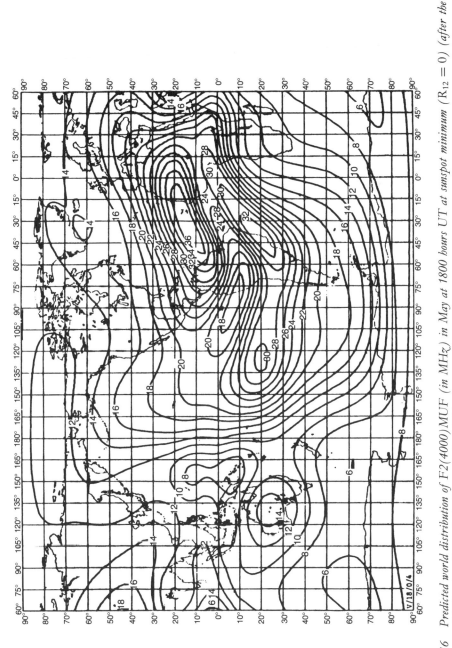

Figure A.16 Predicted world distribution of F2(4000)MUF (in MHz) in May at 1800 hours UT at sunspot minimum ($R_{12} = 0$) (after the CCIR atlas of ionospheric characteristics [60])

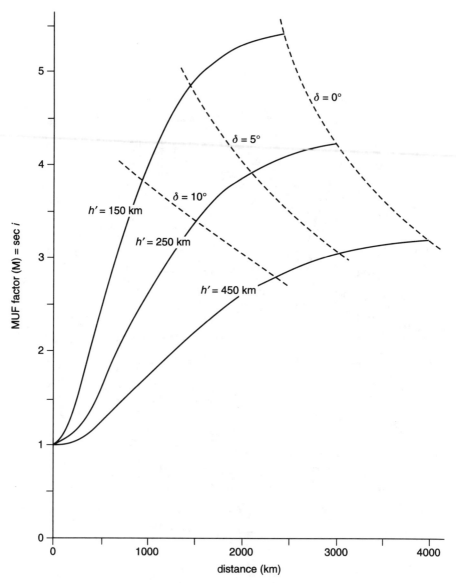

Figure A.17   The MUF factor (M) and the minimum length (d) of a single hop at the MUF for effective heights (h′) of 150, 250 and 450 km

link, often contracted to M, and the MUF factor for a 2000 km link via the E region is often designated $M(2000)E$.

In Figure A.18 the ray that passes from T to R, having a frequency equal to the MUF for the path, leaves T at an angle of elevation δ. Any ray with a higher angle of elevation passes through the E region without reflection although it may be reflected at a

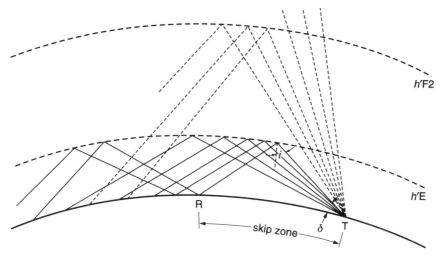

Figure A.18   1E, 2E and 1F2 *propagation modes from a transmitter at* T

higher level, possibly returning to Earth beyond R. Any ray with a lower angle of elevation is reflected to the ground at a greater distance than R; it may be received in the area beyond R or far beyond R, via two-hop propagation after reflection from the ground and a second reflection from the E region. If ground wave propagation is disregarded, a wave at the frequency in question cannot be received between T and R; this zone of silence is called the skip zone.

The MUF for the link between T and R, given above as $5.25 \times f_c E$, is designated the basic MUF. For various reasons it may be found that satisfactory link operation is not obtained at this frequency. For example, it will be seen from Figure A.17 that the ray that links T and R by reflection in the E region has an angle of elevation (δ) at the radio stations of about 2°, and the gain of many HF antennas is small at such a low angle. If this reduces the performance of the link unacceptably, it may be necessary to use a higher angle of elevation and a lower frequency with two-hop E region propagation or, perhaps, reflection from a higher region of the ionosphere with one hop. The MUF applicable to this different but satisfactory mode is called the operational MUF.

However, in practice a system operator will often not have knowledge of critical frequencies in real time (although see the references to real time channel evaluation (RTCE) in Section 6.1.2) and predictions of MUFs are unlikely to be exact, especially when the F2 region is involved. The forecast of solar activity may not be correct. More significantly, the data on which the computation of critical frequencies and effective heights are usually based are averages for periods of one month, whereas ionospheric conditions vary from day to day. It is possible to make allowance for this variability by reducing the calculated operational MUF by an appropriate percentage, producing the optimum working frequency (OWF), estimated to be the frequency which the MUF will exceed on 90% of the days of the month.

To complete this prediction of usable frequencies it is necessary to take into account the lower limit. If the operating frequency is reduced, the level of non-deviative

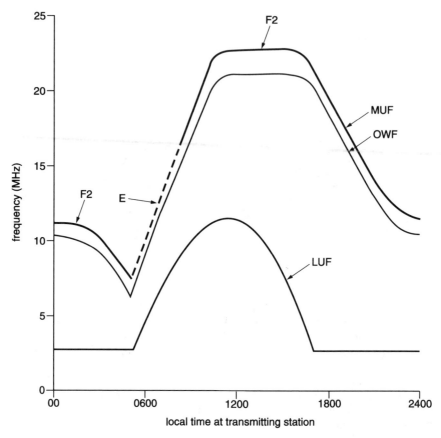

Figure A.19   A sketch of the operational MUF and the corresponding OWF and LUF for a typical long HF link, as a function of local time at the transmitting station

absorption in the D region increases wherever the path of the wave is sun-lit. There is also a general tendency for the level of atmospheric noise, the dominant source of noise at HF, to rise at the receiver with reduction of operating frequency. Thus, for given operating conditions and parameters and for a defined channel performance objective there will be a lowest usable frequency (LUF). Figure A.19 sketches the operational MUF and the OWF and LUF for a typical long HF link. Formal definitions for these various terms are to be found in Recommendation ITU-R P.373-7 [61].

Prediction of MUFs, OWFs and LUFs
The prediction of MUFs, OWFs and LUFs is complex but, given an accurate forecast of solar activity, they can be computed with acceptable accuracy. The method recommended by ITU-R is set out in Recommendation P.533-5 [62]. For explanations of and background information on the method, see Recommendations ITU-R P.1239 [63] and P.1240 [64]. Software and data for the computations that are involved are avail-

able from the ITU. Data for several simpler, if less accurate, methods for calculating MUFs or OWFs are to be found in the literature; see for example Braun [6].

A5 Tropospheric scatter propagation

The signal reaching a receiving antenna by the tropospheric scatter mode is made up of a very large number of very weak elements, each of which has been scattered at a location within the troposphere where there is a discontinuity in the refractive index of the air, each location being visible from both the transmitting antenna and the receiving antenna. The volume of air in which this scattering can occur, called the common scattering volume, is bounded by the upper limit of the troposphere (the tropopause) and the angular range within which the gain of the transmitting and receiving antennas is sufficiently high; see Figure A.20.

Most of the energy that is scattered in this way is deviated from its original path by no more than a few degrees. However, geometrical factors set a minimum angle, shown by θ in the figure, through which energy must be deviated if it is to reach the receiving antenna. Given favourable path geometry and sufficient transmitter power, tropospheric scatter supports reliable communication over distances beyond the reach of line-of-sight propagation, even if allowance is made for refraction and diffraction, out to a maximum distance of about 700 km, extendible to perhaps 900 km in very favourable circumstances. The phenomenon is active over much of the VHF band and upwards, with declining effectiveness, to at least 5 GHz but the frequency range mostly used for fixed systems lies between about 800 MHz and 3 GHz. See Recommendation ITU-R F.698-2 [65].

However, the received signal shows deep and rapid fading. Complex receiving systems are used, typically with eight-fold diversity for the longer links, to minimise the effects of fading. The imperfect coherence of the received signal limits the usable bandwidth of a link; see Recommendation ITU-R F.1106 [66]. Largely empirical methods are given in Recommendation ITU-R P.617-1 [67] for estimating the transmitted e.i.r.p. required to ensure that the signal received with an antenna of given size will be present for an acceptably large percentage of the time.

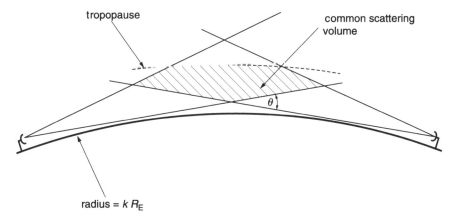

Figure A.20   The scattering volume of a tropospheric scatter link

There is, however, another aspect of tropospheric scatter propagation that must also be considered. Very high transmitted e.i.r.p. is necessary to deliver a usable signal to a distant receiving antenna for say 99.9% of the time but a much smaller e.i.r.p., such as is radiated by many kinds of transmitting station, is sufficient to cause serious interference by this propagation mode for 1.0% of the time over a wide area. For information on interference field strengths by this mode, see ITU-R Recommendation P.452-8 [23].

A6   Meteor-burst propagation

In addition to the showers of meteors, visible from the ground with the naked eye, which enter the atmosphere on occasion and burn up, typically when the Earth crosses the orbit of a comet, the Earth sweeps up a steady stream of so-called sporadic meteors. The sporadic meteors occur in all parts of the world, the number varying diurnally, with a peak around 0600 hours local time, and seasonally, with a maximum in the Northern Hemisphere in July and a minimum in February.

Because of the high velocity with which meteors enter the atmosphere, atmospheric friction is great enough for strong heating to begin at a height of about 120 km and the meteor is usually burnt up by the time it has penetrated to a height of about 80 km. The trail of hot gas which a meteor leaves is intensely ionised and it is capable of deflecting radio waves during its brief lifetime, typically a fraction of a second. If two radio stations are separated by a distance between about 400 and 2000 km, there will often be a brief opportunity for a burst of communication between them by reflection at a meteor trail. The geometry for such a path is much like that shown for tropospheric scatter in Figure A.20.

Not all meteor trails last long enough, are big enough or are suitably oriented to support communication between radio stations that have them simultaneously in view. However, it is found that there is seldom a period lasting more than a few minutes when no usable trail is available to a pair of stations and sometimes there are several per minute. The transmission loss by meteor trail scatter is high but the proportion of usable trails can be increased by raising the transmitter power. The use of wider antenna beams also increases the occurrence of usable trails.

The signal received is somewhat incoherent, and this limits the available bandwidth. The information transmission rate achievable whilst a trail is active is somewhat limited; 8 kbit/s is typical. However, allowing for the time when no trail is visible, an average information transmission rate of a few hundred bits per second is obtainable for messages in digital form operated in a packet mode. Links operating in this mode are reliable and use is being made of them in very extensive, sparsely populated areas lacking infrastructure, typically in arctic regions. Carrier frequencies between 30 and 100 MHz are usable, the optimum for most purposes being around 50 MHz. The factors which determine the carrier frequency range to be preferred are discussed in Recommendation ITU-R P.843-1 [68].

A7   The irregular propagation modes

There are many radio propagation modes other than the five described in Sections A2 to A6. Some, like the whistler mode, which is operative at times at VLF (see CCIR

Report 262-7 [69], are neither useful for communication services nor troublesome as routes for the transmission of interference; these are not considered further here. But other modes, whilst being so intermittent that they cannot be used for regular services, are nevertheless capable of causing troublesome interference when they are active. And some of the regular modes, when active beyond the limits of their normal range of parameters, may cause interference between radio stations that do not interfere under normal conditions. These last two groups of irregular modes are discussed briefly in this section.

A 7.1 Ionospheric cross-modulation

The passage of a powerful radio wave through the ionosphere increases the velocity of movement of free electrons and this in turn increases the ion recombination rate. A wave that is amplitude modulated at audio frequency, passing through the D region or the lower part of the E region, produces variations in the recombination rate, with consequent variations in the transmission loss for the wave, that are a function of the modulation envelope of the wave. The modulation index of the wave is reduced thereby. The effect is greater at the lower modulation frequencies.

However, a second wave, having a different carrier frequency, passing through the same volume of the ionosphere, also experiences the same variation of loss as was generated by the first wave. Consequently, the second wave has the modulation envelope of the first wave imprinted upon it, causing cross-modulation. This phenomenon is sometimes called the Luxembourg effect. This form of interference is significant in particular for amplitude modulated broadcasting emissions at LF and MF. There is an analysis of the process in Recommendation ITU-R P.532-1 [70]. Recommendation ITU-R BS.498-2 [71] examines the impact of the phenomenon on the broadcasting service and provides the results of measurements.

A 7.2 Irregular ionospheric propagation at HF

HF waves are not propagated only by regular reflection by the ionosphere and the ground; they may also be scattered at irregularities on the ground and in the ionosphere, especially the F region. At a time when ionospheric conditions on the great circle path between a transmitting station and a receiving station do not permit communication, scattering at a point off the great circle path may provide a viable though unstable transmission path, of lower absorption or higher usable frequency. See CCIR Report 726-2 [72]. It may also happen that interference is propagated to a receiving station in a similar way, at a time when it would be predicted that regular propagation via the great circle path from the interfering transmitting station would not occur.

A 7.3 F region propagation at VHF

Frequencies above 30 MHz propagated by the tropospheric wave are used mainly for short links, and frequencies below 30 MHz are used mainly for long distance links that depend on the ionospheric wave. However, at the peak of the solar activity cycle during daytime, the F2 region regularly reflects frequencies considerably higher than 30 MHz with hop lengths up to 4000 km. Thus there may be strong interference at some seasons between local systems serving widely separated areas. See Recommendation ITU-R P.844-1 [73].

A7.4 Sporadic E region ionisation

In addition to the regular ionisation in the E region described in Section A4.3, thin clouds of plasma form sporadically in the region for reasons that are not fully understood at present. The symbol E_s distinguishes these sporadic formations from the regular E region ionisation. The E_s ionisation is often more dense than that of the regular E region. The presence of E_s adds to the variability of HF propagation but it is at VHF that its effect is more significant, providing a medium for the transmission over distances up to 4000 km of strong interference for periods lasting several hours at a time, often extending up to 60 MHz and sometimes up to 130 MHz.

Three forms of E_s are recognised. Equatorial E_s forms frequently around the geomagnetic dip equator during the daytime. Auroral E_s tends to form a ring round each geomagnetic pole, principally at night, peak ionisation occurring at a geomagnetic latitude of about 70°; E_s also forms a polar cap within the ring of auroral E_s. Finally, temperate zone E_s is chiefly a summer daytime phenomenon.

Recommendation ITU-R P.534-3 [74] provides a basis for estimating the probability that E_s will occur at any location, with an indication of the highest frequency likely to be affected and the path loss.

A7.5 Ionospheric forward scatter at VHF

Energy is scattered from irregularities in the ion density at the bottom of the E region, at a height of about 85 km. The process is much lossier than normal ionospheric reflection, but it is persistent and, given very high transmitted power, it could provide reliable narrow-band communication links up to about 2000 km long. If it were used in this way, the optimum frequency range would be 30–60 MHz; see CCIR Report 260-3 [75].

Systems making use of this mode were used to a small extent in the 1950s and 1960s. However, such emissions would be a source of interference by the same propagation mode over a wide area. Moreover, the signal will also be propagated by reflection from the F2 region and sporadic E region (E_s) ionisation for a substantial part of the day, particularly at times of high solar activity. These are low loss transmission modes, capable of setting up very powerful interference over large areas thousands of kilometres from the transmitter. Consequently, the utilisation of this propagation mode is strongly deprecated and RR paragraphs S5.44, S8.4 and S8.5, taken together, have the effect of making its use impracticable. Nevertheless, ionospheric forward scatter remains a potential medium for low level interference from transmitters intended for systems designed for conventional propagation modes.

A7.6 Tropospheric ducts

If the rate of change with height (the lapse rate) of the refractive index of the atmosphere should equal -157 parts per million per kilometre, a ray of a radio wave that is initially horizontal will have the same curvature as the Earth's surface. If the ray path is not obstructed, say by hills, the ray will be propagated parallel to the ground for as far as these atmospheric conditions persist. If the rate of change of the refractive index with height should be even more negative, a tropospheric duct may be formed, capable of capturing rays that are not initially horizontal and propagating them round the curved surface of the Earth by successive reflections at its upper and lower surfaces.

Over land, meteorological conditions are not usually favourable to the establishment of extensive surface ducts. Turbulence due to wind and irregularities of the ground tend to prevent uniform stratification of the atmosphere. However, a surface duct, due to an inversion of the normal decrease in the temperature of the air with increasing height, may develop over land at night, when the air in contact with the ground cools more rapidly than the air at higher levels. A shallow surface duct often forms over the sea or a large lake due to the high humidity of the air layer which is in immediate contact with the water. A thick surface duct often forms over a warm sea in the vicinity of land in the evening, when a dry breeze blows off the land. A surface duct formed over land at night may be converted to an elevated duct in the morning when the Sun's heat warms the ground, the duct being raised bodily by convection. Elevated ducts are also formed in anticyclonic weather conditions by a temperature inversion caused by the descent, compression and warming of an air mass over cold low-lying air.

A duct has some of the properties of a lossy waveguide, including a critical frequency, below which transmission in the ducted mode is rapidly attenuated. Propagation by duct at VHF only arises when the layer is unusually thick and the negative lapse rate of the refractive index is very high. It is more common at UHF and frequently occurs at SHF in places where the climate and the topographical situation favour duct formation.

When ducting occurs on paths that would not normally be line-of-sight, a very potent mechanism for trans-horizon interference may be set up. In free space propagation the energy of a wave spreads out in two dimensions orthogonal to the direction of propagation and the transmission loss is related to d^2, the square of the distance. In a duct there may be little spread of energy in the vertical direction and the transmission loss is related to d^n, where n is closer to unity than to two. Consequently the loss for the ducted wave may be considerably less than the free space loss. Recommendation ITU-R P.452-8 [23] gives a method for calculating the transmission loss for ducted waves.

A 7.7 Reflection from aircraft

Signals are reflected passively by aircraft. This can cause sporadic interference between radio stations which are located so that no significant signal passes between them when aircraft are not present. In a similar way a signal reflected from an aircraft that is strongly illuminated by a transmission can cause multipath fading of the reception of the same signal, especially where the direct transmission path to the receiving antenna is obstructed, for example by buildings. Reflections of this kind may be particularly troublesome close to major airports. There are some statistics on the incidence of interference of this kind in CCIR Report 569-4 [76].

A 7.8 Precipitation scatter

Energy is scattered from a radio wave by raindrops and hail, collectively called precipitation or hydrometeors. Some of this scattered energy may reach a receiving antenna that is not directly illuminated by the wave, causing sporadic interference. This kind of scattering differs in its characteristics from other atmospheric scattering phenomena, such as tropospheric and meteor trail scattering; these latter modes deflect radio wave energy through relatively small angles but precipitation scatters energy in all directions. Consequently precipitation interference is not limited to situations where the location at which scattering takes place is on or close to the great circle path between the source of the interference and the receiving station suffering interference.

A body of rain sufficiently intense to cause significant scattering can also be expected to absorb energy from the scattered signal. Both the scattering cross-section of an elemental volume containing rain and the attenuation due to absorption are direct functions of the intensity of the rainfall but they produce opposite effects. The relative importance of the two factors depends on the frequency. A study by Awaka [77] found a minimum in the total transmission loss, giving maximum coupling of interference, at a rain rate near 10 mm/h for typical earth station/terrestrial station propagation geometry at frequencies in the 12–14 GHz range.

It may be concluded that precipitation scatter may be a cause of interference if the main lobes of the two antennas concerned have a common volume in the atmosphere at or below the 0°C isotherm level. Significant interference is unlikely to be coupled between the main lobe of one antenna and the sidelobes of another unless the common volume is close to one station and the gain of the main lobe involved is high.

Precipitation scatter is most likely to be significant when one of the stations is an earth station using a geostationary satellite at a low angle of elevation and the other station is another earth station using the same frequency bands but in the reverse mode or a radio relay station. It may be possible to avoid interference or greatly reduce its incidence by choosing sites for stations that will use the same frequency bands in reverse directions of transmission so that the main lobes of their antennas will not have a common volume within the troposphere. Methods for predicting the extent of interference by this mode are to be found in Recommendation ITU-R P.452-8 [23]. A means to alert Administrations to the possibility of interference by precipitation scatter is incorporated, as Propagation Mode 2, into the coordination procedure for earth stations sharing frequency bands with terrestrial stations; see RR Appendix S7, and Section 5.2.3 above.

A8 Radio noise

A8.1 Introduction

External noise enters a receiving system at the antenna, together with the wanted signal and, perhaps, interference. More noise (internal noise) arises within the receiving system, generated by various mechanisms. All of this noise and interference degrades system performance and it is necessary to design the system so that the ratio, at the input to the demodulator, between the power of the carrier of the wanted signal and the total noise, plus an allowance for interference, is high enough to allow performance objectives to be achieved.

Some, usually most, of the external noise comes from natural sources. However, some external noise is generated elsewhere within the communication system, typically in the transmitter or in intermediate repeaters including satellite transponders. It is part of the system designer's function to ensure that the level of noise generated in-system is low, so that the carrier level at the receiver, and hence the power trans-mitted, which may be relatively costly, can also be low. This also helps to ensure that spectrum is used economically.

In considering the effect of noise on the performance of a system it is often necessary to find the sum of the noise contributions from several sources, internal and external. To do this, it is convenient to express each contribution as a multiple f of the thermal agitation noise (Johnson noise) in a notional resistor replacing the antenna, its impedance matched to the input to the receiver. The noise power available from the resistor within the bandwidth of the receiver, P_s is given by

$$P_s = ktB \text{ W} \tag{A.13}$$

where k is Boltzmann's constant $= 1.38 \times 10^{-23}$ J/K,
 t is the temperature of the resistor (K),
 B is the bandwidth of the pre-demodulator stages of the receiver (Hz).

In most circumstances it is convenient to assume that the temperature of the resistor is 290 K, an arbitrary value for the ambient temperature and denoted by the symbol t_0. Thus, if the noise entering the receiver through the antenna from three separate sources is equal to P_{a1}, P_{a2} and P_{a3}, where

$$P_{a1} = f_{a1} P_s \text{ W} \tag{A.14}$$

and similarly for P_{a2} and P_{a3}, then f_{a1}, f_{a2} and f_{a3} are said to be the noise factors for the three noise sources and the total noise from all three sources is equal to $(f_{a1} + f_{a2} + f_{a3})P_s$. Inputs of interference can be treated in an analogous manner.

It is often convenient to express the noise factor in logarithmic form, when it is called the noise figure and is represented by the symbol F, thus:

$$F_a = 10 \log f_a \text{ dB} \tag{A.15}$$

Noise factors and noise figures, based on a power level of $(kt_0 B)$ where $t_0 = 290$ K, tend to become inconvenient for use for high performance systems above 1 GHz. In such cases it is more usual to use 1 K as the reference temperature of the reference resistor and the noise factor becomes a noise temperature with t as its symbol. Thus, for example,

$$t_a = 290 f_a \text{ K} \tag{A.16}$$

The natural external radio noise sources include:

1   atmospheric noise, radiated from lightning discharges;
2   man-made noise, emitted unintentionally from electrical machinery, electronic devices (including spurious emissions from other radio installations), power transmission lines and the ignition systems of petrol engines; and
3   noise emitted by the environment of the receiving station antenna, thermal noise emitted by the atmosphere and rain, and also noise from extra-terrestrial sources.

However, the dominant noise source and the level of noise from that source varies markedly with frequency; see Figure A.21.

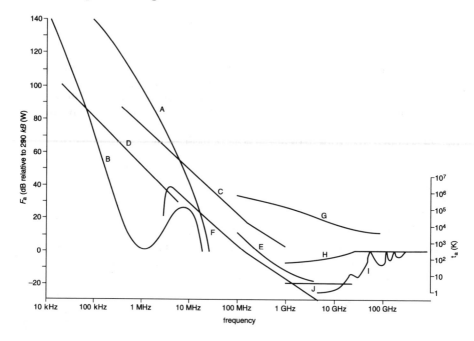

Figure A.21   *Typical external noise levels as a function of frequency, expressed as noise figures*
(F_a) and noise temperatures (t_a) (after Recommendation ITU-R P.372-6
[78])
   A. Atmospheric noise exceeded for 0.5% of the time.
   B. Atmospheric noise exceeded for 99.5% of the time.
   C. Man-made noise, median business area.
   D. Man-made noise, quiet receiving site.
   E. Galactic noise (maximum value).
   F. Galactic noise (typical value).
   G. Solar noise ($\frac{1}{2}°$ beamwidth, directed at Sun).
   H. Sky noise due to oxygen and water vapour (narrow-beam
   antenna, directed horizontally).
   I. Sky noise, as H but antenna directed vertically.
   J. Cosmic background, 2.7 K.

A8.2 Predicted external noise levels

A8.2.1 Atmospheric noise

Atmospheric noise has its origin in lightning discharges, mostly occurring in thunder-storms, many of which are in progress at any time somewhere on Earth. The electro-magnetic energy generated by lightning covers a broad spectrum. The maximum spectral power density occurs below 10 kHz but the amount of power that is produced is very great and it is quite significant at the lower end of the VHF range.

Most thunderstorms take place in a few well-defined geographical areas and atmo-spheric noise is distributed to areas remote from storms by the propagation mechanisms, mostly ground wave and ionospheric wave, that also operate for radio signals. The propagation processes modify the spectral distribution of noise power, attenuating some parts and failing to propagate the higher frequency components over long distances.

Recommendation ITU-R P.372-6 [78] shows predictions of F_{am}, the median atmospheric noise factor at the output of a short vertical monopole over perfectly conducting ground, as a function of frequency, geographical location, time of day and season. There are also data on the statistical distribution of values of F_{am}. No doubt there are variations of F_{am} with the solar activity cycle but these changes have not been quantified yet. See also curves A and B in Figure A.21. It seems likely that F_{am} is dependent to some degree on the directional characteristics of the antenna but it is not clear at present if and how this could be taken into account.

A8.2.2 Man-made noise

Some industrial, scientific and medical apparatus generates radiation with a form that is not greatly different from that of a radiocommunication signal, being mainly concentrated within a relatively narrow band and often relatively coherent in waveform; see Section 4.2.3. Interference from these emissions, while sometimes strong, tends to be concentrated into defined frequency bands and is not considered here. Sensitive communications installations must be located away from such sources of interference.

However, broad-band noise-like electromagnetic radiation is also generated by many kinds of industrial and domestic apparatus, electric traction systems, overhead electric power transmission systems and so on. An important source of man-made noise is the ignition systems of petrol-driven motor vehicles. The level of man-made noise varies considerably from place to place, depending on the level of commercial and industrial activity; where activity is high, man-made noise is likely to be the dominant form of noise at frequencies between 10 and 300 MHz and it is important to choose with care a site for a sensitive radio station if significant constraints on performance are to be avoided.

Figure A.21 shows, in curves C and D, an estimate of F_a from man-made noise for two environments, namely a median business area and a quiet site. There is a more detailed treatment of the subject in Recommendation ITU-R P.372-6 [78]. This is, however, an area of considerable uncertainty.

A8.2.3 Noise from the environment and beyond

Radio noise is received at radio stations on Earth from the ground, from nearby buildings, from the Sun and from many other specific and identifiable sources. Noise of thermal origin is also received from the atmosphere. Noise from some of these sources also enters receivers in satellites. Below 1 GHz these noise entries tend to be overwhelmed by more powerful entries from atmospheric noise due to lightning and man-made noise but most low-noise microwave systems receive noise from several of these environmental sources.

These various sources usually occupy only part of the beam of the antennas of substantial gain that are used for many systems operating above 1 GHz. It is necessary to take this partial exposure into account in estimating the total noise entering the receiver from these various sources. The strength of the radiation from each source can be expressed by a brightness temperature, T_B, and the extent to which an antenna captures noise from the source can be calculated by convolution of the antenna gain pattern and the distribution of T_B due to that source over the sky and the ground visible from the antenna.

It is, however, much simpler and sufficiently accurate for many purposes to define a capture coefficient η corresponding to the proportion of the antenna main beam, within the half-power gain contour that is occupied by the noise source, some suitably small coefficient being assigned to a source which is present only in the sidelobes. Thus $\eta = 1$ would be appropriate for a source which occupied the whole of the main beam at the half-power gain contour, $\eta = 0.1$ would be applied to a source that occupied 10% of the half-power beam area and $\eta = 0.01$ might be appropriate for a small source located in the sidelobes of an antenna with high gain. Then the noise temperature component from that source, t, entering the receiver would be given by

$$t = \eta T_B \, \text{K} \qquad\qquad (A.17)$$

and the total noise temperature from sources of this kind would be equal to the sum of the components, $\sum t$. This approach is used here.

Thermal noise from the atmosphere
Sections A2.2.3 and A2.2.4 discuss respectively, the absorption of energy from a radio wave by liquid water falling as rain and by the gases of the atmosphere. Related to these absorption phenomena there is an emission of thermal noise from the lossy transmission medium. The noise power observed is relatively small, being perceptible only above 1 GHz and only when a receiver with a low noise input amplifier is used, and it is convenient to define it by means of a brightness temperature. It is likely that the capture coefficient η will approach unity for this source if there is no other solid body within the cone of view of the main lobe of the antenna.

If the thermal noise received from a given direction at a given frequency has a noise temperature of t_{at} K and the absorption that a wave from a distant source in the same direction and at the same frequency would suffer is A dB, then

$$t_{at} \approx T_m(1 - 10^{-A/10}) \, \text{K} \qquad\qquad (A.18)$$

where T_m is the temperature of the transmission medium, usually taken to be about 275 K. For large values of A, $10^{-A/10}$ approaches zero and t_{at} approaches T_m. Equation (A.18) tends to overstate the effect of heavy rain, since some of the loss which heavy rain causes to the wave is due to scattering, not absorption. For more data, see section 4 of Recommendation ITU-R P.372-6 [78]. Conversely a radiometric measurement of the noise level received from an antenna directed towards space can be used as a measure of the losses in that path through the atmosphere, allowance being made for noise received from extra-terrestrial sources; see Recommendation ITU-R P.1322 [79].

The directional characteristics of the antenna should be taken into account when estimating t_{at}. Thus, for example, curve H in Figure A.21 is for a narrow-beam antenna directed horizontally; such a beam will receive thermal noise from oxygen and water vapour extending over hundreds of kilometres, a considerable part of it being low in the atmosphere, close to sea level pressure. In consequence, t_{at} approaches T_m over much of the spectrum above 10 GHz. Curve I in the figure is for a vertical beam receiving thermal noise from a column of oxygen and water vapour equivalent to only about 6 km at sea level pressure and $t_{at} = T_m$ only at the frequencies where strong gaseous absorption occurs.

Noise from buildings, the ground and the sea
The ground visible from an antenna on land or from an aircraft flying over land is at the ambient temperature, which can be taken for the present purpose as 290 K, but it does not usually radiate noise as a black body. Thus,

$$T_B = 290\varepsilon \text{ K} \tag{A.19}$$

where ε is the emissivity of the ground. The emissivity varies with the angle of elevation, the moisture content of the soil, the roughness of the surface and, no doubt, the frequency. In addition, the state of polarisation of the noise energy radiated depends to some extent on the angle of elevation. There are some data on the range of typical values of ε in Recommendation ITU-R P.372-6 [78] but 0.75 can be taken as an average value. The situation is similar for an antenna on a ship at sea or on an aircraft flying over the sea. A typical value for the emissivity of the sea can be taken as about 0.5.

Buildings also emit radio noise and, no doubt, their emissivity depends to a considerable degree on the nature of the surface of the walls visible from the antenna. It might be assumed, for example, that the emissivity of a wall sheathed with glass would be appreciably less than unity. However, in the absence of firm data for specific surfaces it would be prudent to assume that ε equals unity.

For ground, sea and buildings it is necessary to evaluate η. For a high gain antenna directed horizontally, η for the ground or the sea is unlikely to exceed 0.4. Buildings are likely to be feeding noise into the sidelobes but the coefficient for even a large building, close to the antenna, is unlikely to exceed 0.1. For a high gain antenna directed at a satellite, none of these coefficients is likely to exceed 0.1. However, for a low gain antenna at street level at a city centre site, a total coefficient of 0.6, divided between ground and buildings might well be appropriate.

Noise from the Sun and Moon
The Sun is a strong source of radio noise. When the Sun is undisturbed, a mean value for T_B is about 10^6 K at 100 MHz, falling to about 10^5 K at 1 GHz, to 10^4 K at 10 GHz and to 6000 K at 100 GHz. The brightness temperature of the quiet Sun varies with the cycle of solar activity and it is considerably increased when the Sun is disturbed.

Much of the radio noise radiated by the Moon comes from the part of its surface that is being illuminated by the Sun, so the Moon's brightness temperature, as seen from the Earth, varies from about 140 K at new moon to about 280 K at full moon, varying little with frequency between 1 and 100 GHz.

Both the Sun and the Moon have an angular diameter of about 0.5° as seen from the Earth, capable of filling the half-power beamwidth of a high gain antenna. Thus, under worst conditions, η equals unity and solar noise interference is very severe. However, the antennas of communication systems do not track the Sun or the Moon. Thus, a high gain antenna of a terrestrial system, typically a line-of-sight link, will receive noise, at worst, for a period of some minutes on perhaps 30 to 40 days per year if and when the Sun or the Moon rises or sets on the azimuth on which the antenna is directed. An earth station receiving from a geostationary satellite suffers similarly while the satellite is in transit across the Sun. For antennas of low gain, η will be much less than unity; noise from the Moon will be negligible; noise from the Sun will be perceptible for periods that are longer than those for high gain antennas but the level will be less severe.

Noise received on Earth from other extra-terrestrial sources

Radio noise is emitted from violent electro-dynamic processes taking place in interstellar gas clouds at the heart of our galaxy. Sky maps in Recommendation ITU-R P.372-6 [78] show that the largest concentrations of noise sources lie in the plane of the Milky Way and above all in the constellation Sagittarius, the direction of the centre of the galaxy. In other directions the brightness temperature, as it would be perceived by an antenna of medium gain, is quite small, typically about 20 K at UHF and much less in the microwave range.

The peak brightness temperature from a relatively large region of Sagittarius is about 3×10^5 K at 10 MHz, falling to about 700 K at 100 MHz and 10 K at 1 GHz. Above 1 GHz T_B is negligible. For an antenna directed towards Sagittarius, galactic noise is perceptible in the HF frequency range when atmospheric noise is low. Galactic noise may be the dominant external noise at VHF and the lower part of the UHF range where man-made noise is low. However, as happens with solar noise, antennas are unlikely to remain permanently directed towards Sagittarius, and a brightness temperature such as those indicated above does not persist all day.

In addition to these relatively broad areas of high brightness temperature in the Milky Way, a number of small, isolated, areas of the sky are sources of a substantial noise flux which extends above 1 GHz. These sources are called 'radio stars'. Small as they are as seen from the Earth, they are much larger than conventional stars although the capture coefficients (η) for even the largest earth station antennas are small. The best known of the radio stars are Cassiopeia A, Taurus A and Cygnus A. None of these sources are close to the Earth's equatorial plane and noise from them is of no operational significance to radio systems. However, these radio stars can be used as signals of known power for measurement of the figure of merit of high gain earth station antennas; see Recommendation ITU-R S.733-1 [80].

Finally, radio noise perceptible from the Earth is emitted by some planets. In particular, although the angular extent of Venus is very small, it has a brightness temperature of about 550 K between 10 and 30 GHz and is preferred to the radio stars for measurements of the figure of merit of high gain earth station antennas above 10 GHz.

Noise received by satellite receivers

The level of noise radiated by the Earth entering the receiver of a geostationary satellite using an antenna beam covering the whole Earth varies according to the satellite's station on the orbit, depending on the proportions of land and sea visible from the satellite. For a satellite over the middle of the Pacific Ocean, seeing mostly sea, the noise temperature will rise from about 130 K at 6 GHz to about 190 K at 30 GHz. For a satellite over Africa, seeing mostly land, the noise temperature will rise from about 195 K at 6 GHz to about 230 K at 30 GHz. For more detailed information, see Figure 9 of Recommendation ITU-R P.372-6 [78]. The level of noise received by a satellite antenna of higher gain can be expected to differ from these figures, depending mainly on the nature of the Earth's surface within the footprint of the antenna beam.

A9 Prediction of radio link performance

A spectrum manager may need to use predictions of propagation loss and noise level for various purposes, typically:

a) to estimate how much transmitter power at a given transmitting station will be required to provide a predetermined standard of channel performance at a receiving station at a given location or in a given service area, given no more than an acceptable amount of interference;

b) to estimate the impact on the channel performance experienced at stations receiving a wanted signal from a transmitter with known characteristics that may be expected to arise from interference from another station with known characteristics; and

c) to estimate whether the power of an emission, considered as interference, would cause a criterion of acceptable interference that had been agreed internationally, to be exceeded.

These three situations are considered below.

It is sometimes important and always desirable for these estimates to be made with precision. In some applications it is quite feasible, even easy, to do so, for example where optical transmission paths for microwave links are involved. It will be evident from what follows that in other applications it would be impracticably laborious to do so or quite impossible in the present state of knowledge; judgement and experience must be used in providing margins of safety in such cases.

Required transmitter power levels
The calculation to determine the required transmitter power (P_t) can be summarised by the expression

$$P_t = L - G_t + F_a + M_i + P_R \text{ dBW} \tag{A.20}$$

where L is the transmission loss (dB),
G_t is the gain of the transmitting antenna (dBi),
F_a is the noise figure of the external noise at the receiver (dB relative to $290 \times k$ watts, where k is Boltzmann's constant),
M_i is a margin for acceptable interference (dB),
P_R is the power required at the receiving location (dBW).

However, in quantifying the terms on the right-hand side of this expression it is necessary to take into account the nature of the service to be provided.

The dominant propagation mode and therefore the transmission loss (L) depends on the operating frequency, the path length, the location of the link and various other factors outlined in Sections A2 to A6. With most propagation modes there is some variation of L with time. If the range of this variation is large it will be necessary to find a value for the loss that will not be exceeded for a percentage of the time that is somewhat less than 100%, since the use of transmitter power high enough to overcome the highest loss conditions may be uneconomical and may use spectrum inefficiently.

The value of transmitting antenna gain (G_t) to be used may not be the antenna's maximum gain. If there is to be only one receiving station, the gain of the transmitting antenna for the ray path that is received, or an estimate of it, should be used. This is a matter of particular concern for ionospherically propagated signals.

The signal power requirement at the receiving location (P_R) is a complex term covering such parameters as receiving antenna gain in the appropriate direction, the receiver noise figure, the pre-demodulator noise bandwidth and the carrier-to-noise ratio required at the input to the demodulator to meet a predetermined standard of channel performance for a predetermined high percentage of the time. This last-mentioned factor will be subject to a wide range of values, depending not only on the nature of the service and performance required but also on the modulation method used, the modulation parameters and whether such techniques as diversity reception or automatic error correction are used to protect the system against transmission impairments. In evaluating P_R it may be convenient to add the noise temperature due to external noise (t_a) to the receiver noise temperature (t_r), instead of using a separate external noise figure (F_a).

Acceptable interference levels will usually be small and M_i can usually be represented adequately by a factor between 0.5 and 1.0 dB, which can be considered as an addition to the noise level.

For a point-to-area service it will often be appropriate to define a set of standard parameters for receiving stations and to relate the channel performance objectives to a predetermined percentage of the receiving locations. Then if the propagation prediction data of Recommendations ITU-R P.370-7 [24], P.528-2 [27] or P.529-2 [28] are applied in determining the transmission loss (see Section A2.2.2), an adjustment will be made automatically to L to reflect those objectives.

The impact of interference on performance
Low level interference has been treated above as equivalent to a small addition to the noise level. However, the level of interference may not be trivial and the impact of interference on the performance of the wanted channel is not a simple function of the interference level; much depends on the nature of the wanted and interfering emissions. Nevertheless, spectrum may be wasted if pessimistic estimates are used in determining whether a specific entry of interference is acceptable. The use of methods for evaluating the impact of interference on performance which are as objective as is feasible is essential in congested frequency bands. Typically, three stages are involved:

a)   calculation of the ratio of power of the wanted emission (C) to each significant interfering emission (I) at the input to the demodulator of the receiver;

b)   conversion of these C/I ratios into a valid measure of the degradation of the wanted signal; and

c)   determination of an acceptable level of degradation for a single interference entry or for all entries in aggregate at a station.

C can be evaluated by the same basic methods as those set out in equation A.20. I can be calculated likewise, given the necessary basic data, and taking into account the irregular propagation modes reviewed in Section A7 as well as the regular modes. However, difficulties will usually arise in determining what values of gain should be assumed for the antennas transmitting and receiving the interference; this latter point is considered further in the next paragraph. If irregular propagation modes are found to be potentially active for the interference path, or if the regular modes are very variable, it may be necessary to predict interference levels exceeded for various percentages of the time, such as 20%, 1% and 0.01% since, to achieve efficient spectrum use, it may be found necessary to accept higher interference for a small proportion of the time than would be tolerated for all or most of the time.

It may not be feasible to determine accurately the gains of the transmitting and receiving antennas for the interference ray path. The direction of the ray at each station can be calculated, but it will usually fall outside the main lobes of the antennas. In particular, for high gain antennas, the gain changes rapidly with frequency and direction in the sidelobes and reliable measurements of the polar diagrams may not exist. This may be a source of considerable difficulty where a transmitter causes interference at a receiver which is under the jurisdiction of another administration. A solution that has been found helpful in the satellite and fixed services when measured antenna polar diagrams are not available is the use of internationally agreed reference radiation patterns for antennas; see for example Recommendations ITU-R S.465-5 [81] and F.699-4 [82]. The polar diagrams of the antennas of geostationary satellites are published for use in coordination with other satellites.

The methods used for evaluating the impact of interference at a given C/I ratio upon the wanted signal depend largely on the nature of the wanted service. For example, for a television broadcast the basic criterion would be the degree of impairment of the received picture perceived subjectively by trained observers; there is an ITU-R handbook on this technique [83]. But the vulnerability to impairment of an analogue television channel using vestigial sideband amplitude modulation has totally different characteristics from that of a digital system with the same picture parameters. The vulnerability to impairment of different digital television systems would also depend on the nature of the modulation system, the encoding algorithms and forward error correction systems used. Subjective tests of this kind having established a relationship between impairment and C/I ratio under precisely defined conditions, acceptable levels of interference power can be defined for use in planning frequency assignments conforming to the same conditions.

Impairment of other kinds of service will have to be judged by other criteria, account also being taken of the technical factors involved. Thus, most interference entering narrow-band analogue channels used for international telephone links is perceived as noise-like and the degradation it causes is judged by the level of the noise relative to test tone level. The criterion that is applied to digital channels used for telephony is the increase in the bit error ratio (BER) which the interference causes. An approach to the calculation of the effect of interference from a single source on the post-demodulator channel noise of an FDM/FM emission is set out in Recommendation ITU-R SF.766 [84]. This involves the definition of an interference reduction factor, B, such that

$$10 \log S/N_i = 10 \log C/I + B \text{ dB} \qquad (A.21)$$

where S is the test tone level in the channel ($= 1 \text{ mW}$) N_i is the interference power in the channel and C/I is as previously defined. B varies in a complicated way with the parameters of the wanted and interfering emissions. Values for B have been determined for a number of specific cases; see, for example, Recommendations ITU-R SF.766 [84] and S.741-2 [85].

International agreement has been reached on the interference limits that would be appropriate for some, although not all, radio services. Agreements of this kind are a necessary part of frequency assignment planning; as an example, the BSS plans for analogue television using frequency modulation that were finally agreed at WARC Orb-88 adopted *C/I* ratios of 31 and 28 dB from interfering emissions operating at the same carrier frequency in Regions 1 and 3 and in Region 2 respectively. An aid of this kind is also valuable for a service that uses frequency coordination in the normal process of selecting operating frequencies. Thus in FSS, and simplifying somewhat, Recommendation ITU-R S.466-6 [86] recommends that interference reckoned as noise entering a channel in an FDM/FM system from all other FSS systems should not exceed 2500 pW0p for more than 20% of any month. And likewise, the interference entering a channel in an FSS system of this kind from terrestrial systems operating in the same frequency band should not exceed 1000 pW0p for more than 20% of any month, nor 50 000 pW0p for more than 0.03% of any month (this short-term figure being included to allow for interference from a radio relay station to an earth station by an irregular low loss propagation mode); see Recommendation ITU-R SF.356-4 [87].

Internationally agreed measures for limiting interference

It has been found useful to apply internationally agreed constraints on the parameters of emissions to ensure that a frequency band can be used efficiently ('technical constraints') and especially to facilitate the use of a band by two services or more ('sharing constraints'). However, it usually happens, especially for sharing constraints, that it is necessary for the constraint to be effective in protecting any kind of wanted emission that is likely to be used in the band from any likely kind of interfering emission. Thus, these constraints are usually expressed in the form of limitations on the e.i.r.p. of the carrier or the spectral density of the emission, regardless of the type of emission. Where the location at which interference becomes significant can be defined with sufficient precision, as for example at the surface of the Earth in cases where a satellite transmitter may interfere with a terrestrial receiver, a limit may be placed on a potentially interfering PFD instead of the e.i.r.p.

A10   References

1 BRUSSAARD, G., and WATSON, P. A.: 'Atmospheric modelling and millimetre wave propagation' (Chapman and Hall, 1995)
2 GIGER, A. J.: 'Low angle microwave propagation' (Artec House Inc., 1991)
3 ALLNUTT, J. E.: 'Satellite-to-ground radiowave propagation' (Peter Peregrinus, 1990)
4 PARSONS, D.: 'The mobile radio propagation channel' (Pentech Press, 1992)
5 DAVIES, K.: 'Ionospheric radio' (Peter Peregrinus, 1990)
6 BRAUN, G.: 'Planning and engineering of shortwave links' (John Wiley and Sons, 1986, 2nd edn)
7 GOODMAN, J. N.: 'HF communications, science and technology' (Van Nostrand Reinhold, 1992)
8 HALL, M. P. M., BARCLAY, L. W., and HEWITT, M. T. (Eds): 'Propagation of radiowaves' (Peter Peregrinus, 1996)
9 'ITU-R handbook on radiometeorology' (ITU, Geneva, 1996)
10 'ITU-R handbook on radiowave propagation information for predictions for Earth-to-space path communications' (ITU, Geneva, 1996)
11 'ITU-R handbook on the ionosphere and its effects on terrestrial and Earth-to-space radiowave propagation from VLF to SHF' (ITU, Geneva, 1997)
12 'The radio refractive index; its formula and refractivity data'. Recommendation ITU-R P.453-6. ITU-R Recommendations 1997 P Series Part 1 (ITU, Geneva, 1997).
13 'Effects of tropospheric refraction on radiowave propagation'. Recommendation ITU-R P.834-2, *ibid.*

14 'Estimation of the actual elevation angle from a station in the fixed service towards a space station taking into account atmospheric refraction'. Recommendation ITU-R F.1333, ITU-R Recommendations 1997 F Series Part 2 (ITU, Geneva, 1997)

15 'Propagation data and prediction methods required for the design of Earth-space telecommunication systems'. Recommendation ITU-R P.618-5, ITU-R Recommendations 1997 P Series Part 2 (ITU, Geneva, 1997)

16 'Propagation data required for the design of Earth-space maritime mobile telecommunication systems'. Recommendation ITU-R P.680-2, *ibid.*

17 'Propagation data required for the design of Earth-space aeronautical mobile telecommunication systems'. Recommendation ITU-R P.682-1, *ibid.*

18 'Propagation data required for the design of Earth-space land mobile telecommunication systems'. Recommendation ITU-R P.681-3, *ibid.*

19 'Effects of multipath propagation on the design and operation of line-of-sight digital radio-relay systems'. Recommendation ITU- R F.1093-l, ITU-R Recommendations 1997 F Series Part 1 (ITU, Geneva, 1997)

20 'Diversity techniques for radio-relay systems'. Recommendation ITU-R F.752-1, *ibid.*

21 'Propagation data and prediction methods required for the design of terrestrial line-of-sight systems'. Recommendation ITU-R P.530-7, ITU-R Recommendations 1997 P Series Part 2 (ITU, Geneva, 1997)

22 'Propagation by diffraction'. Recommendation ITU-R P.526-5, ITU-R Recommendations 1997 P Series Part 1 (ITU, Geneva, 1997)

23 'Prediction procedure for the evaluation of microwave interference between stations on the surface of the Earth above about 0.7 GHz'. Recommendation ITU-R P.452-8, ITU-R Recommendations 1997 P Series Part 2 (ITU, Geneva, 1997)

24 'VHF and UHF propagation curves for the frequency range from 30 MHz to 1000 MHz; broadcasting services'. Recommendation ITU-R P.370-7, *ibid.*

25 'Propagation data for terrestrial maritime mobile services operating at frequencies above 30 MHz'. Recommendation ITU-R P.616, *ibid.*

26 'The prediction of field strength for land mobile and terrestrial broadcasting services in the frequency range from 1 to 3 GHz'. Recommendation ITU-R P.1146, *ibid.*

27 'Propagation curves for aeronautical mobile and radionavigation services using the VHF, UHF and SHF bands'. Recommendation ITU-R P.528-2, *ibid.*

28 'Prediction methods for the terrestrial land mobile service in the VHF and UHF bands'. Recommendation ITU-R P.529-2, *ibid.*

29 'Propagation data for the terrestrial land mobile service in the VHF and UHF bands'. Recommendation ITU-R P.1145, *ibid.*

30 'Digital topographic databases for propagation studies'. Recommendation ITU-R P.1058-1. ITU-R Recommendations, 1997 P Series Part 1 (ITU, Geneva, 1997)

31 'Attenuation in vegetation'. Recommendation ITU-R P.833-1, *ibid.*

32 'Propagation data and prediction methods for the terrestrial land mobile service using the frequency range 30 MHz to 3 GHz. CCIR Report 567-4. Reports of the CCIR, 1990, Annex to Volume V (ITU, Geneva, 1990)

33 'Propagation data and prediction models for the planning of indoor radio-communication systems and radio local area networks in the frequency range 900 MHz to 100 GHz'. Recommendation ITU-R P.1238, ITU-R Recommendations 1997 P Series Part 2 (ITU, Geneva, 1997)

34 'Electrical characteristics of the surface of the Earth'. Recommendation ITU-R P.527-3, ITU-R Recommendations 1997 P Series Part 1 (ITU, Geneva, 1997)

35 'Short distance radio-wave propagation in special environments'. CCIR Report 880-2. Reports of the CCIR, 1990, Annex to Volume V (ITU, Geneva, 1990)

36 'Leaky feeder systems in the land mobile services'. Recommendation ITU-R M.1075, ITU-R Recommendations 1997 M Series Part 1 (ITU, Geneva, 1997)

37 'Specific attenuation model for rain for use in prediction methods'. Recommendation ITU-R P.838, ITU-R Recommendations 1997 P Series Part 1 (ITU, Geneva, 1997)

38 'Characteristics of precipitation for propagation modelling'. Recommendation ITU-R P.837-1, *ibid.*

39 'Rain height model for prediction methods'. Recommendation ITU-R P.839, *ibid.*

40 'The concept of "worst month" '. Recommendation ITU-R P.581-2, *ibid.*

41 'Worst-month statistics'. CCIR Report 723-3, Reports of the CCIR, 1990, Annex to Volume V (ITU, Geneva, 1990)
42 'Attenuation due to clouds and fog'. Recommendation ITU-R P.840-2, ITU-R Recommendations 1997 P Series Part 1 (ITU, Geneva, 1997)
43 'Cross-polarisation due to the atmosphere'. CCIR Report 722-3, Reports of the CCIR, 1990, Annex to Volume V (ITU, Geneva, 1997)
44 'Attenuation by atmospheric gases'. Recommendation ITU-R P.676-3, ITU-R Recommendations 1997 P Series Part 1 (ITU, Geneva, 1997)
45 'Water vapour: surface density and total columnar content'. Recommendation ITU-R P.836-1, *ibid.*
46 'Reference standard atmospheres'. Recommendation ITU-R P.835-2, *ibid.*
47 'Ionospheric propagation data and predictive methods required for the design of satellite services and systems'. Recommendation ITU-R P.531-4, ITU-R Recommendations 1997 P Series Part 2 (ITU, Geneva, 1997)
48 'ITU-R handbook on the ionosphere and its effects on terrestrial and Earth-to-space radiowave propagation from VLF to SHF' (ITU, Geneva, 1997)
49 'ITU-R reference ionospheric characteristics'. Recommendation ITU-R P.1239, ITU-R Recommendations 1997 P Series Part 1 (ITU, Geneva, 1997)
50 'World atlas of ground conductivities'. Recommendation ITU-R P.832-1, *ibid.*
51 'Ground wave propagation curves for frequencies between 10 kHz and 30 MHz'. Recommendation ITU-R P.368-7, ITU-R Recommendations 1997 P Series Part 2 (ITU, Geneva, 1997)
52 'Handbook of curves for radiowave propagation over the surface of the Earth' (ITU, Geneva, 1991)
53 'The phase of the ground wave'. CCIR Report 716-3, Reports of the CCIR, 1990, Annex to Volume V (ITU, Geneva, 1990)
54 'Exchange of information for short-term forecasts and transmission of ionospheric disturbance warnings'. Recommendation ITU-R P.313-8, ITU-R Recommendations 1997 P Series Part 1 (ITU, Geneva, 1997)
55 'Choice of indices for long-term ionospheric predictions'. Recommendation ITU-R P.371-7, *ibid.*
56 'Prediction of sky-wave field strength at frequencies between about 150 and 1700 kHz'. Recommendation ITU-R P.1147, ITU-R Recommendations 1997 P Series Part 2 (ITU, Geneva, 1997)
57 'Prediction of field strength at frequencies below about 500 kHz'. Recommendation ITU-R P.684-1, *ibid.*
58 'Propagation factors affecting systems using digital modulation techniques at LF and MF'. Recommendation ITU-R P.1321, *ibid.*
59 'Implementation of digital sound broadcasting to vehicular, portable and fixed receivers using terrestrial transmitters in the LF, MF and HF bands'. Recommendation ITU-R BS.1349, ITU-R Recommendations 1997 BS Series (ITU, Geneva, 1997)
60 'CCIR atlas of ionospheric characteristics' (ITU, Geneva, 1991)
61 'Definitions of maximum and minimum transmission frequencies'. Recommendation ITU-R P.373-7, ITU-R Recommendations 1997 P Series Part 1 (ITU, Geneva, 1997)
62 'HF propagation prediction method'. Recommendation ITU-R P.533-5, ITU-R Recommendations 1997 P Series Part 2 (ITU, Geneva, 1997)
63 'ITU-R reference ionospheric characteristics'. Recommendation ITU-R P.1239, ITU-R Recommendations 1997 P Series Part 1 (ITU, Geneva, 1997)
64 'ITU-R methods of basic MUF, operational MUF and ray-path prediction'. Recommendation ITU-R P.1240, *ibid.*
65 'Preferred frequency bands for trans-horizon radio-relay systems'. Recommendation ITU-R F.698-2, ITU-R Recommendations 1997 F Series Part 1 (ITU, Geneva, 1997)
66 'Effects of propagation on the design and operation of trans-horizon radio-relay systems'. Recommendation ITU-R F.1106, *ibid.*
67 'Propagation prediction techniques and data required for the design of trans-horizon radio-relay systems'. Recommendation ITU-R P.617-1, ITU-R Recommendations 1997 P Series Part 2 (ITU, Geneva, 1997)
68 'Communication by meteor-burst propagation'. Recommendation ITU-R P.843-1, *ibid.*

69 'ELF, VLF and LF propagation in and through the ionosphere'. CCIR Report 262-7, Reports of the CCIR, 1990, Annex to Volume VI (ITU, Geneva, 1990)

70 'Ionospheric effects and operational considerations associated with artificial modification of the ionosphere and the radio-wave channel'. Recommendation ITU-R P.532-1, ITU-R Recommendations 1997 P Series Part 1 (ITU, Geneva, 1997)

71 'Ionospheric cross-modulation in the LF and MF broadcasting bands'. Recommendation ITU-R BS.498-2, ITU-R Recommendations 1997 BS Series (ITU, Geneva, 1997)

72 'Ground and ionospheric side- and back-scatter'. CCIR Report 726-2, Reports of the CCIR, 1990, Annex to Volume VI (ITU, Geneva, 1990)

73 'Ionospheric factors affecting frequency sharing in the VHF and UHF bands (30 MHz–3 GHz)'. Recommendation ITU-R P.844-1, ITU-R Recommendations 1997 P Series Part 2 (ITU, Geneva, 1997)

74 'Method for calculating sporadic-E field strength'. Recommendation ITU-R P.534-3, *ibid.*

75 'Ionospheric scatter propagation. CCIR Report 260-3, Reports of the CCIR, 1990, Annex to Volume VI (ITU, Geneva, 1990)

76 'The evaluation of propagation factors in interference problems between stations on the surface of the Earth at frequencies above about 0.5 GHz'. CCIR Report 569-4, Reports of the CCIR, 1990, Annex to Volume V (ITU, Geneva, 1990)

77 AWAKA, J.: 'Estimation of rain scatter interference in the case of medium scale broadcasting satellite for experimental purposes (BSE)'. *J. Radio Res. Labs. (Japan)*, 1978, **28**, pp. 23–49.

78 'Radio noise'. Recommendation ITU-R P.372-6, ITU-R Recommendations 1997 P Series Part 1 (ITU, Geneva, 1997)

79 'Radiometric estimation of atmospheric attenuation'. Recommendation ITU-R P.1322, *ibid.*

80 'Determination of the G/T ratio for earth stations operating in the fixed-satellite service'. Recommendation ITU-R S.733-1. ITU-R Recommendations 1997 S Series (ITU, Geneva, 1997)

81 'Reference earth station radiation pattern for use in coordination and interference assessment in the frequency range from 2 to about 30 GHz'. Recommendation ITU-R S.465-5, *ibid.*

82 'Reference radiation patterns for line-of-sight radio relay system antennas for use in coordination studies and interference assessment in the frequency range from 1 to about 40 GHz'. Recommendation ITU-R F699-4, ITU-R Recommendations 1997 F Series Part 2 (ITU, Geneva, 1997)

83 'ITU-R handbook on subjective assessment methodology in television' (ITU, Geneva, 1996)

84 'Methods for determining the effects of interference on the performance and the availability of terrestrial radio-relay systems and systems in the fixed-satellite service'. Recommendation ITU-R SF.766, ITU-R Recommendations 1997 SF Series (ITU, Geneva, 1997)

85 'Carrier-to-interference calculations between networks in the fixed-satellite service'. Recommendation ITU-R S.741-2, ITU-R Recommendations 1997 S Series (ITU, Geneva, 1997)

86 'Maximum permissible level of interference in a telephone channel of a geostationary satellite network in the fixed-satellite service employing frequency modulation with frequency division multiplex, caused by other networks of this service'. Recommendation ITU-R S.466-6, *ibid.*

87 'Maximum allowable values of interference from line-of-sight radio-relay systems in a telephone channel of a system in the fixed-satellite service employing frequency modulation when the same frequency bands are shared by both systems'. Recommendation ITU-R SF.356-4, ITU-R Recommendations 1997 SF Series (ITU, Geneva, 1997)

Index

Administration 1–3
  cooperation between Administrations
    5–6
  see also coordination of frequency
    assignments; International
    Telecommunication Union
  legal basis for spectrum management in
    UK 3–5
  see also allocation of frequencies;
    assignment of frequencies
aeronautical fixed service 120
aeronautical mobile-satellite service (AMSS)
  see mobile-satellite service
aeronautical mobile service (AMS) 5, 26,
  47–9, 165, 175
  frequency allocations and planning
    below 30 MHz 172–4
    117–137 MHz 175, 177–8
    for public correspondence 166, 175
    other available spectrum 178
  see also distress and safety communication
aeronautical radionavigation-satellite
    service (ARNSS) *see* radionavigation
    and radiolocation
aeronautical radionavigation service
    (ARNS) *see* radionavigation and
    radiolocation
allocation of frequencies 2, 26–38
  case for international standardization
    33–6
  international table of allocations 26–33
    amending the table 35–6
  national tables of allocations 37–8
  sharing frequency bands 30–2
    primary and secondary allocations 30,
     33, 76

  see also industrial, scientific and medical
    bands
amateur-satellite service (AmSS) *see* amateur
  and amateur-satellite services
amateur and amateur-satellite services (AmS,
  AmSS) 26, 255
  allocations, sharing and constraints 255–7
  licensing radio amateurs 49–50, 258
  spectrum management for the AmS 258
  spectrum management for the AmSS
    258–9
assignment of frequencies 2, 8, 41–50
  choice of frequencies for assignment
    ad hoc, as for FS at HF 49
    frequency plans 43, 48, 76–7
    non-mandatory coordination 55–7
  delegated management of spectrum 5, 48,
    73–4
  licence fees 4–6, 69–70
    see also spectrum pricing and assignment
    auctions
  non-conforming assignment 50, 78
  see also bandwidth of emissions; cellular
    coverage schemes; radio channel
    planning;

bandwidth of emissions
  frequency tolerance 62–3
  necessary bandwidth 61
  occupied bandwidth 63
  out-of-band emission 63
  see also spurious emission
broadcasting above 30 MHz 137
  adjacent band interference near 108 MHz
    160–1
  allocations 152–5

band-specific sharing issues
  620–790 MHz 154, 222
  11.7–12.7 GHz 154
  40.5–42.5 GHz 128, 155, 202, 224
regulation of terrestrial broadcasting 74,
  139–41
multipoint video distribution systems
  159–60
sound broadcasting
  analogue 160
  digital 161–2
spectrum pricing in broadcasting 139–41
TV frequency assignment planning 152–8
television systems
  analogue 155-6
  digital 158–9
broadcasting-satellite service (BSS) 26, 47–8,
  221–2
allocations 222–4
band-specific sharing issues
  620–790 MHz 154, 222
  1492–1525 MHz 125, 223
  2520–2670 MHz 126, 201, 223
  12 GHz planned bands 195–7, 200–1,
    223, 229, 234–5
  around 17 and 22 GHz 128, 223–4
  40.5–42.5 GHz 155, 202, 224
direct broadcasting by satellite (DBS) 226
  Agreement and plan at 12 GHz 228–32
  assignment of frequencies 226–7
  evolution of plan 232–4
  sharing in downlink bands 234–5
direct-to-home (DTH) broadcasting 217,
  225, 227
feeder link allocations for BSS 201 224
  sharing in planned feeder link bands
    235
regulation of satellite broadcasting 74,
  227–8
sound broadcasting by satellite 236–7
spectrum pricing 74, 227–8
spurious emissions from broadcasting
  satellites 225
broadcasting service (BS) 26, 50 137
regulation of terrestrial broadcasting 138–
  141
see also broadcasting above 30 MHz; HF
  broadcasting; sound broadcasting
  below 2 MHz; tropical
  broadcasting

cellular coverage plans 51–6, 157–8
coordination of frequency assignments 81–2

mandatory, in specified bands 48, 82, 110
mandatory, involving planned BSS 47
mandatory between satellite networks 47,
  81, 83–93
  acceptable levels of interference 105–8
  constraints on emissions 86–7
  current effectiveness of procedure 109–
    11
  due diligence in procurement 91
mandatory, satellite/terrestrial systems
  44–5, 47–8, 81
  constraints to facilitate sharing 100–2
  with GSO networks 93–100
  with non-GSO networks 103–5
  with quasi-GSO networks 102
non-mandatory coordination 49, 55–7

distress and safety communication 7, 76, 187–
  9, 243, 245

Earth exploration-satellite service (EES) 28,
  261
allocations for active sensors 267
allocations for communication links 268–
  71
allocations for passive sensors 265–6
assignment of frequencies 51, 273–4
band-specific sharing issues
  2025–2100 and 2200–2290 MHz 125–6,
    175, 270
  4200–4400 MHz 266, 286
  8025–8400 MHz 87, 126, 271
  10.6–10.68 MHz 127, 266
  15.35–15.4 GHz 105, 201, 244, 266
  18.6–18.8 GHz 127, 199, 267, 278
  25.5–27.0 GHz 128, 272
  29.95–30.0 GHz 87, 199–200, 272
  40.0–40.5 GHz 128, 155, 202, 224, 272
see also passive sensors
electromagnetic compatibility (EMC) 67, 168
emergency position-indicating radio
  beacon (EPIRB) *see* radionavigation
  and radiolocation
emissions, designation of type 45

feeder links 26
fixed-satellite service (FSS) 26, 47, 193–4
applications and allocations
  BSS feeder links 195–6, 201
  commercial FSS C band GSO networks
    194
  commercial FSS Ka band GSO
    networks 196, 198–200

commercial FSS Ku band GSO
networks 195–7
commercial FSS non-GSO networks
196, 200
FSS allotment plan 195, 197–8
government FSS networks 196, 198
MSS feeder links 196, 201
other applications 201–2
band-specific sharing issues
2500–2690 MHz 125–6, 201,223
5091–5250 MHz 105, 201, 244, 286
6700–7075 MHz 105, 194, 197, 201, 244–5
7250–7375 MHZ 198, 241
7450–7550 MHz 198, 271
7900–8025 MHz 198, 241
8025–8400 MHz 126, 198, 271
10.7–12.75 GHz 195–8, 200-1, 223, 229,
234–5
12.75–14.5 GHz 195–7, 200–1, 240, 272,
286–7
15.43–15.63 GHz 105, 201, 244, 265–6,
287
17.7–18.6 GHz 198–201, 223–4, 234–5
18.6–18.8 GHz 199, 267
18.8–19.7 GHz 105, 198–9, 201
19.7–20.2 GHz 87, 199, 240
20.2–21.2 GHz 87, 198, 241
29.1–29.5 GHz 105, 199, 201, 245
29.5–30.0 GHz 87, 199–200, 272
30.0–31.0 GHz 198, 241
37.5–43.5 GHz 128, 155, 202, 224
50.4–51.4 GHz 202, 278
direct-to-home satellite broadcasting 217,
225, 227
factors affecting efficiency of GSO use
202–3
carrier energy dispersal 209–10
earth station antenna characteristics
203–7
polarization isolation 207–8
satellite antenna characteristics 207
uniformity of emission spectrum 209
international spectrum management for
GSO networks 211–12
national regulation and spectrum
management 217–18
non-GSO FSS networks 215–7
spectrum and orbit allotment agreement
113, 212–15
spectrum pricing 217–18
very small aperture terminals (VSATs) 64,
210-11
fixed service (FS) 26

see also fixed service above 30 MHz; fixed
service below 30 MHz
fixed service above 30 MHz
allocations 124–8
band-specific sharing issues
1350–1530 MHz 125, 223
1900–2670 MHz 125–6, 175, 223
8025–8400 MHz 126, 198, 271
10.6–10.68 GHz 127, 266
10.7–12.75 and 17.8–18.6 GHz 127–8, 223,
234–5
18.6–18.8 GHz 127
21.4–22.0 GHz 128, 224
25.25–27.5 GHz 128, 272, 278
31.8–33.4 GHz 130, 269, 278, 287
37.5–42.5 GHz 128, 155, 202
delegated management of spectrum 48
frequency assignment procedure 41–6
spectrum pricing 72–4
systems, line of sight 128–30
high altitude platform systems 131
high density services 128
wireless local loop 131
systems, trans-horizon
meteor burst 131, 324
tropospheric scatter 132–3, 323–4
fixed service below 30 MHz 119
allocations and sharing services 119–20,
147
efficient use of spectrum 123–4
assignment of frequencies 48–9
systems and operating methods 120–4
frequency range designators
for FSS and MSS 193
for general purposes 25
for VHF/UHF broadcasting 152–3

geostationary satellite orbit (GSO)
parameters 67, 82
geostationary satellite network constraints
man-made hazards 67
satellite antenna beam pointing 68
satellite station-keeping 67–8
Global Maritime Distress and Safety System
(GMDSS) 188–9

harmful interference 76
HF broadcasting 137, 148–52
allocations 149–50
applications and systems 148
effective spectrum use 150–1
new modulation systems 148, 151

HF broadcasting (*continued*)
  equitable national access to spectrum 148–9, 151
  frequency assignment 49, 148–9, 152
high altitude platform systems *see* fixed service above 30 MHz

identification of stations and emissions 74–5
industrial, scientific and medical (ISM) bands 32–3, 66
interference and its resolution 46, 75–8
International Telecommunication Union (ITU) 1, 6
  evolution of the Union 7–12
  functions of the Radiocommunication Sector (ITU-R) 17–22
  present organization of the Union 13–7
inter-satellite service (ISS) 28, 202, 277
  allocations 278
  band-specific sharing issues
    24.45–24.75 GHz 278, 287
    25.25–27.5 GHz 128, 272, 278
    32–33 GHz 130, 269, 278, 287
    54.25–58.2 GHz 278
    59–71 GHz 278, 287
ITU-R recommendations and reports, numbering scheme 22

land mobile-satellite service (LMSS) *see* mobile-satellite service
land mobile service (LMS) 26, 165–6
  see also land mobile service above 30 MHz; land mobile service below 30 MHz
land mobile service above 30 MHz 37–8, 165–6, 193–4
  allocations 124–8, 175–6
  assignment of frequencies 48, 49
    regulation of public networks 74, 186–7
    spectrum pricing 186–7
  delegated management of spectrum 48
  sharing at 2025–2100 MHz and 2200–2290 MHz 175
  systems
    low power devices 50, 154–5
    miscellaneous narrowband systems 153–5
    private mobile networks 178–81
    public access networks 181–3
land mobile service below 30 MHz 49, 165–6
  allocations 174
    sharing with tropical broadcasting 147, 174
  Citizens' Band 50, 174

licensing radio stations 2, 45, 69–74
  licence fees 69–74
    see also regulation of radio service provision; spectrum pricing and auctions

man-made noise 66, 331
  see also electromagnetic compatibility
maritime mobile-satellite services (MMSS) *see* mobile-satellite service
maritime mobile service (MMS) 26, 28, 49, 165
  allocations and planning
    below 526.5 kHz 168
    1600–4000 kHz 110, 168–9
    4–30 MHz 110–1, 169–71
    around 156 MHz 176–7
    for on-board communication 177
    other available spectrum 177
  operating methods and systems at VHF 176–7
  operating methods and systems below 30 MHz 166–72
  see also distress and safety communication
maritime radionavigation-satellite service (MRNSS) *see* radionavigation and radiolocation
maritime radionavigation service (MRNS) *see* radionavigation and radiolocation
Master International Frequency Register (MIFR) 44
meteor-burst systems *see* fixed service above 30 MHz
meteorological aids service (MetA) 28, 261
  allocations 271
  sharing issues at 1668.4–1700 MHz 272
  frequencies for meteorological radars *see* radionavigation and radiolocation
meteorological-satellite service (MetS) 28, 47, 261
  allocations 265–6, 271–2
  band-specific sharing issues
    137–138 MHz 268
    1670–1710 MHz 243, 271
    7450–8215 MHz 271
monitoring spectrum use 46
mobile-satellite service (MSS) 26, 47, 239
  allocations for feeder links 201, 243–5
  allocations for MSS 239–43
  band-specific sharing issues (feeder links)
    5091–5250 MHz 105, 201, 244, 286
    6700–7075 MHz 105, 197, 201, 244–5
    15.43–15.63 GHz 105, 201, 244, 265, 287

19.3–19.7 GHz and 29.1–29.5 GHz 105, 199, 201, 245
band-specific sharing issues in MSS bands
bands below 1 GHz 103, 241
1492–1530 MHz 125
1660.0–1660.5 MHz 103–4, 240
7250–7375 MHz and 7900–8025 MHz 241
14.0–14.5 GHz 240
19.7–21.2 GHz and 29.5–31.0 GHz 87, 198, 240–1, 272
other bands above 1 GHz for non-GSO systems 242–3
Commercial GSO networks 245–6
efficient use of orbit and spectrum 199, 246–7
non-GSO systems 64
GMPCS systems 248–50
little LEO systems 247–8
regulation and licensing 247, 249–50
satellite news gathering (SNG) 250-1
spectrum pricing 249–50
mobile service (MS) 26, 28, 49, 124–8, 165–6
see also aeronautical mobile service; distress and safety communications; land mobile service; maritime mobile service

national registers of frequency assignments 45

passive sensors 262–4
see also Earth exploration satellite service; meteorological satellite service, radio astronomy service; space research service
planning of frequency assignments 43, 48, 50–5, 82, 110–13
comparison with ad hoc assignment 112
equitable national access to spectrum 112–13
port operations service 28, 176
prediction of radio link performance 43–4, 105–8, 335–8

radio astronomy service (RAS) 30, 261, 263–4
allocations 264–5
band-specific sharing issues
1660–1660.5 MHz 103–4, 240–1
10.6–10.68 GHz 127, 266
15.35–15.4 GHz 105, 201, 244, 265
22.21–22.5 GHz 267

see also passive sensors
radio channel plans 43, 48, 50–1
radiodetermination-satellite service (RDSS) *see* radionavigation and radiolocation
radiodetermination service (RDS) *see* radionavigation and radiolocation
radiolocation service (RLS) *see* radionavigation and radiolocation
radiolocation-satellite service (RLSS) *see* radionavigation and radiolocation
radionavigation and radiolocation 28
classification of systems by services 28, 281–2
allocations by main usage below 10 GHz
ARNS systems at 960–1215 MHz 285
emergency position-indicating radio beacons (EPIRBs) 189
frequencies for meteorological radars 272, 286
HF position-fixing systems 284
Instrument Landing System (ILS) 160–1, 284
LF medium-range position-fixing systems 283
maritime direction-finding service 284
MF radiobeacons and position-fixing systems 283–4
Microwave Landing System (MLS) 105, 201, 244, 286
OMEGA 283
radio altimeters 286
radionavigation radars 1–10 GHz 285–6
satellite navigation systems 285
search and rescue radar transponders (SART) 189
wind profiler radars 284
allocations and systems above 10 GHz 286–7
band-specific sharing issues above 10 GHz
15.43–15.63 GHz 105, 201, 244–5
24.25–24.65 GHz 278, 287
32–33 GHz 130, 269, 278, 287
59–71 GHz 278, 287
national spectrum management for RDS systems 287
pulse radars and interference 282–3
radionavigation-satellite service (RNSS) *see* radionavigation and radiolocation
radionavigation service (RNS) *see* radionavigation and radiolocation

radio noise 328–30
  atmospheric noise (lightning) 330-1
  atmospheric noise (thermal) 332–3
  environmental noise 331–3
    at satellite receivers 334
  extra-terrestrial noise 333–4
  see also man-made noise
Radio Regulations, ITU 11–2, 14, 17
radio wave propagation 289–90
  free space 290–1
  ground wave 307–9
  ionospheric wave
    cross-modulation in ionosphere 325
    mechanism of reflection 309–14
    propagation by reflection, 2–30 MHz
      317–23, 325–6
    propagation by reflection above
      30 MHz 325–6
    propagation by reflection below 2 MHz
      314–7
    scatter mode 326
  meteor burst propagation 324
  tropospheric scatter propagation 323–4
  tropospheric wave
    absorption by atmospheric gases
      304–5
    effects of obstacles 297–300, 327
    effects of rain, hail etc 301–4, 327–8
    refraction in the troposphere 293–7
    transmission through ionosphere
      305–7
real time channel evaluation (RTCE) at MF
  and HF 49, 123
Regions in frequency allocation 28–9
regulation of radio service provision 3–5, 38,
  47, 70–2
  see also broadcasting; broadcasting-
    satellite service; broadcasting
    service above 30 MHz; fixed-satellite
    service; land mobile-satellite service

safety service 76
satellite link geometry 292–4
satellite system terminology 82–3
search and rescue radar transponder (SART)
  see radionavigation and
  radiolocation
Search for Extra-terrestrial Intelligence
  (SETI) 262
ship movement service 28, 176
sound broadcasting below 2 MHz 137
  frequency allocations 141–2

frequency assignment planning 138,
  145–6
prospective digital systems 142
systems in use 142–5
space operations service (SO) 28, 262
  allocations 268–70
  sharing at 2025–2190 MHz and
    2200–2290 MHz 125–6, 270
space research service (SR) 28, 47, 261
  allocations for active sensors 267–8
  allocations for passive sensors 265–7
  allocations for SR communications
    268–72
  assignment of frequencies 51, 273–4
  band-specific sharing issues for passive
    sensors
    4200–4400 MHz 266, 286
    10.6–10.68 GHz 266
    15.35–15.4 GHz 105, 201, 244, 266
    22.21–22.5 GHz 267
  band-specific sharing issues for SR
    communications
    below 1 GHz 268–9
    2025–2110 MHz and 2200–2290 MHz
      125–6, 175, 270
    7190–7235 MHz 271
    13.75–14.0 GHz 272
    31.0–31.3 GHz 130, 269, 278, 287
    37–38 GHz 128, 202, 224, 272
  see also passive sensors
spectrum manager *see* Administration
spectrum pricing and assignment auctions
  71–4
  see also broadcasting-satellite service;
    broadcasting service above 30 MHz;
    fixed-satellite service; fixed service
    above 30 MHz; land mobile service
    above 30 MHz; mobile-satellite
    service
spectrum utilization efficiency 58–61, 68–9
spurious emissions 64–6
standard frequency and time signal-satellite
  service (TFSS) 28, 262, 273
standard frequency and time signal service
  (TFS) 28, 262, 273

tropical broadcasting 147
tropospheric scatter systems *see* fixed service
  above 30 MHz

wireless local loop *see* fixed service above
  30 MHz
World Telecommunication Policy Forum 17,
  250